INSTRUMENTAL ANALYSIS
OF FOODS

RECENT PROGRESS

VOLUME 2

Edited by

GEORGE CHARALAMBOUS
St. Louis, Missouri

GEORGE INGLETT
Peoria, Illinois

1983

ACADEMIC PRESS, INC.
(Harcourt Brace Jovanovich, Publishers)

ORLANDO SAN DIEGO SAN FRANCISCO NEW YORK LONDON
TORONTO MONTREAL SYDNEY TOKYO SÃO PAULO

INSTRUMENTAL ANALYSIS
OF FOODS

Volume 2

Academic Press Rapid Manuscript Reproduction

Proceedings of the Symposium of the 3rd International
Flavor Conference Held at Corfu, Greece,
July 27–30, 1983

ACADEMIC PRESS, INC.
Orlando, Florida 32887

United Kingdom Edition published by
ACADEMIC PRESS, INC. (LONDON) LTD.
24/28 Oval Road, London NW1 7DX

Library of Congress Cataloging in Publication Data

International Flavor Conference (3rd : 1983 : Corfu,
    Greece)
    Instrumental analysis of foods.

    Proceedings of the 3rd International Flavor Conference,
Corfu, Greece, July 27-30, 1983.
    Includes index.
    Contents: v. 2. Recent progress.
    1. Food--Analysis--Congresses.   2. Instrumental
analysis--Congresses.   3. Flavor--Congresses.
I. Charalambous, George, Date   II. Inglett, G. E.,
Date   III. Title.
TX545.I57 1983   664'.07   83-11756
ISBN 0-12-168902-6 (alk. paper)

PRINTED IN THE UNITED STATES OF AMERICA

83 84 85 86   9 8 7 6 5 4 3 2 1

CONTENTS

CONTRIBUTORS

Numbers in parentheses indicate the pages on which the authors' contributions begin.

Saboor Ahmad (27,33), *Analytical Chemistry Section, Regional Agriculture and Water Research Center, Ministry of Agriculture and Water, Riyadh, Saudi Arabia, c/o USREP/JECOR, APO, New York 09038*

S. Al-Salem (21), *Analytical Chemistry Section, Regional Agriculture and Water Research Center, Ministry of Agriculture and Water, Riyadh, Saudi Arabia, c/o USREP/JECOR, APO, New York 09038*

Dimitrios Apostolopoulos (51), *Department of Food Science, Cook College, Rutgers University of New Jersey, New Brunswick, New Jersey 08903*

C. Ariño (493), *Department of Analytical Chemistry, Facultad de Química, Universidad de Barcelona, Barcelona, Spain*

Itamar Ben-Gera (197), *Wenger International, Inc., Wenger Manufacturing Inc.*

Isario Bertuccioli (409), *Istituto di Industrie Agrarie, Universitá degli Studi, S. Costanzo, I-06100 Perugia, Italy*

André Bossard (307), *Laboratoire de Recherches, Martini & Rossi, Saint-Ouen, France*

M. C. ten Noever de Brauw (335), *Division for Nutrition and Food Research TNO, Institute CIVO-Analysis TNO, Zeist, The Netherlands*

Felix Buccellato (165), *Custom Essence, Inc., Somerset, New Jersey 08873*

D. R. Burgard (173), *The Procter and Gamble Company, Miami Valley Laboratories, Cincinnati, Ohio 45247*

Maria Dolores Cabezudo (357), *Instituto de Fermentaciones Industriales, CSIC, Madrid, Spain*

R. S. Carpenter (173), *The Procter and Gamble Company, Miami Valley Laboratories, Cincinnati, Ohio 45247*

E. Casassas (493), *Department of Analytical Chemistry, Facultad de Química, Universidad de Barcelona, Barcelona, Spain*

Griansak Chairote (119), *Université des Sciences et Techniques du Languedoc, Centre de Génie et Technologie Alimentaires, Place E. Bataillon, F-34060, Montpellier Cédex, France*

William Tsai-Fau Chiu (41), *Taiwan Tea Experiment Station, Taoyuan, Taiwan, R.O.C.*

Jean Crouzet (119), *Université des Sciences et Techniques du Languedoc, Centre de Génie et Technologie Alimentaires, Place E. Bataillon, F-34060, Montpellier Cédex, France*

Cynthia R. Culbertson (187), *The Procter and Gamble Company, Miami Valley Laboratories, Cincinnati, Ohio 45247*

Shun'ichi Dosako (209), *Technical Research Institute, Snow Brand Milk Products Company Ltd., 1-1-2 Minamidai, Kawagoe, Saitama, 350 Japan*

G. Doxastakis (219), *Department of Food Science, Queen Elizabeth College, University of London, London, England*

Michael A. Dressel (243), *Department of Animal Science and Agricultural Biochemistry, University of Delaware, Newark, Delaware 19711*

Lorenz Engel (435), *Bundesforschungsanstalt für Rebenzüchtung Geilweilerhof, D-6741 Siebeldingen, West Germany*

K. I. Ereifej (237), *Department of Food Science and Human Nutrition, Michigan State University, East Lansing, Michigan 48824*

M. Esteban (493), *Department of Analytical Chemistry, Facultad de Química, Universidad de Barcelona, Barcelona, Spain*

I. M. Faruq (21), *Analytical Chemistry Section, Regional Agriculture and Water Research Center, Ministry of Agriculture and Water, Riyadh, Saudi Arabia, c/o USREP/JECOR, APO, New York 09038*

S. G. Gilbert (51), *Department of Food Science, Cook College, Rutgers University of New Jersey, New Brunswick, New Jersey 08903*

Apostolos P. Grimanis (323) *Radioanalytical Laboratory, Chemistry Department, Nuclear Research Center, Demokritos, Athina, Greece*

Helene Hastrich (435), *Bundesforschungsanstalt für Rebenzüchtung Geilweilerhof, D-6741 Siebeldingen, West Germany*

Marta Herraiz (357), *Instituto de Fermentaciones Industriales, CSIC, Madrid, Spain*

M. Jahangir (21), *Analytical Chemistry Section, Regional Agriculture and Water Research Center, Ministry of Agriculture and Water, Riyadh, Saudi Arabia, c/o USREP/JECOR, APO, New York 09038*

I-Ming Juan (41), *Taiwan Tea Experiment Station, Taoyuan, Taiwan, R.O.C.*

George D. Kanias (323), *Radioanalytical Laboratory, Chemistry Department, Nuclear Research Center, Demokritos, Athina, Greece*

G. D. Karaoulanis (519), *Food Technology Institute, Athens, Greece*

Pardul Khan (27, 33), *Analytical Chemistry Section, Regional Agriculture and Water Research Center, Ministry of Agriculture and Water, Riyadh, Saudi Arabia, c/o USREP/JECOR, APO, New York 09038*

Toshiaki Kimura (209), *Technical Research Institute, Snow Brand Milk Products Company Ltd., 1-1-2 Minamidai, Kawagoe, Saitama, 350 Japan*

Werner Knipser (435), *Bundesforschungsanstalt für Rebenzüchtung Geilweilerhof, D-6741 Siebeldingen, West Germany*

Akio Kobayashi (41), *Ochanomizu University, Tokyo, Japan*
Matti Lehtonen (397), *The Research Laboratories of the State Alcohol Monopoly ALKO, SF-00101, Helsinki, Finland*
Pekka Lehtonen (397), *The Laboratory of Rajamäki, Factories of the State Alcohol Monopoly, SF-05200 Rajamäki, Finland*
Peter Liddle (307), *Laboratoire de Recherches, Martini & Rossi, Saint-Ouen, France*
H. Maarse (335), *Division for Nutrition and Food Research TNO, Institute CIVO-Analysis TNO, Zeist, The Netherlands*
Pericles Markakis (237), *Department of Food Science and Human Nutrition, Michigan State University, East Lansing, Michigan 48824*
John M. Mee (21, 27, 33), *Analytical Chemistry Section, Regional Agriculture and Water Research Center, Ministry of Agriculture and Water, Riyadh, Saudi Arabia, c/o USREP/JECOR, APO, New York 09038*
Satoru Mihara (93), *Ogawa & Company Ltd., 6-32-9 Akabanenishi, Kita-ku, Tokyo, Japan*
Gianfrancesco Montedoro (409), *Istituto di Industrie Agrarie, Universita degli Studi, S. Costanzo, I-06100 Perugia, Italy*
Manuel G. Moshonas (137), *U.S. Citrus and Subtropical Products Laboratory, United States Department of Agriculture, Winter Haven, Florida 33883*
Steven Nagy (149), *U. S. Citrus and Subtropical Products Laboratory, United States Department of Agriculture, Winter Haven, Florida 33883*
Osamu Nichimura (93), *Ogawa & Company Ltd., 6-32-9 Akabanenishi, Kita-ku, Tokyo, Japan*
Toshiteru Ohba (375), *National Research Institute of Brewing, Jozo Shikenjo, 2-6-30 Takinogawa, Kita-ku, Tokyo 114, Japan*
E. D. Paneras (519), *Food Science and Technology Department, University of Thessaloniki, Greece*
Vassiliki Pattakou (479), *Chemistry and Technology Laboratory, Cereals Institute, Thessaloniki, Greece*
D. R. Patton (173), *The Procter and Gamble Company, Miami Valley Laboratories, Cincinnati, Ohio 45247*
Donald R. Petrus (149), *U. S. Citrus and Subtropical Products Laboratory, United States Department of Agriculture, Winter Haven, Florida 33883*
S. C. Pillai (463), *CIERS Research and Consultancy Private Limited, 340 Sampige Road, Malleswaram, Bangalore, 560003, India*
María Carmen Polo (357), *Instituto de Fermentaciones Industriales, CSIC, Madrid, Spain*
Adolf Rapp (435), *Bundesforschungsanstalt für Rebenzüchtung Geilweilerhof, D-6741 Siebeldingen, West Germany*
Pascal Ribéreau-Gayon (305), *Director of the Institute of Oenology, University of Bordeaux II, France*
Freddy Rodriguez (119), *Université des Sciences et Techniques du Languedoc, Centre de Génie et Technologie Alimentaires, Place E. Bataillon, F-34060, Montpellier Cédex, France*

Pedro A. Rodriguez (187), *The Procter and Gamble Company, Miami Valley Laboratories, Cincinnati, Ohio 45247*

Galen Rokey (197), *Wenger International, Inc., Sabetha, Kansas*

Makoto Sato (375), *National Research Institute of Brewing, Jozo Shikenjo, 2-6-30 Takinogawa, Kita-ku, Tokyo 114, Japan*

J. Schaefer (335), *Division for Nutrition and Food Research TNO, Institute CIVO-Analysis TNO, Zeist, The Netherlands*

Souleymane Seck (119), *Université des Sciences et Techniques du Languedoc, Centre de Génie et Technologie, Alimentaires, Place E. Bataillon, F-34060, Montpellier Cédex, France*

Philip E. Shaw (137), *U. S. Citrus and Subtropical Products Laboratory, United States Department of Agriculture, Winter Haven, Florida 33883*

P. Sherman (219), *Department of Food Science, Queen Elizabeth College, University of London, London, England*

Takayuki Shibamoto (93), *Department of Environmental Toxicology, University of California at Davis, Davis, California 95616*

Mark R. Silver (243), *Department of Animal Science and Agricultural Biochemistry, University of Delaware, Newark, Delaware 19711*

P. Slump (335), *Division for Nutrition and Food Research TNO, Institute CIVO-Analysis TNO, Zeist, The Netherlands*

Oak B. Smith (197), *Wenger International, Inc., 1 Crown Center, Kansas City, Kansas*

Toshimaro Sone (209), *Technical Research Institute, Snow Brand Milk Products Company Ltd., 1-1-2 Minamidai, Kawagoe, Saitama, 350 Japan*

M. K. C. Sridhar (463), *Department of Preventive and Social Medicine, University of Ibadan, Ibadan, Nigeria*

Keiko Tachiyama (41), *Ochanomizu University, Tokyo, Japan*

Makoto Tadenuma (375), *National Research Institute of Brewing, Jozo Shikenjo, 2-6-30 Takinogawa, Kita-ku, Tokyo 114, Japan*

Kojiro Takahashi (375), *National Research Institute of Brewing, Jozo Shikenjo, 2-6-30 Takinogawa, Kita-ku, Tokyo 114, Japan*

A. C. Tas (335), *Division for Nutrition and Food Research TNO, Institute CIVO-Analysis TNO, Zeist, The Netherlands*

George Thoukis (455), *E. & J. Gallo Winery, Modesto, California 95353*

Hisayuki Toda (93), *Ogawa & Company Ltd., 6-32-9 Akabanenishi, Kita-ku, Tokyo, Japan*

Maria Vassilaki-Grimani (323), *Radioanalytical Laboratory, Chemistry Department, Nuclear Research Center, Demokritos, Athina, Greece*

J. Velísek (335), *Institute for Chemical Technology, Prague, Czechoslovakia*

G. Kasi Viswanath (463), *NCITR/CNTE, School of Engineering and Applied Science, University of California, Los Angeles, California*

E. Voudouris (479), *University of Ioannina, Ioannina, Greece*

Kenji Yamaguchi (93), *Ogawa & Company Ltd., 6-32-9 Akabanenishi, Kita-ku, Tokyo, Japan*

Tei Yamanishi (41), *Ochanomizu University, Tokyo, Japan*

C. Zervos (1), *Pharmaceutical Research and Testing, NCDB, FDA, 200 C Street, S. W., Washington, D.C. 20204*

John P. Zikakis (243), *Department of Animal Science and Agricultural Biochemistry, University of Delaware, Newark, Delaware 19711*

S. S. Zwerdling (173), *The Procter and Gamble Company, Miami Valley Laboratories, Cincinnati, Ohio 45247*

PREFACE

The taste and aroma of foods and beverages remain of the utmost importance to growers, processors, manufacturers of analogs and substitutes of natural substances, brewers, distillers, bakers, confectioners, dairy product manufacturers, and, ultimately, the consumer. Regardless of safety and nutritional characteristics, as well as the other numerous desirable attributes of a successfully marketed food or beverage, acceptable flavor, taste and aroma, is likely to remain its single major asset.

Flavor whether natural, artificial, or a combination, is subject to many vagaries, both biological and nonbiological. Ageing processes, interactions between various components, interactions with packaging materials, storage conditions (time/temperature considerations), all can and do affect the all-important shelf life of foods and beverages.

Changes in taste and aroma of foods and beverages are best determined analytically. Modern instrumentation has achieved great strides and is almost universally employed to this effect—always remembering to compare results with an organoleptic evaluation.

The theme of the 3rd International Flavor Conference held at Corfu, Greece on July 27–30, 1983 was "Instrumental Analysis of Foods and Beverages: Recent Developments." The conference, held under the auspices of the Hellenic Republic Ministry of Agriculture, was cosponsored by the Agricultural and Food Chemistry Division of the American Chemical Society, the Institute of Food Technologists, and The Society of Flavor Chemists, Inc. It was cohosted by the Food Chemistry Department of the University of Ioannina, the Cereal Institute of Thessaloniki, and the Greek Institute of Food Scientists.

Recent findings of over one hundred scientists and food technologists from over twenty countries were reported: food flavor, food quality, food packaging; water, tea and coffee, wine and distilled spirits; dairy products; and fruit juices. They comprise the proceedings of the conference in two volumes, one devoted to the instrumental analysis of foods and the other to that of beverages—published at about the time of the conference.

A great variety of topics was covered, with the emphasis on the most recent developments in instrumental analysis, and both volumes should be most useful as up-to-date, between-two-covers, comprehensive research and technology reports that would save considerable search time.

Thanks are due to all who attended and participated in this conference: to Professor Pascal Ribéreau-Gayon, Director of the Institute of Oenology at Bordeaux, France, and Professor at the University of Bordeaux, for contributing an introduction to the important section on wines and spirits in this volume; and to Academic Press for their unfailing guidance and helpful assistance.

CONTENTS OF VOLUME 1

xvii

INSTRUMENTAL ANALYSIS
OF FOODS

Volume 2

FOOD SAFETY: INTERFACE BETWEEN LEGAL REQUIREMENTS AND ANALYTICAL POSSIBILITIES

C. Zervos

Pharmaceutical Research and Testing
National Center for Drugs and Biologics
U. S. Food and Drug Administration

In modern, developed, industrial societies foods and food items serve social purposes that go far beyond sustenance. Esthetics, nutritional value, economics of production and the obvious or subtle issues known by the collective name of "food safety" are all important aspects of maintaining an adequate food supply (1). Thus, there are many reasons to analyze what we drink and eat. We now analyze the composition of foods, beverages and their ingredients for the purposes of maintaining or improving the esthetic standards of food; maintaining or improving the nutritional value of food; improving the economics of food production; and ascertaining food safety, i.e., that foods, beverages and their ingredients are safe for human consumption.

The scientific and technical aspects of any one of these analytical activities frequently are the exclusive subject of conferences like this one. Consequently, I was uncertain what specific regulatory theme to emphasize.

However, I had a chance to review the abstracts briefly and realized that the theme of this conference is to provide a

ISBN 0-12-168902-6

panoramic and kaleidoscopic view of a very broad subject. Its
purpose is to make aware to all those who deal with instrumental analyses of foods and beverages in any way, of the
problems, frustrations and progress of fellow workers in the
field.

Most of the papers presented in the past few days dealt
with modern techniques of instrumental analysis in the con-
text of food production economics and of the nutritional and
esthetic values of food. A smaller number of papers dealt with
those analytical aspects of food production and processing that
pertain to the need to assess and ascertain food safety. Predictably, this paper will be one of the latter. In what follows I will show you some of the frustrations, the problems
and the progress in analytical regulatory science.

Analytical methods used as tools for assessing and ascertaining food safety must have a number of features that are
not commonly required in food technology. This is so because
regulatory methods must mesh what the law requires with what
analytical science and technology can provide. I will present
and discuss some of these aspects. For background I have
selected a well known food safety problem: the protection of
public health from carcinogenic animal drug residues.

It has been known since ancient times that, besides its
obvious benefits, food engenders risks. It has also been well
understood, accepted and, perhaps more than that, expected
that it is the function of government to engage in activities
designed to manage, ameliorate or eliminate the risks from food
consumption. Hutt, a former Chief Counsel for the U. S. Food
and Drug Administration (FDA), has given a brief but excellent
review of the evolution of food legislation in the western
world (2). Apparently, concern about the safety of food and
the related government activities to ascertain safety, can be
traced back to ancient times.

In the U. S. the basic food law, the current Food, Drug
and Cosmetic Act (to be henceforth refered to as the Act), is
designed to deal, among other things, with the risks inherent
in the production, processing, transport and trade of food
items in a modern industrialized society. The statutory scheme
of the law recognizes that some food-related items are inherent-
ly hazardous and others are not. Accordingly the Act authorizes
the U. S. FDA to engage in a series of risk management activities that are appropriate for the different classes of covered food items. Table I shows the range of authorized risk
management activities for food.

The provisions of the Act that cover food and food items
are cast in the framework of safety. Specifically, the central legal requirement is that to be offered for interstate
commerce all food items covered by the law must be proved "safe".

TABLE I. Mandated Risk Management Activities of the FDA According to Food Item Class[a]

	Animal drugs	Food ingredients	Food contaminants	Food additives	GRAS substances
Premarket clearance	Yes	No	No	No	No
Field inspections and investigations	Yes	Yes	Yes	Yes	Maybe
Response to emergencies	Yes	Yes	Yes	Yes	Yes

[a]Authorized risk-management activity

Safety of course is an elusive concept that defies easy and
universal definitions (3). Because of this it is virtually
impossible for the U. S. FDA to proceed with the task of pro-
tecting the public health from food-borne hazards without developing and adopting: appropriate operational definitions of
safety; procedures that specify how to establish safety; and
specific criteria that indicate when covered food items can be
considered "safe".

During the more than 75 year history of the Act and the
FDA (4) analytical science and analytical methodology have
played an important role in the development of appropriate
operational definitions of food safety and in the establishment of related criteria. This role has been so central that
often one observes a strong tendency on the part of many, even
experts to forget or ignore that scientific disciplines other
than analytical chemistry can be and are used for the purpose.
Be that as it may, the operational definitions and the criteria
of food safety that are based on analytical considerations are
in a sense anchored on one or more of the attributes that characterize analytical methods i.e., accuracy, precision, specificity, and so forth. Therein lies the first if not the largest
source of frustration and problems for those who must mesh legal
requirements and analytical potentialities to develop operational definitions and criteria for food safety. I believe that
most would agree that the number of attributes needed to characterize an analytical method is small, say 4 or 5. However
there is no agreement as to which attributes are fundamental,
which are derived, and how they should be expressed and used
for regulatory work (5). The cause of this confusion is that
analytical methods are developed and used for different purposes. Accordingly, the emphasis of development is placed on
one or the other attribute of an analytical method, depending
on the purpose for which it is developed. Quite often what
might be considered primitive or fundamental attributes of a
method are inconvenient for a specific purpose. In many instances it is more convenient to develop and emphasize derived
attributes. Often through use, the latter assume fundamental
importance in the minds of users with consequent difficulties
in communications even among experts.

In the minds of many the forensic use of analytical methods
is more pristine compared to other uses. Nevertheless, people
in regulatory work are usually faced with a stable of definitions for method attributes of some use, interest and importance. But when it is necessary or preferable to use analytical methodology as the foundation of operational definitions
or criteria of food safety, derived method attributes often
become a hindrance because either do not measure up to the task
or they have obscure origin and meaning. In the case of car-

cinogenic animal drug residues in foods, attempts were made to
peg the operational definition of safety on "method sensitivity",
a derived attribute with obscure origin and meaning that was un-
suitable for the task at hand (6). Those who worked on the case
belatedly realized a need to consider more fundamental attributes. I am getting ahead of myself, though, and I should
start from the beginning of the story to examine in some depth
how analytical methodology in general and specific attributes
of analytical methods in particular played a fundamental and
pioneering role in the development of important operational definitions of safety and of criteria associated with the management of potential risks from animal drug residues in foods.

In the U. S. the case of animal drug residues in foods has
acquired considerable notoriety. It has been discussed at
length and has also been the subject of lengthy litigation and
public policy debate. The outlines, and in some instances the
specific scientific aspects of the case, have periodically been
of great concern to scientists and laymen alike (7). I do not
believe that the case has acquired similar or parallel notoriety
in other countries. I am aware, however, that it has received
some attention because of some recent reports of trade difficulties between countries of the European Economic Community.

At the center of the case of animal drug residues has always been the perennial question of how to deal with the presence of traces of potential carcinogens in foods and drinks,
i. e., how to protect the public health from such substances
adequately without at the same time putting unnecessary and
wasteful restraints on trade, technology, and innovation (8).

The legislative policy about carcinogens in foods, man-
made or natural, is deceptively simple. Even trace amounts of
such substances are bad and are therefore to be avoided whenever and wherever practicable. Table II shows, in laymen's
terms, the specific congressional guidance to FDA concerning
carcinogens in foods according to the class of food related
items.

Traditionally drugs intended for food animals have been
viewed by the FDA in two different ways. One reflects concern for the health and the welfare of the animals themselves.
The other reflects concern for human health which is the corollary to the commonplace notion that drugs administered to
food producing animals are potential sources of drug residues
in food for human consumption. Concern has always been height-
ened by the other simple corollary, that some such residues,
especially those from carcinogenic drugs, might be carcinogens
themselves.

The simplicity of the congressional guidance notwithstanding, handling carcinogens or potential carcinogens in foods has
always been a very confusing issue for the FDA, the Congress,

TABLE II. Congressional Guidance to FDA Concerning Carcinogenic Food Items

Authorized FDA actions	Food ingredients	Food contaminants	Food additives	Animal drugs	GRAS substances
Eliminate avoidable cancer risks	X[a]	X		X	
Proscribe the use of carcinogenic item			X		X

[a] FDA authority not entirely clear.

and for the public health protection establishment of the U.S.
(9). The case of potentially carcinogenic animal drug residues in human foods exemplifies this confusion. After the
Food Additive Amendments to the Act were passed by Congress
in 1958 animal drugs were considered by the FDA to be direct
food additives. Accordingly, the FDA did not permit use of
carcinogenic or potentially carcinogenic substances in food
animals. The prohibition was a direct consequence of the fact
that the amendments contained the now famous or infamous Delaney clause which proscribes the addition to food of any substance found to cause cancer in man or animals.

Soon after passage of the amendments, however, the prohibition of carcinogenic or potentially carcinogenic animal drugs
was seen by many, as unduly restrictive and too severe without
adequately compensating public health protection objectives.
Included among those holding this view was the then Secretary
of the Department of Health, Education and Welfare. From the
available historical record of the Animal Drug Amendments of
1962 one cannot trace back with fidelity the evolution of the
thinking of those who participated in the debate and contributed to the subsequent Congressional decision to make special
provisions for carcinogenic animal drugs. It can be assumed
however that some among the principle actors believed that the
analytical methods of the time could tell whether and when the
residues of a drug had "disappeared" from the flesh of a treated food animal. For instance, writing to Congress on the matter, the then Secretary of Health, Education and Welfare said:

> "There is one respect to which the anti-cancer proviso
> has proved to be needlessly stringent as applied to the
> use of additives in animal feed. For example in the
> case of various animals raised for food production cer-
> tain drugs are used in animal feed which will leave no
> residue in the animal after slaughter or in any food
> product obtained from the living animal and which are
> therefore perfectly safe for man."(10)

Continuing, the Secretary offers a proviso to the Delaney
Clause of the 1958 Amendments which anchors animal drug safety on the ability of analytical methods to determine the presence or absence of drug residues in products derived from food
animals. In support of this proviso the Secretary states that:

> "Under the amendment the assay methods applicable in
> determining whether there will be residues shall be those
> prescribed or approved by us by regulation." (10)

This proposal was finally adopted and became the famous DES
proviso of the anti-cancer clauses in the Food Additive Amendments of 1958. Conceptually, the Secretary's approach appears

to be straight forward at first sight; but its simplicity is
deceptive. Before proceeding to see the difficulties it en-
gendered I will examine briefly whether any among those that
made the proposal and those who accepted it had any idea of
what lay ahead.

The record contains no indication that the involved scientists harbored any concerns about the implications of the Secretary's proposal. It appears, however, that from plain experience with past regulatory activities at least one person
had reservations about it. Representative Sullivan objected
to it during the House debate and it is significant to consider
the reasons for her objection. She reminded her colleagues in
the House that the use of DES, a known carcinogen, in chickens
had been regarded as safe because at first it was found to leave
no residue in human food. Later, however, such use had to be
proscribed by FDA because a better and more "sensitive" method
of analysis had been developed and could detect DES residues in
chicken meat (11). Thus, even though scientists were apparently unconcerned, there was at least one person, who, because of
a commonplace experience could not make the logical jump from
the experimental observation: "no residue could be found" to
the legally required conclusion: "there is none there".

Even though her skepticism was founded on empirical knowledge and common sense rather than theoretical consideration
of the potentialities of analytical methods Representative
Sullivan was skeptical about the ease with which such potentialities could be meshed with the requirement of the law.

As it is oftentimes the case the skeptic's doubts were set
aside ostensibly on the grounds of a higher principle. In the
case of the DES proviso it was the need to be equitable to all
producers of DES implants for growth promotion use in food animals. Apparently, some of them had obtained FDA approval prior
to the date of passage of the 1958 amendments and their product were therefore "grandfathered". Others wished to begin
producing implants after the date and therefore came under the
provisions of the new amendments. The need for equity was
found greater than the need to recognize the limitation of
analytical science. Accordingly, the skeptics were given the
empty and perhaps irrelevant assurance that the full vigor of
consumer protection afforded by the Delaney Clause itself would
indeed be preserved and the DES proviso became law.

There is a lesson of course in the history of the DES proviso to the Delaney Amendments of the Act. It concerns the
interface between science and the law. It is I believe: that
the scientific and legal paradigms are essentially incompatible,
the former yielding only tentative inferences of well defined
scope while the latter requires sweeping and firm conclusions;
and that, for whatever reason, ignoring, or papering over, this

essential incompatibility is a sure invitation to regulatory difficulties.

But the latter part of this lesson will become apparent from what happened after the DES proviso was adopted.

Let us begin by noting that the proviso does two things in terms of determining and ascertaining the safety of food animal drugs. First, it defines "safety" operationally in terms of the results of chemical measurement. Specifically it declares that safe conditions of use of an animal drug shall exist if analysis by a method approved by the Secretary, shows no residues of the drug in products from treated food animals.

Second, the DES proviso implicitly specifies a criterion for deciding whether the conditions of safe use of a carcinogenic animal drug are actually met. This criterion is the lowest concentration of residues of the drug in the food animal products that can be measured by the method approved by the Secretary. Essentially this criterion constitutes a practical definition of "Zero" and on that definition hinges FDA's interpretation of the proviso.

At first sight the approach seems perfectly good. Adoption and use of an analytical method as an anchoring device for the operational definition of safety appears to be a handy solution to a very complex policy problem. It meets a basic requirement that public health protection efforts by FDA be based on facts. Science is the source of facts. Chemical and biochemical analyses are based upon sciences. They provide concrete scientific facts. What better choice then, than analytical science to serve as the base that supports regulations and regulatory activities designed to protect the public health from risks in the food supply? What actually happened with this choice after passage of the DES proviso to the 1958 Food Additive Amendments to the Act is described succinctly in the sensitivity of the method (SOM) regulation published by the FDA in 1979 (12). It states:

"The enactment in 1962 of the so-called DES proviso to the Delaney Clause has been a source of continuing controversy. There is no unanimity on the proper interpretation of the proviso and the legislative history of the proviso summarized above does not lay to rest all doubts."

A fraction of this controversy should be disregarded because lawyers will always argue among them what Congress really meant and what it didn't when it passed a specific piece of legislation. Also it should be assumed that the DES proviso was not meant by Congress to be a dead letter, i. e., that the Congress meant for the FDA to approve the safe uses of carcinogenic animal drugs. After that it becomes instructive to consider FDA's interpretations of the proviso, their characteristics, which among them was adopted, when, and for what reasons.

After passage of the proviso FDA had the task to find a way
of slecting the analytical methods on which to peg the operational definition of safety for carcinogenic or potentially
carcinogenic animal drugs. There are three alternatives:

Alternative #1: At the time of each decision the FDA could
choose the best available method of analyzing food animal products for residues of the substance being considered for veterinary uses. The Agency could then define the legal requirement
"no residues" or "zero residues" to mean the lowest limit of
reliable measurement of the method and anchor on it the operational definition of safety for the substance.

Alternative #2: Periodically the FDA could survey the field
of analytical science in general or specifically the science of
analysis of drug residues in food animal products. It could
determine the analytical potentialities therein and anchor the
operational definition of safety for all carcinogenic animal
drugs on a definition of "zero" that corresponds with what can
be achieved by, the best practical analytical technology; the
best available analytical technology; or the best attainable
analytical technology.

Alternative #3: Finally the FDA could choose a cancer risk,
say one in one million, it would consider acceptable; take into
account that different substances have different potentials to
induce cancer; and anchor the operational definition of safety
on a definition of "zero" that is either: appropriate for the
most potent carcinogen known; or appropriate for the carcinogenic potential of the substance under consideration (13).

The choice before the FDA has not been easy. Obviously,
the Agency has always wished to choose the best among the alternatives. But the overall quality of each and their relative
merits depend on a number of factors. Most important among
them are:

. *Residual Risk,* i.e., whether and to what relative degree
the alternative, if adopted, would diminish the risk of human
cancer from carcinogenic drug residues in food animal products.

. *Equity,* i. e., whether, under the alternative, the strin-
gency of the requirements for collection of scientific information in support of safety is commensurate with the potential
carcinogenic risks from a substance considered for veterinary
uses.

. *Administrative Efficiency,* i. e., whether, under the alternative, administration of the proviso would require resources
in appropriate proportions with the FDA budget and commensurate
with the resources required by the other missions and objectives
of the Agency.

. *Legality*, i. e., whether the alternative is in any way contrary to the provisions of the Act or other applicable statutes.

. *Definitional Clarity*, i. e., whether and to what relative degree the operational definitions and the criteria that derive from the alternative are scientifically and technically obvious.

. *Technical and Scientific Difficulty*, i. e., the relative complexity of the science and the technology needed to support the alternative.

. *Cultural Barriers*, i. e., whether a significant fraction of the population will perceive the alternative to run counter to deeply rooted ethical standards and to the mores of society; and

. *Side Effect*, i. e., whether the alternative will have predictable side effects (beneficial or deterimental consequences other than those intended by the proviso).

Table III shows in a matrical form how the three alternatives score with respect to each of the factors.

The FDA chose from the alternatives on two occasions. Soon after 1962 when the DES proviso became law the FDA seems to have chosen alternative #2. In the early 1970s this choice became increasingly untenable and the FDA changed to alternative #3b. In neither occasion did the Agency use the explicit matrical approach suggested by Table III. However, during the change from alternative 2 to 3b there was implicit consideration of most of the factors listed in Table III and explicit consideration of residual risk, equity, legality and technical complexity.

The first alternative was apparently rejected by the FDA from the start. It is hard to document exactly why and how much though originally went into the decision to do so as a matter of official policy. Unfortunately, prior to the early 1079s the FDA did not explain its decisions in a manner that became customary by 1973 when it began publishing detailed explanatory preambles to policy statements, formal decisions, and proposed or final rules. Once can imagine, however, that the reasons for rejecting it did not change substantially with time and that they are, therefore, similar to those published in 1979 (14) when it was agreed that the first alternative makes no sense because it does not take into account differences in cancer risks posed by the different substances. It was noted earlier that analytical methods measuring a particular substance in a matrix of other substances are developed for different purposes. According to the purpose, the emphasis is placed into one or the other specific attribute of the method. The lowest limit of reliable measurement and specificity are the aspects of most concern to scientists developing methods to measure the concentration of potentially carcinogenic residues in food animal products. From the public health pro-

TABLE III. Implementation of the Deproviso of the FD&C Act[a]

| | | | Character |
Alternatives	Residual risk	Equity	Administrative efficiency
Alternative #1			
Chose the best available method	High	Low	High
Alternative #2			
Define common "zero" according to:			
(a) best practical technology	2a 1[b]		
(b) best available technology	2b 2a	Moderate	Moderate
(c) best attainable technology	2c 2b		
Alternative #3			
Define "zero" appropriate for:			
(a) most potent carcinogen	2a 2b	Moderate	Moderate
(b) the carcinogenicity of the special drug	3a 2c	High	Low

[a] The three alternative definitions of "zero residue" and

[b] The meaning of the expression 2a. 1 is that overall resi
1. Analogous expressions have similar meaning.

[c] Tentative, events might prove otherwise.

istics

Legality	Cultural barriers	Definitional clarity	Technical complexity	Side effects
No obvious reason to question (Courts might decide other-wise)	Unknown	High	Low	Some
			Medium	
	Unknown	Moderate	High	Some
			High	
	Unknown	Moderate	High	None[b]
	Some	Low	High	

their characteristics.

dual risk from alternative 2a is less than that from alternative

tection perspective this is understandable. The FDA wishes to
know with certainty the concentration of the specific residues
in food. When it can no longer measure them the Agency wishes
to know the maximum concentration that could go into food undetected. However methods ordinarily developed by animal drug
manufacturers have as their most likely purpose the analysis
of the content and the purity of drugs in bulk and in dosage
forms. Very rarely are they concerned with the lowest limit
of reliable measurement. Most often the development emphasis
goes to the attributes of precision, accuracy and speed of
measurement and not on matching the lowest limit of reliable
measurement with carcinogenic risks which are acceptable.
Thus, without the technology forcing features envisioned by
alternative #2 or #3, the kind of methods that would be available for the control of animal drugs residues in food animal
products would likely run afoul of the requirements for risk
reduction and equity envisioned by any reasonable interpretation of the DES proviso.

Alternative #1 predictably fares very well with respect to
all the other factors except one: side effects. One could
visualize two competing firms selling each a different carci-
nogenci animal drug approved for similar indications on the
basis of available analytical methods for the determination of
their residues in food animal products. One could also visualize these two competitors trying to gain advantage in the market by improving each other's analytical methods and demanding
from FDA the actions required by the law. Such events would
likely place the FDA in an extremely precarious position. In-
cidentally, if there is doubt that such events could happen
they would disappear upon examination of the imaginative uses
entrpreneurs have found for the Freedom of Information Act (16).

Alternative #2 appears to have been the FDAs first choice
after the proviso became law. At first glance it is reasonable
to anchor the operational definition of safety for animal drugs
on a definition of "zero" based on the potentialities of analytical science. Also, in 1962, when the FDA first faced the
need to make a choice, the prevailing ideas about quantitative
risk assessments for carcinogens were to say the least embryo-
nic. Thus, absent the risk assessment tools which are necessary for alternative #3, alternative #2 is the only one the
FDA could choose after rejecting alternative #1.

Alternative #2 defines "zero residues" on the basis of the
current status of analytical science. There are three ways
this can be done. The FDA, for instance, can choose for the
definition the best practical analytical technology at the time
of decision (alternative 2a). The required lowest limit of
reliable measurement would be set at a level that would be
attainable by anyone in the business.

Alternately the FDA could base the definition of zero on what it considers the best available technology at the time of decision (alternative 2b). Presumably the determination of what is best available technology would require proper studies of the field of analysis and some very difficult decisions. Obviously certain animal drug producers would be required to upgrade their analytical capabilities.

Finally, the FDA could choose for the definition the best attainable analytical technology at the time of decision (alternative 2c). In that instance the lowest limit of reliable measurement required would be determined by the potentialities of analytical science at the time of decision and not by the average or the best a specific industry can offer. Besides proper studies of the field of analysis determination of what is best attainable technology would require some very difficult judgments on the part of the FDA. The consequence of selecting alternative #2c would be that all or nearly all animal drug producers would be required to upgrade their analytical capabilities.

It is very difficult to discern which of the three variants of alternative 2 did the FDA select after the proviso became law. However, the operational definition of zero for carcinogenic residues in food animal products hovered over 2 parts per billion until the middle 1970s. A retrospective review of the analytical potentialities of the period between 1960 and 1975 suggests that 2.00 ppb was neither the best practical nor the best attainable level. It mus therefore be assumed that the choice was made on what appears to have been the best available analytical technology.

The reasons for the selection of 2 ppb are not clear. Undoubtedly the selection had a technology forcing effect in that it compelled animal drug producers to develop methods with emphasis on the lowest limit of reliable measurement instead of other method attributes. Consequently, rightly or wrongly, it created the public perception that risks from residues of carcinogenic drugs in food animal products were controlled by the FDA adequately. For that reason alternative #2 remained an acceptable way of implementing the DES proviso for over 10 years as it was perceived to be advantageous relative to alternative #1 the only other feasible alternative until the early 1970s.

It is likely, although one cannot be certain, that the two were considered comparable with respect to all other factors except perhaps definitional clarity, technical difficulty and side effects.

It was noted earlier that "zero" was defined to be 2.00 ppb or less without an official clarification of this definition at the time the standard was set. The Agency did not opine whether 2.00 ppb was the best practical, the best available

or the best attainable technology of the time and why. As a
consequence the standard was criticized as obscure at least
during the early 1970s. Predictably there were also complaints
about the complexity of the technology and science required by
a technology forcing standard. However side effects grew to
be the most serious defect of alternative #2. Beginning with
the early to middle 1960s general scientific and technological
progress propelled analytical science into giant leaps forward. Comparable progress in the area of residue analysis was
undoubtedly motivated by the 2.00 ppb requirement, i.e., the
technology forcing standard of the FDA.

The best known side effect of this progress was continuous
pressure on the FDA both from external and internal sources to
revise the standard to comport with the continuous improvements
of analytical science. Justification for the pressure was the
wording of the proviso which made the use of a carcinogenic
animal drug illegal if its "residues" were "found" in or on
food animal products. There is ample evidence that for some
time the FDA felt helpless because analytical methodology development was too rapid to permit rational regulation of carcinogens in animal drugs (16). Thus, with the development of
our notions about comparative chemical oncology and the con-
comitant evolution of means for quantitative risk assessments
the FDA began to look at alternative #3 as a cure for its carcinogenic animal drug problems.

Alternative #3 either the a or b version can theoretically
tailor the residual cancer risk from the use of carcinogenic
drugs in food animals to be consistent with some level that is
considered acceptable. In that respect it has a telling advantage over alternatives #1 and #2.

With respect to equity alternative 3a appears neither better
nor wors than any of the variants of alternative 2. However,
alternative 3b appears theoretically superior to all others
because it implies that the research and development requirements in pursuit of an appropriate analytical method are dictated by the proposed uses of a substance and by its carcinogenic potential.

Compared to alternatives #1 and #2 the administrative efficiency of alternative 3 would appear to suffer somewhat in
that it requires the FDA to develop and maintain adequate ex-
pertise in a variety of fields.

None of the alternatives appears to contradict the legal
requirements in the proviso, other parts of the Act or other
applicable statutes. Thus, none has an obvious advantage in
that respect relative to the others. However, whether that is
true remains to be established by the Courts of Law.

There has been substantive cultural resistance to alternatives 3a and 3b which can be viewed as setting tolerances for
carcinogens. It stems from widely held notions that one can

always tell carcinogens from non-carcinogens and all one need
do to eliminate risks is to ban the use of the carcinogenic
animal drugs whose residues are found in food animal products.

With respect to relative clarity of definition alternative
#3a is not better than the variants of alternative #2. How-
ever, the complexity of alternative #3b notwithstanding, its
definitional clarity is by far the better in the lot.

With the preceding discussion I hope to make you aware that
it is not as easy as it might appear at first glance to mesh
what science can produce and what the health protection laws
usually require.

I also hoped to convince at least some of you that regulatory science does not end with the preparation of a paper or a
report and that it is much more than that.

For better or worse the FDA has changed, at least in practice, the operational definition of safety required for the
implementation of the DES proviso of the Act. The change has
required a major scientific effort on the part of the Agency.
It is not yet possible to know whether the change will be good
i.e., whether it will result in increased public health protection from carcinogenic drug residues in foods without dis-
rupting the continuity and the vigor of the food supply. The
omens thus far are not encouraging. They don't augur that
past problems will go away or that excessive and unnecessary
rancor between industry, the FDA and consumerist will diminsh.
Quite the contrary the fact that the change was proposed in
1973 and has not been made officially final yet would suggest
that rational regulation of carcinogenic animal drugs has become more difficult with the passage of time and with scientific progress.

The preceding discussion covers only a part, albeit an important part of the scientific problem of the DES proviso of
the Act. The problem has other parts. Among them is the conversion of the lowest limit of reliable measurement, i.e., the
key attribute of the approved analytical method, into a criterion or legal standard for distinguishing adulterated from
safe food. This conversion represents another example of the
difficulties one might encounter in the interface between legal
requirements and analytical possibilities. Time however does
not permit me to describe and analyze it here.

REFERENCES

1. Look for instance at Chapter 10 of "Changing Attitudes and
    Lifestyles Shaping Food Technology in the 1980s" by
    Marilyn Chou in *Critical Food Issues of the Eighties,* M.
    Chou and D. P. Harmon, Eds. Pergamon Policy Studies,

Pergamon Press, New York, Oxford, Toronto, Sydney, Frankfurt, Paris 1979.

2. *The Basis and Purpose of Government Regulation of Adulteration and Misbranding of Food.* P. B. Hutt in FD&C Law Journal, Vol. *33*(10), p. 505 (1978).

3. Scanning the book *"Of Acceptable Risk. Science and the Determination of Safety"* by W. W. Lowrance (Wm. Kaufmann, Inc., Los Altos, CA, 1976) one can glean at the elusive nature of the term "Safety". Lowrance for example is correct in stating that judging safety is equivalent to judging the acceptibility of risks which is an essentially normative function (p. 75). His definition of safety however as "a judgment of the acceptibility of risks" (p. 8) is patently wrong however. No one would call war or mountain climbing "safe" even though societies and individuals respectively judge the risks rherein acceptable.

4. *A Legislative History of the Federal Food, Drug and Cosmetic Act* by Mary Nell Lehnhard, Congressional Research Service, Library of Congress, TX501B, 73-174-ED. *The Protectors. The Story of the Food and Drug Administration* by H. E. Neal, Julian Messner, New York 1968.

5. For example in *Evaluation of Analytical Methods Used for Regulation (J. Assoc. Off. Anal. Chem., 65*(3), 525 (1982)) W. Horwitz discusses a number of the important scientific characteristics of analytical methods. In their nature these characteristics parallel the method attributes discussed by the FDA in *"Chemical Compounds in Food Producing Animals; Criteria and Procedures for Evaluating Assays for Carcinogenic Residues. Federal Register, 44,* 17070 (March 20, 1979). They are however considerably different from the performance criteria discussed by W. Horwitz in *"The Inevitability of Variability in Pesticide Residue Analysis"* in Advances in Pesticide Research, Part 3, H. Geissbuhller (Ed.) Pergamon Press, Oxford and New York 1979.

6. *Chemical Compounds in Food Producing Animals; Criteria and Procedures for Evaluating Assay for Carcinogenic Residues, Federal Register, 44,* 17070 (March 20, 1979) Section VIII. D.5.

7. For example see: *Diethylstilbestrol; Withdrawal of Approval of New Animal Drug Applications; Commissioner's Decision,* in *Federal Register, 44,* 54835 (September 21, 1979; *A Quantitative Evaluation of Estrogens (including DES) in the Diet,* T. H. Jukes in the Amer. Statistician, *36,* 273 (1982), *Regulatory History of DES,* P. B. Hutt, in the American Statistician, *36,* 267 (1982); and *FDA's Ban of DES in Meat Production,* C. Zervos and J. V. Rodrick, in the American Statistician *36,* 278 (1982).

8. See "Study of the Delaney Clause and other Anticancer Clauses" Hearings Before a Subcommittee of the Committee on Appropriations House of Representatives 93[d] Congress Secon Session, Part 8 and specifically the Remarks by Dr. Tepper, p. 43.

9. *Public Policy Issues in Regulating Carcinogens in Foods* by P. B. Hutt in FD&C Law Journal *33*(10), 541 (1978). *Regulation of Cancer Causing Food Additive - Time for a Change?* A report to Congress of the United States by the Comptroller General, United States General Accounting Office, December 11, 1981. HRG-82-3.

10. House of Representatives Report No. 86-1761 (H. R. 7624) Committee on Interstate and Foreign Commerce, 86th Congress, 2nd Session, 1960.

11. Congressional Record, *108,* 19916; September 27, 1982 (Remarks by Representative Sullivan).

12. See Ref. 6 Section I.B.

13. See Ref. 12 Section V.D.

14. See Ref. 12 Section V.B.1.

15. Hearings before the Subcommittee on the Constitution of the Committee on the Judiciary, U. S. Senate 97th Congress 1st Session (Serial No. - J-97-50 and specifically remarks by Senator Dole (Vol. 1. p. 2).

16. See Ref. 8 pp 2-8.

ANALYSIS OF CARBONATES AND BICARBONATES IN BOTTLED WATERS BY AUTOTITRALIZER

John M. Mee[1]
M. Jahangir
I. M. Faruq
S. Al-Salem

Analytical Chemistry Section
Regional Agriculture & Water Research Center
Ministry of Agriculture & Water
Riyadh, Saudi Arabia

I. INTRODUCTION

Carbonates $(CO_3^=)$ and bicarbonates (HCO_3^-) are important anions which may contribute up to 85% of the inorganic salts in total dissolved solids of the bottled waters.

In water analysis, $CO_3^=$ and HCO_3^- can be measured either by conventional indicator titration method, or potentiometric titration technique dependent on pH or hydrogen ion activity or alkalinity of the test water samples.

From a given water sample, both alkalinity $(CO_3^=$ and HCO_3^- of lime, magnesia, sodium and potassium, etc) and pH interact and manipulate water chemistry to certain extent which may significantly affect the water quality, taste, flavor and the 'body' of the drinking water (1).

This paper reports the use of an automated titration system for rapid determination of $CO_3^=$ and HCO_3^- in domestic and imported bottled waters. Assessments of the autotitralizer system used in water analysis was reported elsewhere (2).

---

[1]Mailing address: USREP/JECOR/USDA, APO NEW YORK 09038

Instrumental Analysis of Foods
Volume 2

21

II. RESULTS AND DISCUSSIONS

Fig. 1 (A & B) shows the instrumental calibration curves established for $CO_3^=$ and HCO_3^- analysis by the Autotitralizer instrument (Fisher M-381) from which linearity was attained within each working concentration range of the anions. The end point of titration for $CO_3^=$ and HCO_3^-, when against standard 0.01 N sulfuric acid, was pre-selected and set at pH 8.3 for $CO_3^=$ and pH 4.5 for HCO_3^-. Throughout the auto-analysis, pH values were automatically printing out for the test samples, and also during the first and second end points of titration.

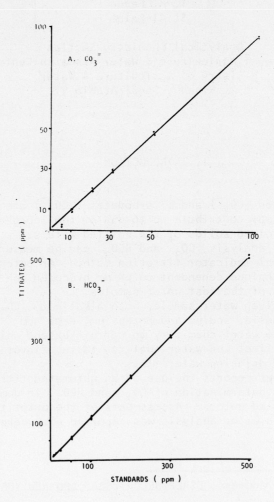

Fig. 1. Calibration curves by Autotitralizer.

Fig. 2 illustrates the autotitration profiles for $CO_3^=$ and HCO_3^-. Thus, the determination limit for $CO_3^=$ is at 10 ppm (mg/L) level and for HCO_3^- is 5 ppm.

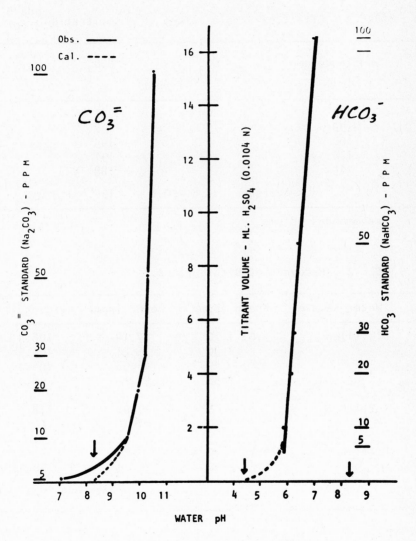

Fig. 2. Autotitration profiles of $CO_3^=$ and HCO_3^- ions.

   The effect of sample volume on titrant values was studied. It was noticed that the autotitralizer requires at least 75 ml of the test sample for the immersion of the electrode and the stirrers during the potentiometric determination.

   Table 1 below recommends the use of 100 ml water sample for each analysis:

TABLE 1. Effect of Sample Volume on Autotitration.

Sample Volume (ml)	$CO_3^=$ (ppm)	HCO_3^-
75	6	201
90	5	197
100 *	5	195
110	4	191
125	4	188
Average	4.8 ± 0.8	194.4 ± 5.2

*Recommended volume.

TABLE 2. Recovery of Bicarbonates.

Water Sample	Known (ppm)	Found (ppm)	Recovery(%)
Distilled	5.03 + 20	27.14	108.4
	+ 30	35.21	100.5
	+ 40	46.98	104.3
	+ 50	54.44	98.9
Bottled A	17.80 + 5	27.02	118.50
	+ 50	67.15	99.04
Bottled B	24.75 + 50	73.33	98.10
	+100	124.11	99.49
		Average	104.1 ± 6.4

Table 2 presents the HCO_3^- recovery data. The autotitration system provides the range of HCO_3^- recovery from 98 to 119% which yields an average of 104% recovery by addition method or spiking technique from the increments of 5 to 100 ppm.

The precision and error of the automatic titration can also be discussed briefly here. Below 10 ppm of $CO_3^=$, the standard deviation was 28% and percentage of error was 85% based on triplicate analysis. In case of HCO_3^-, below 5 ppm the standard deviation was about 10% and error 9%.

TABLE 3. Analysis of Bottled Water for $CO_3^=$ and HCO_3^-.

Sample (domestic & imported)	$CO_3^=$ (ppm)	HCO_3^-
Nissah (S.A.)	—	170
Oasis (S.A.)	—	278
Taiba (S.A.)	—	90
Sohat (Lebanon)	—	24
Dreikiche (Syria)	—	135
Gulfa (UAE)	—	146
Apollinaris (FRG)	—	1677
Peters Val (FRG)	—	960
Perrier (France)	—	364
Penda (U.S.A.)	—	164

Table 3 summarizes the determination of $CO_3^=$ and HCO_3^- in selected domestic and imported bottled waters. Apparently $CO_3^=$ ion was not detectable due to the nature of water pH at neutral or acidic range.

The automatic titration system appears attractive for our water quality protection programs specially for the bottled water available in the Kingdom.

REFERENCES

1. Faruq, I.M. and Mee, J.M. in "Water Resources Development in the Kingdom of Saudi Arabia" Vol 2, P. 364, 1982. Ministry of Planning, Kingdom of Saudi Arabia.
2. Mee, J.M., Al-Salem, S., Jahangir, M. and Faruq, A. in "Pittsburgh Conference, No. 216" 1983. Atlantic City, U.S.A.

RAPID ANALYSIS OF ORGANIC MATTERS IN BOTTLED WATER BY INFRA-RED SPECTROSCOPY

John M. Mee[1]
Pardul Khan
Saboor Ahmad

Analytical Chemistry Section
Regional Agriculture & Water Research Center
Ministry of Agriculture & Water
Riyadh, Saudi Arabia

I. INTRODUCTION

Organic matters in bottled water beverages can cause taste effects as well as physiological effects. Analytical methods vary from the conventional evaporation-burning off-difference by weight to sophisticated chromatographic techniques with modern instruments-data systems such as GC, HPLC and GC-MS. Each method has its own merits and shortcomings.

We have developed a simple and rapid screening test for the detection and, for the best, identification of organics in drinking and bottled water using infra-red spectroscopy.

The analysis is based on $CHCl_3$ extraction and the '$CHCl_3$ Extractables' are scanned by IR spectrophotomer or FT-IR through 200 to 4000 cm^{-1} wavenumber and the absorption bands are recorded.

This paper deals with the nature of '$CHCl_3$ Extractables' via IR spectral data obtained from the drinking and bottled water samples, with or without prior heat treatment on the water samples.

---

[1] Mailing address: USREP/JECOR/USDA, APO NEW YORK 09038.

Instrumental Analysis of Foods
Volume 2

27

II. RESULTS AND DISCUSSIONS

Fig. 1 shows the infra-red spectrum in terms of wavelength or wavenumber(frequency) in relation to other known electro
magnetic spectra such as X-ray, UV, etc., having energy higher
than that of Near IR; and Radio, TV, etc., having energy lower
than that of Far IR.

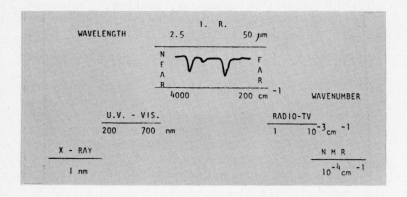

Fig. 1.   Region of Infra-red and Electromagnetic
           Spectrum.

Fig. 2 shows the typical IR spectra of a distilled water
sample taken from $CHCl_3$ extracts of water at ambient temperature and of identical water sample upon heating at 110 $^\circ$C for
one hour prior extraction. IR band of 1350 cm^{-1} was the results of thermal activation of the organics in water.
     The addition of extra IR band upon heating the water at
110 $^\circ$C/1 h is further demonstrated in Fig. 3 from which two
IR bands (1600 and 3680 cm^{-1}) became significantly intensified
due to the thermal activation of organics in bottled water.
     Fig. 4, on the other hand, shows the loss of 1700 cm^{-1}
due to the thermal reaction by heating at 110°C/1 h the second
drinking water.

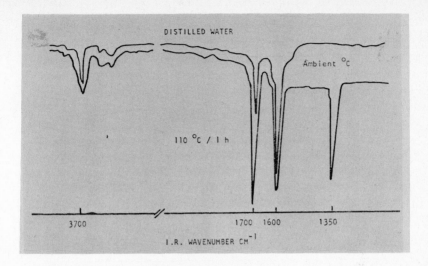

Fig. 2.   Infra-red Spectra of Distilled Water Showing
the Addition of 1350 cm⁻¹ Band by Thermal
Activation.

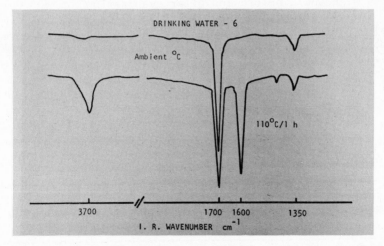

Fig. 3.   Infra-red Spectra of Drinking Water Showing the
Addition of 1600 and 3680 cm⁻¹ Bands by Thermal
Activation.

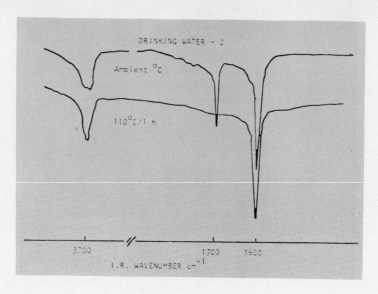

Fig. 4.   Infra-red Spectra of Drinking Water Showing the
          Deletion of 1700 cm^{-1} Band by Thermal Activation.

We further studied the volatility of the organics in
water. We collected eight composite samples (N=8) with a wide
range of salinity (0.6 to 9.1 mMho / cm) and extracted the organics with CHCl$_3$ before and after evaporation at 180 °C.

Table 1 summarizes the results in which about 50% of the
total organics were considered as volatile and the balance
composition could be a combination of synthetic (by thermal
activation) as well as natural occuring non-volatile organic
compounds.

TABLE 1. Volatility of Water Organics-Of-Interest.

Organics/Water	Av. Intensity (1600 cm^{-1})	Composition (%)
Total (Ambient °C)	0.67	100
Non-Volatiles (180 °C)	0.33	49
Volatiles [1]	0.34	51

[1] Total - Non-Volatiles, by difference.

TABLE 2. Library Search for Potential Water Organics by PU SP 3-080 IR-Data System (Water A).

Organics / Water A	Probability Score (1600 and 3680 cm^{-1})
Hydrocarbon:	
Heptane	92
2-Methyl-Butane	92
Octane	92
2-2-Dimethyl-Butane	91
Chlorinated Hydrocarbon:	
CCl_4	91

TABLE 3. Library Search for Potential Water Organics by PU SP 3-080 IR-Data System (Water B)

Organics / Water B	Probability Score (1350, 1600 and 1700 cm^{-1})
Hydrocarbon:	
Butane	92
Hexane	87
Heptane	87
Nonane	87
Hexadecane	87
Chlorinated Hydrocarbon:	
CCl_4	89
Alcohol:	
Decanol	88
Aromatic or Unsaturated N:	
Benzonitrile	87

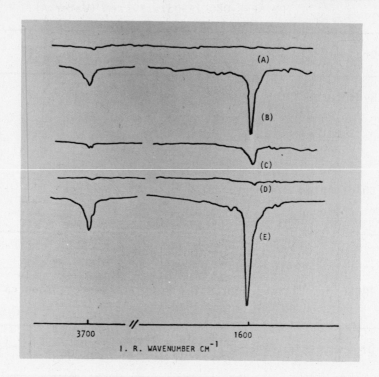

Fig. 5.   Infra-red Spectra of Solvent Background (A);
          CHCl₃ Extracts of Bottled Water, 3 X (B, C, D);
          and ³CHCl₃ Extract of CCl₄ Spiked Water Residue(E).

To search for organics, we applied the extracts via a FT-IR instrument equiped with IR data system PU 3-080.

Table 2 shows the water sample of 2 major IR bands (1600 and 3680 cm⁻¹) and likewise, Table 3 provides data from another bottled water having 3 major IR bands (1350, 1600 and 1700/cm). The print-out data of Table 2 and 3 provide us some information on volatile organics and probability of their presence in the test samples. To verify the FT-IR data, we spiked CCl₄ to the water residue after CHCl₃ extraction and washings, Fig. 5 (E). Thus, identical IR spectra are matched between Spectrum (B) and (E) in Fig. 5.

Non-volatile organics may be traced by different library search dependent on the availability of I.R. data base. On the other hand, we have applied a Direct Mass Spectrometric method for rapid identification and quantification of water non-volatile organics (see this conference, 1983).

APPLICATION OF DIRECT MASS SPECTROMETRY FOR RAPID ANALYSIS OF ORGANICS IN WATER BEVERAGES

John M. Mee[1]
Pardul Khan
Saboor Ahmad

Analytical Chemistry Section
Regional Agriculture & Water Research Center
Ministry of Agriculture & Water
Riyadh, Saudi Arabia

I. INTRODUCTION

Instrumental analysis of trace organics by GC and/or GC-MS is commonly used in environmental chemistry and clinical bio-chemistry. For monitoring water organics, however, the need for a rapid and direct assessment of most, if not all, organic compounds which may be present in the bottled or potable water sample is not readily met by a conventional sample preparation and chromatographic separation, which is laborious and time-consuming; and in many cases of routine analysis, depending on the purpose and objective, the GC-MS results may still be lack-ing of analytical assurance simply due to the diversity of pol-lution, the types and kinds of contaminants, and the limitat-ions of GC operational conditions.

The invention of a micro-analytical technique for the bio-logical active compounds and the principle of handling complex physiological samples amenable to a low resolution MS instru-ment suggests a suitable alternative for organic analysis in water and since chromatographic separation is not required a complete analysis is possible within minutes of analysis time (1-6).

---

[1]Mailing address: USREP/JECOR/USDA, APO NEW YORK 09038.

33

This paper reports the analytical findings on water organics by 'direct mass spectrometry' (DMS) analysis for rapid measurements of organics-of-interest. The quantitative assessment of DMS methodology for the ultimate analysis of water pollutants via an internal standardization technique will be the subject of other communication.

II. RESULTS AND DISCUSSIONS

It may be useful, in the first place, to understand the instrument characteristics and capabilities between GC/MS and DMS.

Table 1 compares the practical aspects for both instrumentation according to objective in which DMS (chemical ionization) is versatile in operation and rapid for specific chemical separation and measurement. In many occassions, separation, detection and quantification of certain compounds could be rapidly resolved by DMS-CI instead GC/MS (1-6).

Table 2 shows the operational requirements and results of DMS-CI designed for organics monitoring program. It should be emphasized, however, that a temperature programmable solid probe with a special data management system is essential features for this purpose.

TABLE 1. Instrument Capability for Water Quality Monitoring

Characteristics	GC/MS	DMS-CI
Separation	Tedious	Rapid
Identification and Quantification	Yes	Yes
Operation	Limited by GC	Versatile

We have monitored the organic molecules in a variety of potable/bottled water samples as well as city and well waters. Examples are illustrated as follows:

Fig. 1 shows the total ion current (TIC) profiles of drinking water (A to C) as compared to that of distilled water. Similar source of potable water yielded identical TIC profiles (A and B) for the total organic content (evaporated) near 100 °C at 1 Torr. However, sample C showed a different pattern at

TABLE 2 Direct MS Operation and Data Availability

1. Selective extraction with organic solvent(s).
2. Analysis by molecular volatility.
3. Operation by temperature programming scans.
4. Data on molecular weight.
5. Profiles of TIC and SIC and integrated values.
6. Computer display, search and print-out.
7. Statistical treatment and data management.

100-200 $^{\circ}$C, and the TIC significantly extended to 300 $^{\circ}$C zone indicating the pollutant intensity as well as the characteristics of "high boiler" organics. On the other hand, the distilled water sample displayed only a narrow TIC distribution at lower temperature or near 100 $^{\circ}$C.

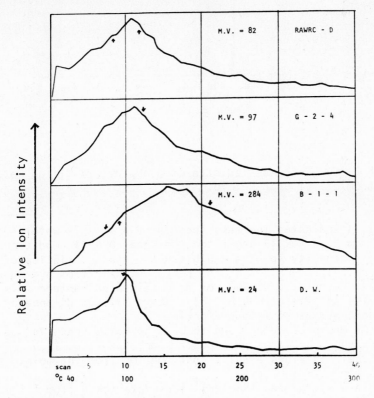

Fig. 1. Direct mass spectrometry (CI) of total organics in water as expressed by total ion current (TIC) during spectral scanning at indicated temperature.

Fig. 2 domenstrates the typical organic compounds found in distilled water sample showing protonated molecular ions (MH)$^{+}$ taken at 100°C from DMS-CI scan No. 10. It is of interest to note that the relative clean spectral profile of TIC provided only few (MH)$^{+}$ peaks such as at m/e 257 and 391 which are commonly recognized as plasticizers.

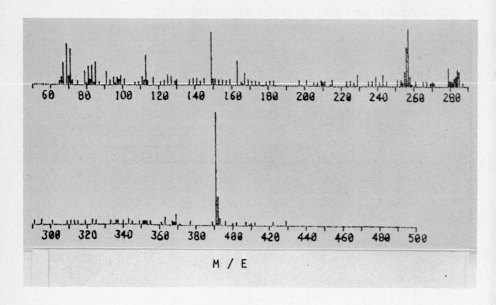

Fig. 2. Organic mass spectra of a typical distilled water taken from DMS-CI Scan NO. 10.

Fig. 3 indicates the mass spectra of a potable water (B) taken from Scan NO. 13 under the identical DMS-CI conditions. The high-intensity spectra exhibited the characteristics of methylene units, -CH$_2$- , of the petrochemicals from hydrocarbons m/e 71, 85 .... to 393, or mono-saturated pantene (M.W. 70) series as well as m/e 69, 83 .... to 391, or di-unsaturated derivatives of isoprene series.

Water organics were found to be "thermally reactive" molecules as shown by infra-red spectroscopic profiles (7,8). DMS-CI again provides a quick confirmation from the molecular distribution point of view.

Table 3 summarizes the (MH)$^{+}$ ions data taken from TIC scan at 350 ^{o}C for both potable water wamples with or without heating at 110 ^{o}C /1 h, in the sealed glass tubes. These organic ions m/e are useful markers which may provide a clue to the mystery of the water resource or of contamination origin.

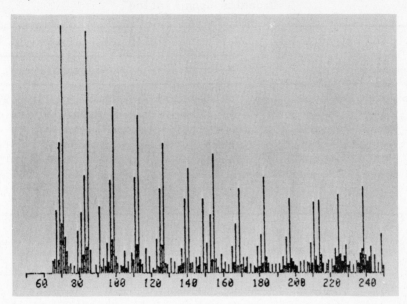

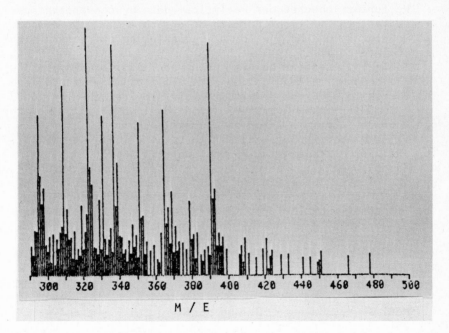

M / E

Fig. 3. Direct mass spectra (DMS-CI) of a potable water sample B, Fig. 1.

TABLE 3. Direct Mass Spectrometry of Water Extracts for Organics upon Heating[a]

CHCl$_3$ Extractables in Sample	Water Organics[b], $(MH)^+$, Probe 350°C	
	Matchable m/e	Unmatchable m/e
1. Potable R	67,69,71,73,75 83,113	85,87,91,93,95,99 110,117,127,<u>141</u>, <u>145</u>,147,149
2. Drinking Well D	67,69,71,77,79 81,83,250	96,106,114,<u>124</u>,<u>141</u> 223
3. Drinking Well M	67,69,79,81,83	65,85,<u>93</u>,99,113,117 149,<u>248</u>,<u>250</u>

[a]Room Temp. vs. 110 °C/1 h. in sealed glass tubes for water samples.

[b]Underlined m/e values to be peaks only in heated samples.

We further checked our analytical data by converting spectral data into Empirical Formula of probable chemical compound via molecular ion $(MH)^+$ information (assuming on chemical ionization mechanism)..

Table 4 presents a partial prepared list, for example, from which one can imagine the possibilities if we could have the access to a "FT-DMS" which rapidly provides chemical data via a computer-based library information in assessing the organic contaminants in water beverages.

It is of particular interest to see the $(MH)^+$ at m/e 117, for example, our selective ion current (SIC) profile actually displayed three (3) separated mass spectral ion peaks representing clearly the different volatilities of the molecular species which were present in the test sample. Accordingly, the SIC profile of $(MH)^+$ at m/e 117 represents at least three or more organic compounds of identical molecular weight 116.

REFERENCES

1. Mee, J. M., Halpern, B. and Korth, J., U.S. Patent 4,224, 031.
2. Mee, J. M., Korth, J. and Halpern, Biomed. Mass Spèctrom. 4:178 (1977).

TABLE 4. A Partial list of Predicted Organics in Drinking
Water Samples Detected by DMS-CI [a]

$(MH)^+$ Peak, m/e	Empirical Formula	Probable Compounds
69	C_5H_8	Isoprene
71	C_5H_{10}	Pantene
73	C_4H_8O	Methylethyl Ketone
85	CH_2Cl_2	Methylene Chloride
87	$C_4H_{10}O$	Diethyl Ketone
95	C_6H_6O	Phenol
117	$C_5H_8O_3$	Methyl Acetoacetate
	$C_6H_{12}O_2$	Diacetone Alcohol
	$C_7H_{16}O$	Heptanol
:	:	:
:	:	:

[a] Probe temp. 350 $^\circ$C.

3. Mee, J. M. and Halpern, B., in "Recent Developments in Mass
   Spectrometry in Biochemistry and Medicine, Vol. 1" (A. Fri-
   gerio, ed.) p. 291 and p. 321. Plenum, London, 1977.
4. Mee, J. M., in "Quantitative Mass Spectrometry in Life Sci.
   II" (R.R. Roncucci, C. van Petephem, ed.)p. 175 and p. 227.
   Elsevier Amsterdam, 1978.
5. Mee, J. M., American Laboratory, 5:55 (1980).
6. Mee, J. M., in "Direct Mass Spectrometry of Body Metaboli-
   tes - Quantiative Methodology and Clinical Application.
   U.S. Library of Congress, 1982.
7. Khan, P., Mee, J. M. and Ahmad, S. in "Proceedings of the
   1st Sym. on Water Resources Development in the Kingdom of
   Saudi Arabia, Riyadh, 1982."
8. Mee, J. M., Khan, P. and Ahmad, S. in "Proceedings of the
   3rd International Flavor Conference, Corfu, Greece, 1983."

FORMATION OF POUCHONG TEA AROMA
DURING WITHERING PROCESS

Tei Yamanishi, Akio Kobayashi, Keiko Tachiyama
Ochanomizu University, Tokyo, Japan

I-Ming Juan, William Tsai-Fau Chiu
Taiwan Tea Experiment Station
Taoyuan, Taiwan, R.O.C.

ABSTRACT

Pouchong(Pauchung) tea is a kind of semi-fermented Chinese tea and is known for its characteristic floral aroma, which develops by solar-withering followed by indoor-withering while being turned over at adequate intervals.

The aroma concentrates were prepared from tea leaves at several stages of withering process and analyzed by capillary GC-MS in which the gas chromatograms were drawn by TIC and TMIC, and also by capillary GC-FID with a computing integrator.

In comparison of aroma compositions among nine tea samples of different withering stages, remarkable variations were recognized and found that solar-withering and indoor-withering were very effective to produce the aroma compounds responsible to typical pouchong tea flavor such as jasmine lactone, nerolidol, indole, benzyl cyanide and some other compounds.

I. INTRODUCTION

Tea is roughly classified into three categories according to its manufacturing process ; (1) fermented tea or black tea, (2) semi-fermented tea and (3) non-fermented tea or green tea. Semi-fermented tea involes pouchong tea and oolong tea.

Semi-fermented tea, originally produced only in Fukien, China, was introduced by the Fukienese immigrants into Taiwan in 1796 and called "Oolong" by Fukienese. In old days, oolong tea of common grade is mixed with jasmine or gardenia flowers prior to the final firing process and called "scented tea" . But in Taiwan, it was called "pauchung jasmine tea" which is not a pure kind of pouchong tea as a Chinese adage goes that only common tea requires scenting. The pure pouchong tea or pouchong tea, originated from Bohea area of Fukien, is one of the highest quality semi-fermented teas(1), and not scented with any flowers.

General manufacturing process of pouchong tea is shown in Fig.1. Degree of fermentation of pouchong tea is lighter than that of oolong tea. As shown in Fig.1, the fresh tea leaves are subjected first to solar-withering a short time and then withered indoors while being turn over at adequate intervals. This special withering method is essential to develop the characteristic flowery note.

The aroma components of pouchong tea has been investigated previously(2) and found that nerolidol, jasmine lactone, methyl jasmonate, indole, benzyl cyanide and linalool oxides were major contributory constituents to the characteristic elegant floral note.

In the present work, investigation of the characteristic aroma formation during the special withering process was carried out.

II. EXPERIMENTAL

A. Materials

In order to find out the effects of solar-withering and turn over treatment during indoor-withering on the aroma formation, nine different grades of withering were applied for tea leaves of var. Chin-shin-oolong as shown in Table I.

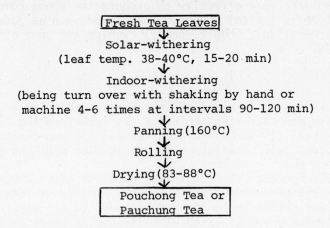

Fig. 1. Manufacturing process of pouchong tea.

TABLE I. Samples for Investigation of the
Effects of Solar-withering and Turn
Over during Indoor-withering

Sample	A*					B*			
	1	2	3	4	5	1	2	3	4
Solar-withering	0	0	min 7	min 17 (st.)	min 37	st.	st.	st.	st.
Turn over times during indoor-withering	0	st.	st.	st.	st.	0	1	3	4 (st.)

Note; st. means standard.
   * Fresh tea leaves for sample A and B were plucked
   on September 25th and 26th, 1982 respectively.

B. Preparation of aroma concentrate

   Every 100g of tea sample was mechanically powdered and
immediately mixed with 5ml of deionized water in the 2 liters
flask of a rotary evaporator, equipped with a rotation speed
regulator, connected with a condenser and two traps which
were cooled in freezing mixture(-15∿-18°C)and dry ice-acetone
mixture(-78°C) respectively. Distillation was conducted at
65 70°C, 25-30mm Hg under rotation speed 140 rpm. When the
material in the flask was almost dry, another 500ml of water
was added and continued distillation. Thus, about 700ml of
distillate was obtained. The combined distillate was saturated with sodium chloride and extracted with ether. After drying over sodium sulfate, the ether extract was concentrated
by distilling off the ether at 38∿40°C. The concentrate had
an intense aroma characteristics of individual tea samples.

C. Instrumental analysis

   1. Gas Chromatography. A Shimadzu GC-7A gas chromatograph
equipped with a flame ionization detector(FID) connected to
a computing integrator was used for quantitative analysis to
acquire all profiling data on 9 different tea samples. Analytical conditions were as follows;
   Column: 0.28 mm i.d.X 30 m length glass SCOT, liquid
         phase FFAP.

Column temperature: programmed from 60 to 180°C, rate
                    2°C/min
Detector and injection temperature: 200°C
Carrier gas: Nitrogen, 1 ml/min(0.8kg/cm^2) at column
                    inlet, Spilit ratio 1:50

2. Combined Gas Chromatography and Mass Spectrometry
   (GC-MS). Two GC-MS instruments were used for identifi-
cation of the components, as well as for a part of quantita-
tive analysis.
   a. Hitachi 063 GC combined with Hitachi RM-50GC mass
spectrometer. Analytical conditions were as follows;
   Column: 0.25 mm X 50 m fused silica, WCOT, liquid phase
           Carbowax 20M
   Column temperature: programmed from 60 t0 180°C, rate
                    2°C/min
   Carrier gas: Helium
   Electron voltage: 20 eV
   Ion accelerating voltage: 3 KV
   Temperature of ionization chamber: 180°C
   Injection temperature 200°C
   b. Hitachi 663 GC combined with Hitachi M-80 mass spect-
rometer connected with computer M-003. Analytical conditions
were as follows;
   Column: 0.35 mm X 50 m silica SCOT, liquid phase FFAP
   Column temperature: programmed from 70 to 210°C, rate
                    3°C/min
   Carrier gas: Helium
   Injection temperature: 250°C
   Interface temperature: 280°C
   Mass range: 25 to 400
   Scanning speed: 4 sec.

III. RESULTS AND DISCUSSION

   Figure 2 shows a typical gas chromatograms of sample A-4
recorded by total ion current(TIC) and total maximizing ion
current(TMIC) and of sample A-2 recorded by TMIC. Peak identification is shown in Table II.
   Quantities of individual compounds in each tea sample
were calculated based on the yields of aroma concentrate
and peak area percentages. In order to find out the effects
of solar-withering and turn over treatment during indoor-
withering, the quantities of main aroma components were compared among nine samples as shown in Table III and Table IV
respectively.

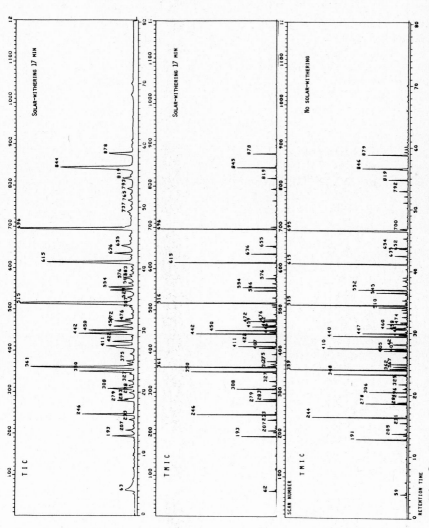

Fig. 2. Gas chromatograms of A-4 (upper and middle) and A-2 (bottom).

TABLE II. Identification of Aroma Constituents of
Sample A-2 and A-4

Peak scan No.*	Compound, identified
88	1-Penten-3-ol
112	1-Pentanol
126	Trimethylbenzene
132	Tridecane
135	cis-2-Penten-1-ol
145	2-Methylhept-2-en-6-one
149	Hexanol
162	cis-3-Hexen-1-ol
174	Tetradecane
191	Linalool oxide I (furanoid)
205	Linalool oxide II (furanoid)
231	Benzaldehyde
244	Linalool
278	3,7-Dimethyl-1,5,7-octatrien-3-ol
296	Phenylacetaldehyde
306	cis-3-Hexenyl hexanoate
325	4-Nonanolide
330	Ethyl hexanoate
336	Heptadecane
348	Linalool oxides III & IV (pyranoid)
359	δ-Farnesene
367	Methyl salicylate
410	Geraniol
421	Benzyl alcohol
440	Phenethyl alcohol
447	Benzyl cyanide
458	cis-Jasmone
487	Phenol
505	Cadinenol (tentative)
514	Nerolidol
519	Octanoic acid
552	cis-3-Hexenyl benzoate
575	Phenethyl hexanoate (tentative)
582	5-Decanolide
603	δ-Cadinol & Methyl palmitate
614	Jasmine lactone
652	Methyl jasmonate
654	Isoeugenol
696	Indole
700	Coumarin

Note: * Peak scan No. corresponds to that of GC in Fig.2.
Retention time corresponds to 4 sec. X scan No.

TABLE III. Effect of Solar-withering

Main aroma compounds	10^{-2} mg% in A-1	Ratio to A-1				
		A-1	A-2	A-3	A-4	A-5
cis-3-Hexen-1-ol	1.2	1	1.1	1.1	1.4	1.2
cis-3-Hexenyl butyrate	0.9	1	1.4	2.4	0.8	3.7
cis-3-Hexenyl hexanoate	2.1	1	2.5	3.5	2.7	3.9
cis-3-Hexenyl benzoate	3.2	1	3.2	2.9	3.2	2.3
Linalool	27.5	1	0.8	1.0	o.9	1.0
Linalool oxide I	5.5	1	1.3	1.5	1.2	1.4
Linalool oxide II	4.1	1	0.7	0.8	0.6	0.8
Linalool oxide III & IV	6.1	1	1.5	1.6	2.5	4.1
3,7-Dimethyl-1,5,7-octa-trien-3-ol	6.4	1	1.7	1.6	1.9	2.9
Geraniol	31.4	1	1.0	1.0	1.1	1.4
Benzyl alcohol	10.0	1	1.6	1.8	2.8	3.6
Benzaldehyde	1.4	1	1.1	1.3	2.1	3.2
Phenethyl alcohol	14.6	1	1.6	1.8	3.3	2.9
Phenylacetaldehyde	2.3	1	4.5	4.5	5.3	4.2
Methyl salicylate	2.6	1	1.5	1.6	1.9 }	2.6
⍺-Farnesene	19.6	1	2.3	2.4	1.8 }	
Nerolidol	19.3	1	3.9	3.4	5.0	3.5
cis-Jasmone	6.1	1	2.1	1.7	1.8	1.8
Jasmine lactone	16.4	1	5.9	4.4	5.3	3.8
Methyl jasmonate	0.8 }	1	4.2*	3.7*	4.6*	5.1*
Isoeugenol	2.1 }	1				
Benzyl cyanide	8.2	1	2.5	3.3	5.9	4.8
Indole	68.2	1	4.2	3.9	6.0	5.9

Note : * From gas chromatograms recorded by TMIC, amount of methyl jasmonate was about 1/2.5 and 1/3 of isoeugenol in A-2 and A-4 respectively.

Yields of the aroma concentrate : A-1(3.5mg%) A-2(8.8mg%), A-3(8.1mg%), A-4(10.6mg%), A-5(10.3mg%)

TABLE IV. Effect of Turn Over during Indoor-withering

Main aroma compounds	10^{-2}mg% in B-1	B-1	B-2	B-3	B-4
		—Ratio to B-1—			
cis-3-Hexen-1-ol	1.5	1	0.8	1.1	1.1
cis-3-Hexenyl butyrate	3.0	1	0.6	1.0	1.4
cis-3-Hexenyl hexanoate	3.6	1	0.9	2.2	2.5
cis-3-Hexenyl benzoate	3.3	1	1.1	2.5	2.3
Linalool	29.6	1	0.8	0.9	0.8
Linalool oxide I	0.5	1	5.3	9.5	12.8
Linalool oxide II	3.1	1	0.7	0.8	1.0
Linalool Oxide III & IV	4.6	1	0.4	0.8	0.5
3,7-Dimethyl-1,5,7-octa-trien-3-ol	10.2	1	0.4	0.9	1.5
Geraniol	28.3	1	1.0	1.1	1.0
Benzyl alcohol	12.4	1	0.6	1.0	1.0
Benzaldehyde	1.4	1	0.7	1.2	1.3
Phenethyl alcohol	14.4	1	0.6	1.3	1.4
Phenylacetaldehyde	7.5	1	0.4	1.7	1.9
Methyl salicylate	4.6	1	0.4	0.8	0.5
ᴋ-Farnesene	22.3	1	0.6	1.3	1.8
Nerolidol	21.0	1	1.5	2.9	3.0
cis-Jasmone	7.0	1	1.5	1.6	1.4
Jasmine lactone	22.2	1	1.0	3.1	3.2
Methyl jasmonate ⎫ Isoeugenol ⎭	2.4	1	2.4	2.6	2.5
Benzyl cyanide	16.3	1	0.4	1.4	1.4
Indole	171.9	1	1.0	1.7	1.7

Note : Yields of the aroma concentrate : B-1(4.8mg%),
B-2(4.2mg%), B-3(7.9mg%), B-4(8.1mg%).

As seen in A-2, A-3, A-4 and A-5(Table III), linalool oxides(pyranoid), 3,7-dimethyl-1,5,7-octatrien-3-ol(oxydized product of linalool), geraniol, benzyl alcohol and benz-aldehyde showed   straight increases by solar-withering, while cis-3-hexenyl esters, ⋌-farnesene, nerolidol, jasmine lactone and indole showed irregular changes. The latter three compou-nds, as well as phenethyl alcohol, phenylacetaldehyde and benzyl cyanide decreased by excessive solar-withering(A-5).

⋌-Farnesene and nerolidol were predominant components in A-1(without any withering process). When the concentration of ⋌-farnesene was reduced by solar-withering, the concentra-tion of nerolidol increased correspondingly. This seemed to be an interesting phenomenon from aspect of their chemical structures.

α-Farnesene $C_{15}H_{24}$ Nerolidol $C_{15}H_{26}O$

In comparison between A-1 and A-2, remarkable increases of following compounds were recognized,i.e. cis-3-hexenyl hexanoate, cis-3-hexenyl benzoate, cis-jasmone, benzyl cyanide, ⋌-farnesene, nerolidol, jasmine lactone, isoeugenol(mixed with methyl jasmonate), phenylacetaldehyde and indole. Especially the latter five compounds increased greatly. This means that indoor-withering is very effective to develop the compounds which contribute to the characteristic aroma of pouchong tea, even in the case of no solar-withering.

When tea leaves were subjected to standard solar-wither-ing prior to indoor-withering, times of turn over treatment had not so large effect on increasing the components except linalool oxide I(furanoid), nerolidol and jasmine lactone, as seen in Table IV. It seemed that three times of turn over treatment is enough to develop the major aroma components.

IV. CONCLUSION

In summary, in comparison of A-1 with A-4, the combina-tion of standard solar-withering and indoor-withering with three times turn over treatment was very effective to develop the major components which contribute to aroma characteristics of pouchong tea. Indoor-withering while being turn over at adequate intervals seemed to be more important than solar-withering.

ACKNOWLEDGMENTS

The authors wish to thank Mr.T.Nakata and Mr.K.Shizuku-
ishi of Naka Works, Hitachi Ltd., for GC-MS analysis using
a Hitachi M-80-M-003.

REFERENCES

1. Wu,C.T., Bull of Taiwan Tea Experiment Station 94,179
    (1980).
2. Yamanishi, T., Kosuge, M., Tokitomo, Y. and Maeda, R.,
    Agric. Biol. Chem. 44, 2139 (1980).

WATER SORPTION OF COFFEE SOLUBLES
BY INVERSE GAS CHROMATOGRAPHY

Dimitrios Apostolopoulos
S. G. Gilbert

Food Science Department
Rutgers University
New Brunswick, N.J. 08903

I. ABSTRACT

Water sorption isotherms can be obtained using the pulse
method of Inverse Gas Chromatography. This method assumes
that equilibration between the water in the mobile phase and
water in the stationary phase is attained instantaneously and
hysteresis is not involved. However, this is not always justified. Where time dependent relations of diffusion and
hysteresis are involved, the data reflects a nonequilibrium
status.

A continuous transport of humidified gas through the stationary phase is used to produce a frontal chromatogram of a
desired step-height. Pulses made at the equilibrium plateau
may then provide more accurate equilibrium sorption isotherms.

This combined method, referred to as step and pulse
method, has been developed and used in this work to study
water sorption by coffee solubles within the region of low
moisture. The water sorption isotherms of coffee solubles
determined at 25, 35 and 45°C show that water sorption is
favored by low temperatures.

The mode of change of the thermodynamic parameters: Free
energy (ΔG), Enthalpy (ΔH) and Entropy (ΔS), with increased
step-height water vapor pressure and/or uptake, indicate an
energetically favorable water sorption on the most active
sites, followed by a sorption on active sites of less binding
energy resulting in a less structured water-coffee system.

Desorption isotherms obtained by Frontal chromatography
show hysteresis compared to the sorption isothermal data.

II. INTRODUCTION

Instrumental Analysis of Foods
Volume 2

51

Water is a major factor in shelf life of foods. To obtain comprehensive data on the water relations of a food system for any temperature, the water sorption isotherms must be determined.

Water sorption isotherms describe the amount of water present in a solid at equilibrium with a specific water vapor pressure.

The elution or pulse method of inverse gas chromatography, which employs injections of water, has been used to study the sorption of water on a variety of food materials. This method assumes that equilibration between the water in the mobile phase and the water in the stationary phase is attained instantaneously and hysteresis is not involved. However, this is not always justified. Where time dependent relations of diffusion and hysteresis are involved, the data reflect a nonequilibrium status.

It was reasoned that these drawbacks of the pulse method could be eliminated by introducing a continuous transport of humidified gas through the stationary phase to produce a frontal chromatogram of a desired step-height and establish a constant water vapor concentration plateau. Injections of different amounts of water made on the equilibrium plateau could generate pulse data which provide more accurate sorption isotherms by more closely approaching the equilibrium state.

If the humidified gas is switched off and the stationary phase flushed with dry carrier gas, a desorption chromatogram could be obtained which would permit calculation of desorption isotherms.

When desorption isotherms are compared with the sorption isothermal data obtained by frontal sorption chromatography, formation about hysteresis can be obtained.

In addition, thermal studies of the sorption process can be used for evaluations of the thermodynamic quantities of free energy, enthalpy and entropy.

The combined method, referred to as "elution on a plateau" (Conder and Purnell, 1969) or "step and pulse method of inverse gas chromatography," was used to study water sorption by coffee solubles within the region of bound water, where caking and flavor retention are of considerable practical importance (Karel,1975)(Peleg et al., 1973).

A practical advantage which the IGC offers is the rapidity with which equilibrium sorption data are obtained, as opposed to the lengthy equilibration time frequently required by the static methods.

The objectives of the study were as follows:
1. To develop a rapid method which would provide accurate water sorption isotherms.
2. To use the method to study the sorption of water on coffee solubles, within the low moisture region, which is of great

importance to the shelf life of this dry product.
3. To determine the thermodynamic parameters of the water-coffee interaction and
4. To investigate any hysteresis effects involved in the water sorption-desorption of coffee solubles.

III. LITERATURE REVIEW

A. Inverse Gas Chromatography

A method used for the determination of sorption isotherms is inverse gas chromatography (IGC). IGC is a form of chromatography, where an unknown stationary phase is investigated by its interaction with known low molecular weight probe molecules, present in the mobile phase. (Gilbert, 1983)

When water sorption is studied, the probe molecules are water vapor. Depending on the way the probe molecules (water vapor) are introduced in the mobile phase, IGC can be described as:

   i) Frontal inverse gas chromatography
  ii) Elution or pulse inverse gas chromatography and
 iii) Step and pulse inverse gas chromatography

Each version of inverse gas chromatography will be discussed in detail.

a. Frontal Inverse Gas Chromatography

Basically this method involves the introduction and transport of a continuous stream of sorbate vapor at a constant concentration, through a column packed with sorbent, by means of an inert, nonsorbable carrier gas.

After the sorbate breaks through the stationary phase a frontal chromatogram is developed. When the sorbent in the column is saturated, the detector response reaches a maximum step-height and levels off, establishing a plateau. (Fig. 1)

The step-height of a plateau represents an equilibrium between the sorbate in the mobile phase and the sorbate in the stationary phase. The equilibrium can be verified by the composition of the ingoing mixture.

If the supply of sorbate to the column is stopped after equilibrium is established, and the column is flushed with dry carrier gas so that after a certain period of time the concentration of the sorbate in the outgoing mixture begins to decrease, a characteristic desorption chromatogram is developed. (Fig. 1)

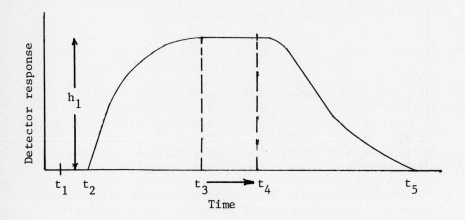

FIGURE 1.   Frontal chromatography

where:

t_1 = start of feeding sorbate vapor in the column.
t_2 = sorbate vapor breaks through the column
t_3 = equilibrium is established between the sorbate vapor in the mobile and stationary phase.
t_4 = stop of feeding sorbate in the column.
t_5 = all of the sorbate is driven off the stationary phase.
h_1 = step-height

This frontal method is unaffected by non-ideal distortion because the technique depends not on the shape, but on the area bounded by the chromatogram, which is not influenced by kinetic bond broadening process. (Conder and Purnell, 1969). This makes frontal analysis especially attractive for determination of sorption/desorption isotherms, thus permitting a direct check on the possibility of hysteresis in the finite concentration region. (James and Phillips, 1954), (Eberly, 1961), (Eberly and Kimberlin, 1961), (Stock, 1961), (Greenstein and Issenberg, 1967), (Conder, 1968), (Huber and Gettse, 1971), (Hussey and Parcher, 1973), (Kalinichev, et. al., 1978), (Kucera et. al., 1981).

The sorption and desorption isotherms can be derived from the frontal and desorption chromatograms by analysis of the effluent stream as a function of time.

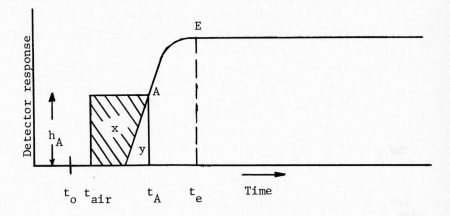

FIGURE 2.   Frontal Analysis

where:

t_o = start of feeding sorbate in the column.

t_{air} = emergence of an unsorbed species (air)

x = area before sorbate vapor breaks through the column and is related to the amount sorbed at time t_A

y = area under the front curve and is related to the amount of sorbate desorbed at time t_A

h_A = step-height corresponding to point A

t_e = equilibrium is established between the sorbate vapor in the mobile and stationary phases

Figure 2 illustrates the method of transformation of a breakthrough curve with a diffuse front to sorption isotherm. According to Glueckauf (Cerny, 1970), the equilibrium amount sorbed at a point A is given by equation:

Eq. 1

$$a = \frac{n^o \bar{c} - \mu_c}{g}$$

where:

a = the equilibrium amount sorbed at point A

n^o = the total number of moles of sorbate and carrier gas fed to the column

$\bar{c}$ = the concentration of sorbate in outgoing mixture at time t_A (point A)

μ_c = the number of moles of sorbate retained in the column when its outgoing concentration is $\bar{c}$

g = denotes grams of sorbent in the column

The product $n^o\bar{c}$ is equal to the product of the sum of areas X and Y and the volume rate of flow,W. The number of moles of sorbate retained in the column (μ_c) is equal to the product of the area Y and the volume rate of flow, W.

The equilibrium amount sorbed corresponding to the concentration $\bar{c}$, at point A on the frontal chromatogram, is therefore equal to the product of the area X and the volume rate of flow, W.

Following that equation (1) can be written:

Eq. 2
$$a = \frac{(X + Y)\ W - YW}{g} = \frac{XW}{g}$$

where: W = the flow rate of the carrier gas

As it is shown from the graph of Figure 2, X has the dimensions $(t_A - t_{air}) \cdot h_A$, and is a nonlinear function of h_A with respect to time. Therefore, it is difficult to quantify X directly. The sum $X + Y = C_F$ can be readily determined since it represents a rectangle with dimensions $(t_A - t_{air}) \cdot h_A$, both of which are directly measurable, since h_A is a characteristic point of t_A.

Thus equation (2) develops as follows:

Eq. 3
$$a = \left(\frac{X}{X + Y}\right) \frac{W \cdot C_F}{g}$$

Substituting $(t_A - t_{air}) \cdot h_A$ for C_F, equation 4 is derived:

Eq. 4
$$a = \left(\frac{X}{X + Y}\right) \frac{W(t_A - t_{air})}{g} h_A$$

From a calibration curve constructed independently by injecting different amounts of sorbate vapor in an analytical column, h_A can be expressed in terms of sorbate concentration $(\bar{c})$. The conversion of height to sorbate concentration can be done using the relationship:

Eq. 5 $\qquad h_A = C_A \cdot \text{Slope}_{\text{calibration}} = \bar{c}$

The molar value $\bar{c}$ can be substituted for the value h_A and equation (4) can be written as modified by Apostolopoulos-Gilbert (Gilbert, 1982).

Eq. 6 $\qquad a = \left(\dfrac{X}{X + Y}\right)\dfrac{W(t_A - t_{air})}{g} \cdot \bar{c}$

The partial pressure, P_A, of the sorbate in the gas phase at point A, is a function of the step-height $P_A = f(h_A)$ and can be deduced from the calibration curve using the relationship:

Eq. 7 $\qquad P_A = \bar{c} \; RT$

By repeating the outlined procedure for different values of $\bar{c}$ or different points on the frontal chromatogram, the complete sorption isotherm can be constructed.

The transformation of a diffuse desorption chromatogram to a desorption isotherm is illustrated in Figure 3.

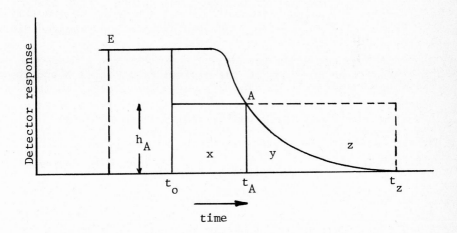

FIGURE 3.   Desorption branch analysis.

where:

t_o = stop of feeding sorbate in the column.

t_z = all of the sorbate has been driven off the stationary phase.

$(x+y)$ = sum of areas x and y which represents the amount of sorbate still remaining in the column at time t_A.

z = area which is related to the amount of sorbate, desorbed at time t_A.

According to Glueckauf the equilibrium amount sorbed at point A, corresponding to an outgoing concentration $\bar{c}$ of sorbate, is given by equation:

Eq. 8
$$a = \frac{n\bar{c} + \mu_{\bar{c}}}{g}$$

where:

n = the number of moles of carrier gas fed to the column after the feeding of sorbate had been stopped and for a period of time equal to $t_A - t_o$

$\mu_{\bar{c}}$ = the number of moles of sorbate retained in the column when its outgoing concentration is $\bar{c}$ (point A)

g = denotes grams of adsorbent in the column

The quantity $n\bar{c}$ is equal to the product of the area X and the volumetric flow rate, W. The parameter $\mu_{\bar{c}}$ is equal to the product of the area Y and the volumetric flow rate, W.

Following that equation (8) can be written in the following form:

Eq. 9
$$a = \frac{XW + YW}{g} = \frac{(X + Y)W}{g}$$

From Figure 3 it can be seen that X has the dimensions: $(t_A - t_o) \cdot h_A$ and Y: $(t_Z - t_A) : h_A$, but Y is a nonlinear function of h_A and therefore it is difficult to quantify X + Y directly.

However, the sum $X + Y + Z = C_F$ can be readily determined since it represents a rectangle with dimensions $(t_Z - t_A) \cdot h_A$, both of which are directly measurable.

Thus equation (9) can be rewritten as follows:

Eq. 10
$$a = \left(\frac{X + Y}{X + Y + Z}\right) \frac{W\ C_F}{g}$$

By substituting $(t_Z - t_o) \cdot h_A$ for C_F, the above equation becomes:

Eq. 11
$$a = \left(\frac{X + Y}{X + Y + Z}\right) \frac{W(t_Z - t_o) \cdot h_A}{g}$$

Using the relationship in equation (5) the step-height, h_A can be converted to concentration, $\bar{c}$, and by substituting $\bar{c}$ for h_A in equation (11), it can be rewritten as modified by Apostolopoulos-Gilbert (Gilbert, 1982).

Eq. 12
$$a = \left(\frac{X + Y}{X + Y + Z}\right) \frac{W(t_Z - t_o)\ \bar{c}}{g}$$

The partial pressure, P_A of the sorbate in the gas phase at point A of the desorption chromatogram is a function of the step-height $P_A = f(h_A)$ and is calculated similarly to the method used for sorption isotherms.

By repeating the procedures outlined for a number of chosen outlet concentrations or for a number of different points on the desorption chromatogram, the complete desorption isotherm is constructed.

b. Elution or Pulse Inverse Gas Chromatography

This method involves the injection of a known amount of volatile probe molecules (sorbate) into a column packed with the material of interest. The injected probe molecules move in the direction of the carrier gas with an initial velocity, U_O. The molecules of the injected sorbate collide with the surface of the sorbent. If there is no interaction the velocity U_o remains the same throughout the column. Whereas, if interaction is taking place, the velocity will be retarded and the area response and the shape of the pulse will be affected to a certain extent depending on the strength and nature of the interaction. Wilson (1940) first examined the quantitative relation of equilibrium distribution between the stationary and mobile phase to the equilibrium chromatogram. (Gluckauf 1947, 1949) devised a method of deducing the equilibrium sorption isotherm. In the theory of equilibrium chromatography, it is assumed that the existing conditions virtually rule out diffusion and kinetic forms of broadening. This equilibrium (but not ideal) chromatography corresponds to the following equation

for material balance in an elementary layer as described by
Kiselev and Yashin, (1969):

Eq. 13 $-u_0 v \left(\dfrac{dc}{dx}\right)_t = v \left(\dfrac{dc}{dt}\right)_x + V_a \left(\dfrac{dca}{dt}\right)_x$

where:

    u_o = linear velocity of the carrier gas

    v = volume of the gas phase in the column

    V_a = volume of the adsorbed layer

    c = concentration of the probe in the gas phase

From equation 13 Kiselev and Yaskin (1969) derived equation (14):

Eq. 14 $a = 1/m \displaystyle\int_o^c V_c d_c$

where:

    a = uptake of probe molecules by the sorbent

    m = mass of the sorbent

    V_c = retention volume of sorbate at concentration c

Equation 14 can be expressed in terms of factors obtainable directly from a chromatogram to give:

Eq. 15 $a = \dfrac{m_p s_p}{ms}$

where:

    m_p = mass of probe molecules

    s = area of chromatographic peak (Y)

    s_p = the sum of the areas X and Y, as shown in Figure 4.

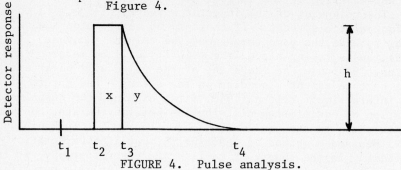

FIGURE 4.   Pulse analysis.

where:

t_1 = injection of sorbate in the mobile phase.
t_2 = emergence of unsorbed species (air)
t_3 = emergance of sorbate peak
x = prepeak area
y = peak area
h = peak height

The concentration, c, corresponding to an uptake, a, is given by equation:

Eq. 16
$$c = \frac{m_p \, q \, h}{sw}$$

where:

q = recorder chart speed
w = flow rate of carrier gas (corrected for pressure drop along the column)

The partial pressure of probe, P, can be calculated from equation:

Eq. 17
$$P = \frac{m_p \, q \, h \, R \, T}{sw}$$

where:

R = universal gas content
T = column temperature

The elution or pulse method of inverse gas chromatography has been applied to the study of water sorption by food systems, such as proteins (Coelho, et. al., 1979), sucrose and glucose (Smith et. al., 1980), and starches (Smith, 1982). It may be of interest to mention that the same method has been successfully used to investigate the surface and bulk properties of polymers (Braun and Guillet, 1976), polymer-migrant interactions in packaging materials, (Kinigakis, 1979), (Senick and Sanchez, 1979), (Orr, 1979), (Gilbert et. al., 1979), (Apostolopoulos, 1980), polymer degradation (Bereskin et. al., 1977), activity coefficients (Varsano and Gilbert, 1971), (Khalil, 1976), diffusion coefficients (Fuller and Giddings, 1965), and oxidation of well defined substances (Evans and Newton, 1976).

c.   Step and Pulse or Elution on a Plateau

The method of step and pulse or elution on a plateau is a combination of the two techniques described previously. A front is developed and after equilibrium is established so as to set up a concentration plateau of given height invariant with time and distance along the column, different amounts of sorbate are injected into humidified carrier gas at the column inlet. These injections provide increases in elution peaks lodged at the equilibrium plateau, as shown in Figure 5. (Conder, 1968), (Valentin, 1976).

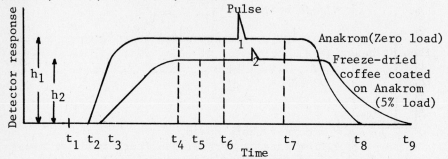

FIGURE 5.   Diagram of sorption-desorption chromatograms
by pulse and step chromatography for two
load levels of reactor columns.

where:

t_1 = start of feeding sorbate vapor in column packed
either with anakrom or freeze-dried coffee
coated on anakrom.

t_2, t_3 = sorbate vapor breaks through the columns packed
with anakrom and freeze-dried coffee coated on
anakrom, respectively.

t_4, t_5 = equilibrium is established between the sorbate
vapor in the mobile and stationary phase in
both columns.

t_6 = injection of sorbate pulses in either column.

t_7 = stop of feeding sorbate in the columns.

t_8, t_9 = all of the sorbate vapor has been driven off
the stationary phase.

From the chromatogram (Figure 6), the sorbate uptake and the partial pressure corresponding to the equilibrium plateau, can be calculated using equations (6) and (7).

The variation of the retention time and/or the area response of the pulses with different step-heights, different amounts of sorbate injected and temperature are used to calculate the additional sorbate uptakes and partial pressure provided by the pulses. The Kiselev-Yashin procedure is used for analysis of the pulse data. Equations (18) and (19) presented in the following pages provide for calculation of the total uptake, a_t, and total partial pressure, P_t, corresponding to a specific step-height and a pulse of a defined size.

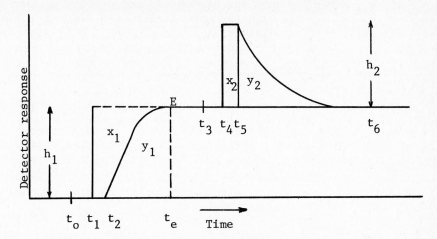

FIGURE 6.   Step and pulse analysis.

where:

t_o = start of feeding sorbate vapor in the column.

t_1 = emergence of an unsorbed species (air).

t_2 = sorbate vapor breaks through the column.

t_e = equilibrium has been established between the sorbate vapor in the mobile and stationary phase.

t_3 = injection of sorbate in the mobile phase.

t_4 = emergence of unsorbed species (air)

t_5 = emergence of sorbate peak

t_6 = sorbate introduced by pulse has been eluted of the stationary phase.

x_1 = area before sorbate breaks through the column

y_1 = area under the front curve.

x_2 = prepeak area

y_2 = peak area

h_1 = step-height

h_2 = peak height

Eq. 18 $$a_t = a_{eq} + a_p$$

Eq. 19 $$P_t = P_{eq} + P_p$$

where:

a_{eq} and P_{eq} = the sorbate uptake and partial pressure corresponding to equilibrium plateau

a_p and P_p = the additional sorbate uptake and partial pressure provided by the pulse

In other words, adding together the uptakes and the partial pressures of the sorbate, as calculated from the frontal analysis and the pulses injected on the plateau, the sorption isotherms of the sorbent are determined.

The step and pulse method has been applied by Valentin (1976) to the determination of isotherms of n-pentane dissolved in squalene and of n-hexane sorbed on graphitized carbon black. In both cases the isotherms derived were in excellent agreement with those obtained static methods.

B. Thermodynamic Aspects of Sorption

A better insight into the water sorption process can be obtained from the calculation of the thermodynamic parameters involved in this process.

For instance, such thermodynamic parameters may be related to theoretical interpretations of isotherms, indicate the mode of sorption of the water molecules or give information about the configurational modifications of the sorbent during the course of sorption. (Bettelheim and Velman, 1957).

a. Calculation of Thermodynamic Parameters

Sorption data at more than two temperatures are required for the calculation of the thermodynamic parameters: Free energy (ΔG_s), Isosteric heat or Enthalpy (ΔH_s) and Entropy of Sorption (ΔS_s), (Iglesias, et. al., 1975), (Iglesias, et. al., 1976), (Chirife, et. al., 1977).

i) Free Energy of Water Sorption

Water molecules sorbed on a solid set up an equilibrium with water vapor over the solids, which can be represented by equation:

Eq. 20   Solid(P=0) + n moles of H_2O(at P) = Solid with n moles of H_2O(at P)

where:
> P = the partial pressure of the water vapor (in the gas phase) at a given temperature T
> n = the number of moles of water sorbed, which depends on P

The differential Gibb's Free Energy change can be obtained from equation 21.

Eq. 21

$$\frac{dG}{dn} = \Delta G_s = RT\ln \frac{P}{P_o} = RT\ln A_w$$

where:
> ΔG = the change in integral Gibb's Free Energy
> R = the gas constant
> T = the temperature
> $x = P/P_o$ represent the relative vapor pressure of the water
> P_o = the pressure of the water vapor in an atmosphere saturated with water

(Rockland, 1969),(Wurster, 1979),(Iglesias et. al., 1976)

ii)   Isosteric Heat or Enthalpy of Sorption

The isosteric heat or enthalpy of sorption can be calculated using Clausius – Clapeyron equation at a fixed moisture content, m.

Eq. 22

$$\left(\frac{d\frac{1}{n}P}{dT}\right)_m = \left(\frac{\Delta H_s}{RT^2}\right)_m$$

This equation is thermodynamic in origin and is valid only for a thermodynamically reversible process.   (Gregg, 1951). Equation (22) can be modified to give:

Eq. 23

$$\left(\frac{d\ln P}{d\frac{1}{T}}\right)_m = \left(\frac{\Delta H_s}{R}\right)_m$$

Considering ΔH_s as the average isosteric heat or enthalpy of sorption over the temperature range studied with fixed m values integration of equation (23) results in the working expression:

Eq.24

$$\ln P = \frac{\Delta H_s}{R}\left(\frac{1}{T}\right) + Const.$$

Thus, a plot of lnP versus 1/T at a given moisture content may result in a straight line with slope $\Delta H_s/R$.

where: P = the partial pressure of water vapor

T = temperature

R = universal gas constant

ΔH_s = isosteric heat or enthalpy of sorption

The thermodynamic parameter ΔH_s, is referred as isosteric heat or enthalpy of sorption because it can be calculated either from sorption isotherms or sorption isosteres (Iglesias and Chirife, 1976).

iii)   Entropy of sorption

The entropy of sorption can be calculated by using the values obtained from equations (21) and (24) in the relationship:

Eq. 25

$$\Delta S_s = \frac{\Delta H_s - \Delta G_s}{T}$$

b.   Thermodynamic considerations

i)   Free energy of sorption

The thermodynamic requirement for water sorption is that the free energy of the solid-water system be less negative than for the individual components in the pure state. Thus the free energy provides a criterion of whether water sorption is a spontaneous or non-spontaneous process, depending on the sign of the ΔG values. Negative ΔG values denote spontaneous sorption and the more negative the quantity the stronger the tendency toward sorption. Conversely, positive ΔG values are indicative of a non-spontaneous process. Consequently, the free energy of sorption, ΔG, is an indicator of the favorability of the sorption process.

The free energy change of sorption, G_s, is also a measure of the relative number of moles of water and solid involved in mixing.

ii)   Isosteric heat or enthalpy of sorption

When vapor is sorbed by a solid, heat is always given off; in other words, sorption is an exothermic process. Correspondingly, if vapor is pumped out of a sorbent, heat is absorbed and the sorbent becomes cooler; desorption is an endothermic process. The quantity of heat released (or absorbed) during sorption is referred to as isosteric heat or enthalpy of sorption, ΔH_s.

Since heat is a form of energy, enthalpy changes can be considered as a measure of the potential energy changes occurring in the system. Many of these energy changes result from binding of the system components which occurs upon mixing. Consequently, enthalpy changes can be associated with the binding or repulsive forces of the system, depending on the sign of the ΔH_s values. Negative ΔH_s values denote the existence of binding forces as opposed to repulsive forces indicated by positive ΔH_s values.

iii)   Entropy of sorption

Entropy(s), is a thermodynamic state function associated with the number of different equilibrium energy states or spatial arrangements in which a system may occur. Once the conditions for a system are specified, that is, temperature and pressure, the entropy function is defined. Changes in the conditions can cause the structure of the system to become less regular or less confined and make the entropy increase or decrease. The entropy change ΔS can be used to characterize and determine the degree of randomness or disorder occurring in the system as compared to its original state. The order-disorder concept in relation to entropy changes is very useful for interpreting such processes as dissolution, crystalization and/or swelling, which usually occurs during water sorption by food products (Iglesias et al., 1976).

iv)   Enthalpy/Entropy compensation

The changes occurring in the thermodynamic parameters during the water sorption process, can be combined in the form:

Eq. 26
$$\Delta G_s = \Delta H_s - T\Delta S_s$$

From a close examination of the above equation, it is apparent that the free energy function depends on both enthalpy and entropy changes. Therefore, the entropy contribution, represented by the quantity $-T\Delta S_s$ may increase or decrease the spontaneousness sorption process. When ΔS_s is positive, meaning that the system becomes more random, the term $-T\Delta S_s$ makes a negative constribution to ΔG_s and increases the spontaneity of the sorption process.

When ΔS_s is negative, the term $-T\Delta S_s$ becomes positive which results in less negative ΔG_s values which decreases the spontaneity of sorption.

Temperature is another important consideration. Both ΔH_s and ΔS_s are in principle capable of changing with temperature. In practice, however, the only quantity in equation (26) that changes markedly with temperature is $-T\Delta S_s$, which in turn can affect the spontaneity of the sorption process (Brown, 1977), (Labuza, 1980).

C. Hysteresis

For many systems, sorption is completely reversible so that the saturation at any pressure is independent of the fact that equilibrium was reached after a decrease or an increase in pressure.

Frequently, however, it is found that when a sorbent containing sorbed vapor is subjected to diminishing pressure, the desorption isotherm fails to retrace the path of the isotherm which formed when the pressure was increased. The sorbent then holds more vapor for a given pressure on the curve of desorption than on that of sorption. In such cases, all or part of the desorption isotherm is above the sorption isotherm, indicating a resistance to desorption. This behavior is known as hysteresis (Hassler, 1963).

The presence of hysteresis indicates that the system, though reproducible, is not in true equilibrium. It is doubtful whether either the sorption or desorption isotherm represents conditions closer to equilibrium.

Moisture sorption hysteresis has important theoretical and practical implications in foods. The theoretical implications range from general considerations of the irreversibility of the sorption process to the question of validity of thermodynamic functions derived therefrom. The practical implications deal with the effects of hysteresis on chemical and microbiological deterioration, and with its importance in low and intermediate moisture foods (Kapsalis, 1981).

A variety of hysteresis loop shapes can be observed in foods. Wolf et al. (1972) found wide differences in magnitude, shape, and extent of hysteresis of dehydrated foods, depending on the type of food and the temperature. Variations could be grouped into three general types:

i) In high-sugar-high pectin foods, hysteresis occurs mainly in the monomolecular layer of water region.

ii) In high protein foods, moderate hysteresis begins in the

capillary condensation region and extends over the rest of
the isotherm to zero water activity.

iii) In starchy foods, a large hysteresis loop occurs with a
maximum within the capillary condensation region.

A great number of theories have been advanced to explain
sorption hysteresis. The interpretations proposed are based
on the structure of the sorbent (Arnell and McDermett, 1957).
Hysteresis on nonrigid solids deals with interpretations
related to changes in structure, as these changes hinder pene-
tration and egress of the sorbate.

For example, Kamya and Takahashi (1979) suggested that
during sorption water may penetrate in the more accessible
regions of a material, and break the intercahin hydrogen bonds.
During desorption the reformation of the dissolved bonds is
hindered irreversibly by the presence of sorbed water and as a
result sorption hysteresis occurs.

In general hysteresis seems to be the net result of
reinforcing or competing variables between sorbent and sorbate.
From the stand point of the sorbate, important factors in
hysteresis are the hydrogen-bonding ability, the amount sorbed
and the molar volume (Benson and Richardson, 1955).

Sorbates of greater hydrogen-bonding ability develop
large hysteresis loops, but those of little or no hydrogen-
bonding ability give very small loops. Sorbates that are
sorbed in greater amounts cause greater hysteresis loops
(greater total deformation) than sorbates sorbed in small
amounts.

Important in hysteresis is the molar volume of the
sorbate. Larger molar volumes give greater hysteresis loops
because there is greater network deformation per sorbed
molecule of sorbate. On the other hand, the increasing bulk
of the sorbate molecule may result in decreased sorption due
to too large local deformation, where sorption becomes energetically unfavorable, due to the increased work required by the
larger deformation.

Bettleheim and Ehrlich (1963) attributed hysteresis in a
swelling polymer to mechanical constraints contributed by the
elastic properties of the material and on the ease with which
the polymer swells. Polymer with a large sorptive and swelling
capacity, have polymer chains weakly bound with water and lead
to a small hysteresis loop. When sorptive and swelling
capacity is small, as in a tightly bound matrix with strongly
bound water, hysteresis is large.

IV.   EXPERIMENTAL PROCEDURES

A. Apparatus

An apparatus was built to conduct measurements by the "step and pulse" method. This apparatus consisted esssentially of a carrier gas supply, a gas stream humidifier, a flow control system and a reactor connected in series to a gas chromatograph opitimized for water vapor assay. A block diagram of the apparatus is given in Figure 7.

1. Carrier gas supply

Helium gas was drawn from cylinders, dried and purified by passing through moisture and oxygen traps installed in the gas lines. Two gas lines ensured that the apparatus would be operable. One supplied the reactor and the gas chromatograph with high purity carrier gas and the other supplied the humidifier with a gas stream for producing humidified gas.

2. Gas stream humidifier

The step and pulse method requires that a stream of water vapor be continuously introduced into and transported by the gas phase. Accurate results can be assured only when using a humidified gas stream in which the water vapor concentration is very stable. A gas stream humidifier is such a piece of equipment. The device should equilibrate easily and rapidly at any setting, operate in a broad range of concentrations, and deliver a very stable mixture. The gas humidifier assembled and installed in the apparatus, consisted basically of two stainless steel cylinders connected in series ant attached to the carrier gas line by a 1/8" copper tube. The cylinders were partially filled with distilled water and immersed in a constant temperature (30°C) bath thermostated at ±0.5°C. An accurately controlled stream of helium gas was drawn from the gas supply at a low flow rate, bubbled through the water in one cylinder and mixed with the head space of the other one to generate a gas stream saturated with water vapor.

The saturated gas stream formed in the humidifier was flowed via an outlet to a mixing chamber, where it was diluted with the main stream of dry carrier gas at a defined low mole fraction. The humidified carrier gas at a constant water vapor concentration could be introduced into either of the reactor columns. The water vapor concentration of the humidified carrier gas entering the reactor column was determined by the flow rate of the gas stream flowing through the humidifier and the water vapor pressure in the humidifier, which was directly

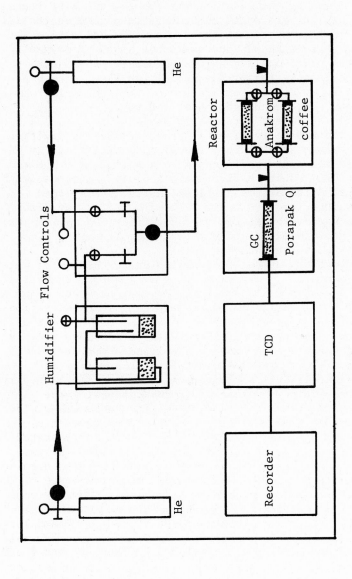

FIGURE 7. Schematic diagram of gas chromatographic reactor used for step & pulse method

related to the temperature of the humidifier water bath and the
ratio of mixing between the saturated gas stream from the
humidifier and the dry carrier gas.

The concentration of the humidified carrier gas could be
adjusted by changing the gas flow through the humidifier, the
temperature of the water bath and the ratio of mixing.

A flow control system equipped with the appropriate flow
and pressure regulators and the automatic heater immersed in
the water bath provide all controls necessary to regulate the
water vapor concentration of the humidified gas. Care had to
be taken not to raise the temperature of the humidifier water
bath too much, since condensation of the vapor in the gas lines
connecting the different parts of the apparatus could result.

Insulation and heating, of all the wet gas lines with
heating tape was essential to prevent condensation, which
together with temperature fluctuations along the humidified
gas route, could result in fluctuations of the water vapor
concentration, drift of the baseline, noise, which could lead
to inaccurate results.

Vapor condensation inside the humidifier, was avoided by
flushing it between experiments with a slow stream of gas
which was vented to the atmosphere.

The humidifier was refilled periodically through a
removable closure.

3. Flow control system

The flow control system consisted of a set of pressure
and flow regulators which could control and stabilize the gas
flow through the humidifier and enable mixing of the saturated
gas stream with the dry carrier gas at a defined low mole
fraction. Prevention of water vapor condensation was assured
by placing the whole system in a controlled temperature box
heated with circulating hot air.

4. Reactor

A system of two columns was housed in the oven of the
reactor. One was packed with Anakrom Q (zero coffee loading)
and was used for calibration and the other was packed with
freeze-dried coffee on an inert support at 5%. A system of on
and off valves permitted the flow of dry carrier gas or
humidified gas through only one column at a time depending on
the experiment run.

Temperature control in the reactor oven was maintained by
submerging the column system in a glycol bath which kept the
column temperature constant within $\pm 1^{\circ}C$. To assure a uniform
temperature throughout the columns, tubing spacers, were

placed between the hot (125°C) injection port of the reactor and the column outlet and the tubing connecting the reactor with the gas chromatograph. The reactor was also equipped with a hot (125°C) injection port, which could be used to inject pulses of water vapor in the mobile phase, to increase the water vapor pressure of the system, whenever this was desirable.

5. Gas chromatograph

A Varian 3700 research gas chromatograph was used to assay water vapor. The chromatograph was equipped with a quad filament thermal conductivity detector cell and an analytical column. A column packed with Porapak Q was used to buffer pressure surges which occurred when the humidifier was switched on, to detect any air diffusing through the system and to prevent any coffee particles from reaching the detector. Temperature control in the chromatographic oven was mantained by circulating air. The temperature was constant to within $\pm 0.5^{o}$C.

The operating conditions were as follows:

Column temperature:	180°C
Detector temperature:	250°C
Injection port temperature:	230°C
TCD current:	151ma
Carrier gas, flow rate:	60cc/min

B. Materials

1. Freeze-dried coffee, packaged in a hermetically closed jar was purchased from a supermarket. The brand was "Maxim" a product of the Maxwell House Division, General Foods Corporation White Plains, N.Y.

2. Inert support, Anakrom Q 60/80 mesh (Diatomaceous earth-special acid washed and silanized) supplied by Analabs, Inc., N. Haven, CT.

3. Porapak Q (Polymerizing ethylvinyl) benzene and divinyl benzene), gas chromatographic column packing material, manufactured by Water Associates Inc., Milfor, Mass. and supplied by Analabs, Inc.,

4. Distilled water.

5. Potassium nitrate (KNO_3) in the form of crystals (certified A.C.S.) was supplied by Fisher Scientific Co.

C. Methods

a. Selection of loading level

The assumptions involved in the theory of equilibrium
chromatography require that peak broadening be minimized and
diffusion diminished in order to ensure chromatographic
ideality. These requirements ruled out using freeze-dried
coffee in its original bulk form. The coffee had to be
diluted with an inert support material. Furthermore, it was
reasoned that by coating the freeze-dried coffee on diatomaceous earth as a thin layer would reduce the diffusional
resistance of the material and minimize hysteresis. The
selection of the level of coffee loading was based on
optimizing conditions. It was established through preliminary
experiments that sufficient coated coffee was needed to differentiate the retention time of injected water from that of
an unsorbed species (air). If too much of loading was used,
too much peak broadening would occur. Peak broadening was
increased dramatically with loadings of 7.5 - 10% (w/w), while
peaks obtained at 5% (w/w) were sufficiently narrow to be
acceptable.

b. Freeze-dried coffee coating on Anakrom Q 60/80 mesh by lyophilization

The amount of freeze-dried coffee and anakrom needed for
coating was determined by the desired coffee loading 5% (w/w).
These materials were weighed, using an analytical balance with
an accuracy of ±0.1 mg. Then the weighed freeze-dried coffee
was dissolved with cold distilled water (75ml) in a sealable
container. After the coffee dissolved, a preweighed amount of
solid inert support was mixed with the coffee solution. The
mixture was stirred vigorously to assure complete and homogeneous impregnation of the anakrom support with coffee
solubles. After 2-3 hours the container with the sample was
transferred to a low temperature (-50°C) glycol bath (shell-
freezer), where the mixture was solidly frozen. The slush-
freezing technique employed and the rotation of the container
in the bath, allowed for continuous mixing and uniform distribution of the coffee solids on the anakrom particles, throughout
the frozen layer formed on the inside of the container. Sub-
limation of the ice phase and further drying of the matrix was
accomplished by a freeze-dryer (Labconco-12), operating under a
high vacuum (10 microns Hg) and at a condenser temperature of
-75°C.

The dryness of the sample was determined by valving off
the vacuum pump and measuring the vapor pressure of the drying
chamber. A pressure rise of 10 microns or less was considered

as the termination point for the drying process. As soon as
the vacuum was broken the container was sealed immediately to
prevent any moisture absorption of the dried material from the
atmosphere. The resulting powdery material was very homogeneous
and uniform, and was either used without sieving to pack the
column immediately or stored in a freezer (-12°C) until used.

2. Preparation of the columns

Aluminum tubing (3' x ¼" O.D.) was used as a column.
Packing was done with the aid of a vacuum pump and a mechanical
vibrator. The amount of packing in the column was determined
by weighing the columns and the container holding the stationary phase before and after packing. The descriptions of the
packed columns are as follows:

i)   Column packed with Porapak Q, amount of packing: 7.6g

ii) Column packed with Anakrom Q, amount of packing: 5.0g

iii) Column packed with 5% freeze-dried coffee coated on
     Anakrom Q, percent loading: 5% w/w

A different column was prepared and used to determine
each sorption isotherm. Thus the amount of packing material
ranged from 5 to 5.5g.
The amount of coffee in the column, as calculated, varies
from 0.25 to 0.30g.
In order to remove most of the moisture (without affecting the structural properties of coated coffee) that might
have been picked up during the process of packing, all the
columns were conditioned for 48 hours by passing dry helium
through them, before being used experimentally.

3. Measurement of the carrier gas flow rate and pressure

The flow rate of the carrier gas was measured at room
temperature (25°C) at the outlet of the gas chromatograph
column using a soap bubble flow meter and a stop watch.
The inlet and outlet pressures of both reactor and
chromatograph columns were measured using a mercury manometer
with a range up to 100cm Hg. These flow rate and pressure
measurements allowed for calculation of the pressure drop in
the system and correction of the flow rate of the carrier gas,
using the James and Martin (1952) equations:

Eq. 27 $$J = 3/2 \frac{(Pi/Po)^2 - 1}{(Pi/Po)^3 - 1}$$

Eq. 28 $w = W_n \cdot \dfrac{T}{T_r} \, J$

where:

  J = James and Martin compressibility factor

  Pi = pressure at column inlet

  Po = pressure at column outlet

  w = corrected carrier gas flow rate

  W_n = measured carrier gas flow rate

  T = column temperature

  T_r = room termperature

4. Calibration curve construction

Saturated salt solutions of potassium nitrate (KNO_3) were prepared in 120cc vials, sealed properly and placed in a controlled temperature chamber at 30°C. The solutions were left in the incubator for one week to equilibrate and establish a 92% relative humidity, as given by the tables (Rockland, 1960).

Using a gas tight syringe different volumes (0.1-0.5 ml) of head space were withdrawn from the vials containing the KNO_3 solution and injected in the analytical column of the gas chromatograph, while the responses were recorded. The gas chromatograph operating conditions are reported on page 23

Based on the Kiselev-Yashin procedure the amounts of water injected with the different volumes of salt solution head space were converted to a concentration of water in the carrier gas or water vapor partial pressure. By plotting concentration or partial pressure versus peak height response, the concentration and partial pressure calibration curves were constructed (Figure 8). These calibration curves were used to calculate the water vapor concentration or water vapor partial pressure of the humidified gas coming out of the reactor columns.

5. Development of the frontal and desorption chromatograms

By switching on the humidified gas and flushing the column with dry carrier gas after equilibrium was established, the sorption and desorption chromatograms of three predetermined step-heights (0 - 3.0mm Hg) were obtained at three temperatures (25°C, 35°C and 45°C). Each step-height was run for both columns, the anakrom and the column packed with coated coffee, in order to correct for any anakrom sorption if

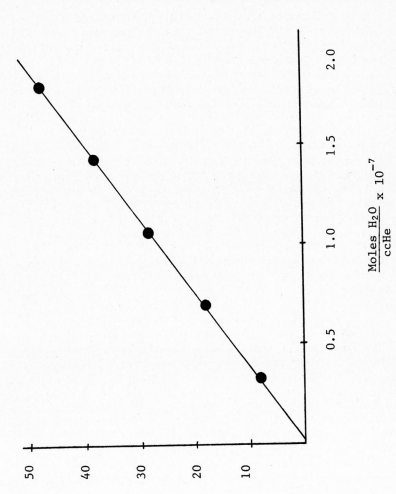

FIGURE 8. Calibration curve.

Moles H_2O $\times$ 10^{-7} / ccHe

necessary. Calibration for the emergence of an unsorbed
species (air) was done for both columns by injecting a small
volume of air prior to and after each isothermal step-height
run.

6. Injection of water pulses

Different amounts of water (0.5μl - 7μl) were injected
at the equilibrium plateau of the steps, at three temperatures
(25°C, 35°C and 45°C) and in both columns, the anakrom and the
column packed with coated coffee.

A Unimetrics 1μl capacity syringe and Hamilton 5 and 10μl
capacity syringes were used for the injections. Each injection
was repeated three times at each step and temperature setting.
An average of the three injections was used for the calculations.

Pulse data obtained for the anakrom column was used to
correct for any water sorption by the inert support material.
In a similar way, calibration for the emergence of an unsorbed
species (air) was done for both columns prior to and after
each isothermal water injection series.

7. Integration of the response areas

The response areas of the chromatograms were integrated
using a planimeter (L. A. Scientific Co.).

V. RESULTS AND DISCUSSION

A. Sorption Isotherms

The water sorption isotherms of freeze-dried coffee
coated on anakrom, are presented in Figures 9, 10 and 11.

The isothermal data were plotted as water uptake, a,
versus water vapor partial pressure, P. The use of pressure
instead of water activity, A_w, for sorption isotherms of more
than one temperature, follows from the inherent temperature
dependence of the term A_w in food systems.

As it can be seen from the graphs, water sorption
isotherms for coffee solubles were determined at 25°C, 35°C
and 45°C and at three step-heights equivalent to a water
vapor pressure of 0, 1.4 and 2.97mm Hg. The values for water
uptakes obtained were from 5 to 100 x 10^{-3} g of water/g of
coffee solids. The water vapor pressures covered, range from
0.15 to 15.5 x 10^{-3} atmospheres or in terms of water activity,
the A_w values range from 0.005 to 0.25.

Figure 9 represents the water sorption isotherms of coffee
solids coated on anakrom, as calculated from the pulse data
obtained at a step-height of zero water vapor partial pressure.

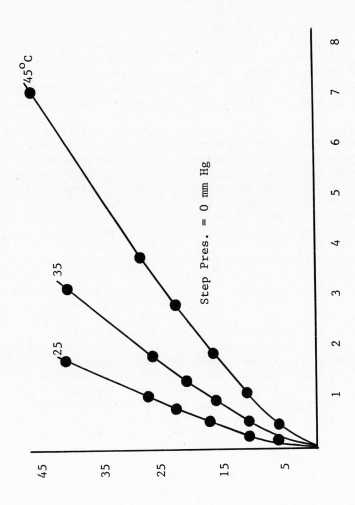

FIGURE 9. Sorption isotherms of 5% freeze dried coffee coated on anakrom Q.

Figures 10 and 11 show the water sorption isotherms calculated from the frontal and pulse data obtained at two different step-heights, equivalent to 1.4 and 2.9mm Hg of water vapor pressure, respectively. The shape of the water sorption isotherms in this work, is very much dependent on the step-height water vapor partial pressure. The sorption isotherms of the zero step-height are of a convex non-linear shape, as opposed to those of higher step-heights which exhibit a concave shape. This is probably due to the approach of true equilibrium conditions as the step-height is increased.

Increase of the step-height water vapor pressure introduced in the system, changes the shape of the elution peaks. Furthermore, peaks with highly diffused trailing become less diffused and as a result the distribution of the isotherms is different.

Also, the slope of the sorption isotherms increases quite steeply as the water vapor pressure increases with the increase of the step-height and the injection of larger amounts of water at the equilibrium plateaus. This can can be explained by the fact that at higher water vapor pressures structural changes may occur in the system which result in exposure of an increased number of active sites, higher water uptakes, and as a consequence the sorption isotherms start curving upwards.

From an examination of all the sorption isotherms obtained, it is apparent that water sorption on coffee solubles is favored by low temperatures.

Water uptakes obtained for freeze-dried coffee coated on anakrom, by the step and pulse method were higher than those reported in the literature (Hayakawa et al., 1978) for bulk freeze-dried coffee which had been determined by the static method.

The difference in water uptake between coated and bulk freeze-dried coffee can be explained by the fact that coated has a larger surface area with a greater number of active sites compared to bulk freeze-dried coffee. Preliminary data obtained in this work for both types of coffee by the step and pulse method showed that coated coffee sorbed higher amounts of water than bulk freeze-dried coffee under the same conditions.

B. Free Energy of Sorption, ΔG_s

The free energy (ΔG_s) of water sorption by coffee solids, was calculated using equation (25). The thermodynamic parameter, ΔG_s, represents the tendency of the coffee-water system for interaction.

The free energy (ΔG_s) values of the water mixing with coffee were to be negative in the step-height, uptake and

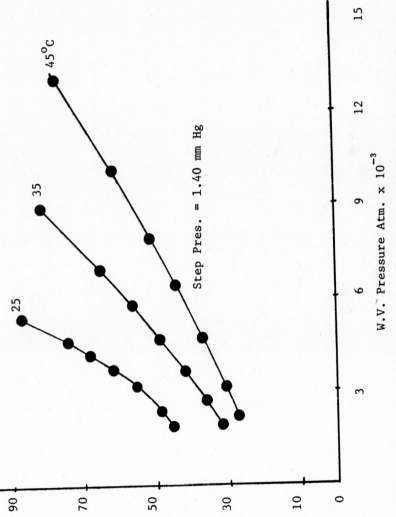

FIGURE 10. Sorption isotherms of 5% freeze dried coffee coated on anakrom Q.

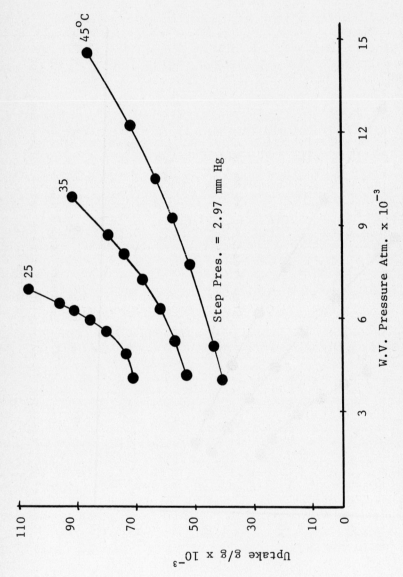

FIGURE 11. Sorption isotherms of 5% freeze dried coffee coated on anakrom Q.

temperature range covered. The negative values indicate that the water-coffee interaction is an energetically favorable process.

ΔG_s values obtained at 35°C and at three different step-height water vapor partial pressures are presented as a function of the water uptake in Figure 12.

In all cases, the free energy, ΔG_s values initially were -3.5 kcal/mole for the lowest uptake values, and became less negative with increased amounts of sorbed water and step-height water vapor pressures. This indicated that the driving force toward mixing in the water-coffee system was decreasing and reaching sorption equilibrium faster.

C. Isoteric Heat or Enthalpy of Sorption, ΔH_s

The ΔH_s values for the water sorption of coated coffee, were determined using equation (24). Actually by plotting the natural logarithms of the water vapor pressures of the 25°C, 35°C and 45°C isotherms at fixed water uptakes versus the inverse of temperature (°K) values, straight lines with slopes equal to $\dfrac{\Delta H_s}{R}$ were obtained. The slopes of these lines were used subsequently for the calculation of the ΔH_s values.

A plot of the ΔH_s values calculated for different water uptakes as a function of the step-height water vapor partial pressure, is shown in Figure 13.

The ΔH_s values obtained for the water sorption process were negative within the step-height, water uptake and temperature range covered. These negative ΔH_s values indicated the existence of attractive forces in the water-coffee system, which in turn governed the binding of water on the coffee solids.

ΔH_s values were approximately -12 kcal/mole for the very low water uptakes and zero step-height water vapor partial pressure. As the step-height and the amount of sorbed water increased ΔH_s values became less and less negative until they leveled off at about -7 kcal/mole.

An explanation of this $-\Delta H_s$ decrease offers itself when it is recalled that the sorbent may exhibit sorption active sites with different binding energies.

Thus, the water molecules may be sorbed preferentially onto points where the attractive forces are strongest; followed by sorption on active sites of less and less activity. Eventually a point is reached where all remaining sites have the same binding capacity.

D. Entropy of Sorption, ΔS_s

Entropy of sorption, ΔS_s values were calculated using equation (29).

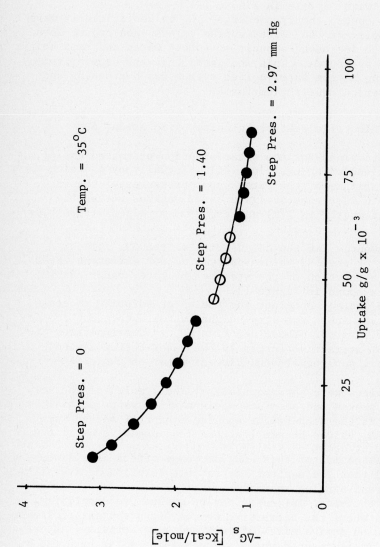

FIGURE 12. Free energy of water sorption on 5% freeze dried coffee coated on anakrom Q as a function of step height–V. pressure

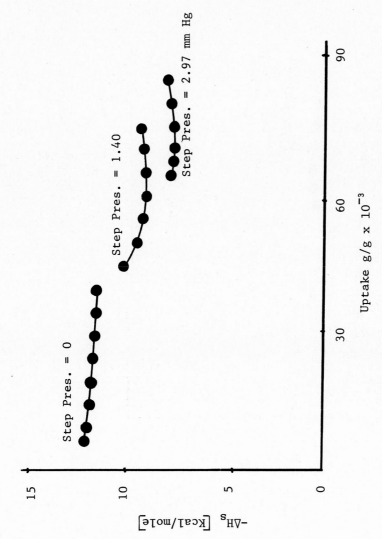

FIGURE 13. Enthalpy of water sorption on 5% freeze dried coffee coated on anakrom Q as a function of step height – V. pressure

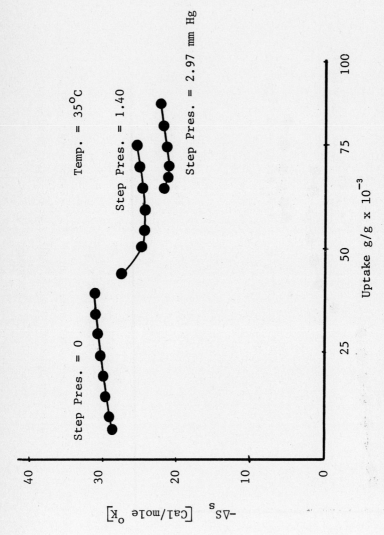

FIGURE 14. Entropy of water sorption on 5% freeze dried coffee coated on anakrom Q as a function of step height – V pressure

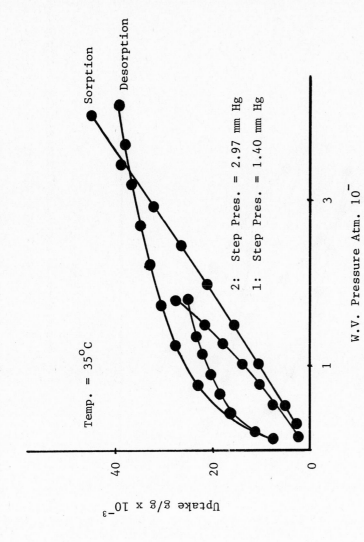

FIGURE 15. Sorption-desorption isotherms of 5% freeze dried coffee coated on anakrom Q.

Hysteresis as a function of step height-V pressure

87

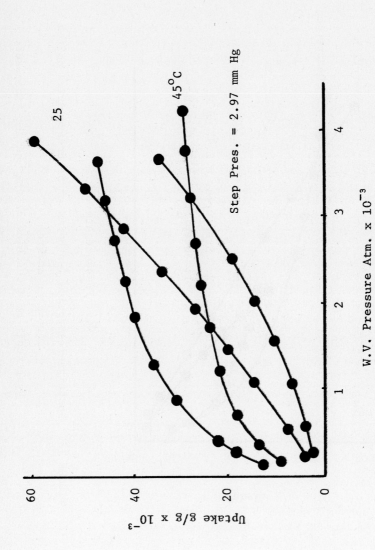

FIGURE 16. Sorption-desorption isotherms of 5% freeze dried coffee coated on anakrom Q. Hysteresis as a function of temperature

A plot of the ΔS_s values calculated at different water uptakes as a function of the step-height water vapor partial pressure at a temperature of 35°C, is shown in Figure 14.

ΔS_s values are negative (-30 Cal/mole) at zero-step-height and become slightly more negative as the water uptake increases up to 25g H_2O/g solid x 10^{-3}. This indicated that the water-coffee system became more structured at low water uptakes. However, as the water uptake increased with increased step-height water vapor pressure, ΔS_s values became less negative (-20 Cal/mole), and the system eventually became more disordered.

E. Hysteresis

Desorption isotherms obtained by frontal chromatography showed hysteresis compared to sorption isotherms. Hysteresis loops were scanned over a water vapor pressure range of 0 - 3.5 x 10^{-3} atmospheres and at temperatures of 25°C, 35°C and 45°C, (Figures 15 and 16).

The occurrence of hysteresis demonstrates that the coffee-water system, though reproducible, is not in true equilibrium and as a result the sorption-desorption process is irreversible. Hence, there is an entropy production in the system that indicated the degree of irreversibility, including structural changes occurring in the sorbing solid matrix. The extent of these structural changes seemed to be a function of the amounts of sorbed water. The higher the amounts of sorbed water at higher step-height water vapor pressures, the greater the hysteresis loops observed (Figure 15).

Hysteresis is also a temperature dependant phenomenon, as is shown in Figure 16. Temperature increases make the hysteresis loop shift towards lower water uptakes. This can be explained by the lower energy inputs required for desorption at higher temperatures.

VI. BIBLIOGRAPHY

Apostolopoulos, D. V., 1980. The interaction between PVC and VCM using I.P.G.C. - Thermodynamic and structural considerations. M.S. Thesis, Rutgers University, Food Science Dept., New Brunswick, N. J. 08903.

Arnell, J. C. and H. C. McDermot, 1957. Sorption hysteresis. In "Surface Activity." Vol. 2. Edited by J. H. Schulman, p. 3. Butterworths, London.

Bain, J. W., 1935. An explanation of hysteresis, in the hydration and dehydration of gels. J. Am. Chem. Soc. 57. 699.

Benson, S. W. and B. A. Richardson, 1975. Reversibility of water vapor sorption by cottage cheese whey solids. J.

Dairy Sci., 58, 25.

Bettelheim, F. A. and D. H. Velman, 1957. Pectic substances-
    water. Part II. Thermodynamics of water vapor sorption.
    J. Polym. Sci., Vol., XXIV, 445.

Braun, J. M. and J. E. Guillet, 1976. Study of polymers by
    I.G.C. Advances in polymer science, Vol. 21, 108.
    Springer - Verlag.

Brown, T. L. and H. E. LeMay, 1077. Free energy, entropy and
    equilibrium. In Chemistry. Edited by T. L. Brown and E.
    H. LeMay, p. 429. Prentice-Hall, Inc.

Cerny, S., 1970. Theory of adsorption on active carbon. In
    "Active carbon, manufacture, properties and applications."
    Edited by M. Smisek and S. Cerny, 71-154. Elsevier
    Publishing Co.

Coelho, U. J., 1978. The use of I.P.G.C. in determining the
    water activity of collagen in the bound water region. M.S.
    Thesis, Rutgers University, Food Science Dept., New
    Brunswick, N. J. 08903.

Conder, J. R., 1968. Physical measurement by gas chromato-
    graphy. Progress in gas chromatography. Advances in
    analytical chemistry and instrumentation. Vol. 6, 209.

Conder, J. R. and J. Purnell, 1969. Theory of frontal and
    elution techniques of thermodynamic measurement. Trans
    Faraday Soc., 13, 337.

Dole, M. and A. D. McLaren, 1947. The free energy, heat and
    entropy of sorption. J. Am. Chem. Soc., 69, 50.

Eberly, P. E., 1961. Measurement of adsorption isotherms and
    surface area by continuous flow method. J. Trans Faraday
    Soc., 65, 1261.

Eberly, P. E. and C. N. Kimberlin, 1961. High temperature
    adsorption on silica-alumina and melybdena-alumina
    catalysts by use of flow techniques. J. Trans Faraday
    Soc., 57, 1169.

Evans, M. B. and R. Newton, 1976. I.G.C. in the study of
    polymer degradation. Part I oxidation of squalene as a
    model for the oxidative degradation of natural rubber.
    Chromatographia, 9, 11.

Gilbert, S. G., 1982. New methods for storage life determina-
    tion. Journal series, N. J. Agricultural Experimental
    Station, Cook College, Rutgers University, Food Science
    Dept., New Brunswick, N. J. 08903.

Gilbert, S. G., 1982. Inverse gas chromatography. Journal
    series, N. J. Agricultural Experimental Station, Cook
    College, Rutgers University, Food Science Dept., New
    Brunswick, N. J. 08903.

Greenstein, G. and P. Issenberg, 1967. Adsorption of Volatile
    Organic compounds in dehydrated food systems. I.
    Measurement of sorption isotherms at low water activities.

Paper presented at the synposium on dehydration. 27th annual I.F.T. meeting Minneapolis, Minnesota. May 17, 1967.

Gregg, S. J., 1951. Surface chemistry of solids, p. 15. Reinhold Publishing Co.

Hassler, J. W., 1963. Elementary aspects of adsorption by active carbon. In "Activated carbon." Edited by J. W. Hassler. p. 15. Chemical Publishing Co., Inc., New York, N. Y.

Hayakawa, K. I., J. Matas, and M. P. Hwang, 1978. Moisture sorption isotherms of coffee products. J. of Food Sci., Vol. 43, 1026.

Huber, J. F. K. and R. G. Gerritse, 1971. Evaluation of dynamic gas chromatographic methods for determination of adsorption and solution isotherms, J. Chrom., 58, 137.

Hussey, C. L. and J. F. Parcher, 1976. A modified gas chromatograph for thermodynamic measurements by frontal chromatography. J. Chrom. 92, 47.

Iglesias, H. A., J. Chirife, and P. Viollaz, 1976. Thermo-dynamics of water vapor sorption of the sugar beet root. J. Food Technol. 11, 91.

Iglesias, H. A. and J. Chirife, 1976. Isosteric heats of sorption. Part I. Analysis of the differential heat curves. Leben. Wiss. Tech. 9: 2-116.

Iglesias, H. A. and J. Chirife, 1976. Isosteric heats of water vapor sorption on dehydrated foods. Part II. Hysteresis and heat of sorption comparison with BET theory. Leben. Wiss. Tech., 9: 2-123.

James, D. H. and C. S. G. Phillips, 1954. The chromatography of gases and vapors. Part III. The determination of adsorption isotherms. J. Chem. Soc., 1066.

Kalinichev, A. I., Y. A. Pronin, K. V. Chmutov, and N. A. Goryacheva, 1978. Theory of non-linear frontal chroma-tography. J. Chrom. 152, 311.

Kapsalis, G. J., 1981. Moisture sorption hysteresis. In "Water activity" Influence of food quality." Edited by L. E. Rockland and G. F. Stewart, p. 143. Academic Press, Inc.

Karel, M., 1975. Retention of organic amp including flavor during freeze-drying. Principles of food science. Part II. Physical principles of food preservation. Edited by O. R. Fennema, p. 379. Marcel Dekker Inc., New York and Basel.

Khalil, H., 1976. A gas chromatographic study on the effect of co - and ter - polymerization on the interaction of polyacrylonitrile with selected low molecular compounds. Ph.D. Thesis, Rutgers University, Food Science Dept., New Brunswick, N. J. 08903.

Kinigakis, P., 1979. The migration of VCM from food grade PVC into selected food stimulants: Equilibrium partition and thermodynamic considerations. M.S. Thesis, Rutgers University, Food Science Dept., New Brunswick, N. J. 08903.

Kiselev, A. Y. and Y. A. Yashin, 1969. "Gas adsorption chromatography." Plenum Press, New York, N. Y.

Kucera, P., S. A. Moros and A. R. Mlodozeniec, 1981. Differential frontal analysis of carboxylic acids. J. Chron., 13, 742.

Labuza, T. P., 1980. Enthalpy/entropy compensation in food reactions. Food Technology, Feb. 1980, p. 67.

Moreyra, R. and M. Peleg, 1981. Effect of equilibrium water activity on bulk properties of selected food powder. J. of Food Sci., Vol. 46, 1918.

Peleg, M., C. H. Mannheim and N. Passy, 1973. Flow properties of some food powders. J. Food Sci., Vol. 38, 959.

Pietsch, W. B., 1969. Adhesion and agglomeration of solids during storage, flow and handling. J. of Engin. for Industry, 435.

Rockland, L. B., 1969. Water activity-storage stability. Food Technology, Vol. 23, 1241.

Senich, E. and I. C. Sanchez, 1979. I.G.C., a method for studying polymer migrant interactions in polyolefin packaging materials. Org. Coat. Plast. Chem. 41, 345.

Smith, D. S., C. H. Mannheim, and S. G. Gilbert, 1981. Water sorption isotherms of sucrose and glucose by I.G.C. J. Food Sci., 46(4): 1051.

Smith, D. S., 1982. The thermodynamic and structural aspects of water sorption by starches using I.G.C. Ph.D. Thesis. Rutgers University, Food Science Dept., New Brunswick, N. J. 08903.

Stock, R., 1961. Determination of surface area by gas chromatography. Anal. Chem. Vol. 33, No. 7, 967.

Valentin, P., 1976. Determination of gas-liquid and gas-solid. Equilibrium isotherms by chromatography: I. Theory of the step and pulse method. J. of Chrom. Sci., Vol. 14, 56.

Valentin, P., 1976. Determination of gas-liquid and gas-solid. Equilibrium isotherms by chromatography: II. Apparatus, specifications and results. J. of Chrom. Sci., Vol. 14, 136.

Wolf, M., J. E. Walker and J. G. Kapsalis, 1972. Water vapor sorption hysteresis and dehydrated foods. J. Agr. Food Chem. 20, 1073.

CHEMICAL STUDIES ON TROPICAL FRUITS

Kenji Yamaguchi
Osamu Nishimura
Hisayuki Toda
Satoru Mihara

Ogawa and Co., Ltd.,
6-32-9 Akabanenishi, Kita-Ku, Tokyo, Japan

Takayuki Shibamoto[1]

Department of Environmental Toxicology,
University of California, Davis, CA 95616

I. INTRODUCTION

More than six hundred species of tropical and subtropical
fruits are grown in the world. Nearly twenty-five species
are sold commercially today. Tropical fruits have attracted
peaple since ancient times. Their unique flavors and bright
colored appearance give an exotic atomosphere to a table.
Recently, many soft drinks using tropical fruit flavors have
also appeared on the market in addition to the fresh fruits
themselves. To accommodate the increasing demand, the production of tropical fruits has been increasing steadily. The
market for tropical fruit is as yet limited, but the potential
is there.
In this study, the four most common tropical fruits-mango,
papaya, passion fruit, and avocado were subjected to chemical
analysis using modern analytical instruments (GLC, GC/MS, 1H
NMR, 13C NMR, IR, UV). A newly developed fused silica capillary column was used for analysis.

---

[1]To whom all inquiries should be addressed.

93

II. EXPERIMENTAL

A. Sample Preparations

All fruits were obtained commercially from the places listed in Table I. Peel and seeds were removed from the fresh fruits and edible parts were homogenized with 1.5 l of de-ionized water using a homogenizer (the amounts used are shown in Table I). The homogenized samples were water distilled under reduced pressure (40 mm Hg) in a nitrogen stream at 40° C using a water bath. An acetone/dry ice trap was attached in front of the vacuum pump. The distillation was continued until 1.5 l of distillate was obtained. Each distillate was extracted with 150 ml of dichloromethane using a liquid-liquid continuous extractor for 12 hours. The extract was dried over anhydrous sodium sulfate and the solvent was removed in vacuo. The volatiles obtained were stored at 5° C for further experimentations.

In the case of passion fruits, the processed puree and juice were diluted with 1.5 l of deionized water, then water distillated under the same conditions as the fresh fruits.

The extracts prepared from fresh tropical fruits and processed tropical fruits were further concentrated under a nitrogen stream. The yields of concentrated volatiles obtained from each fruit are shown in Table I.

B. Analysis of Volatiles

Identification of volatile constituents of the tropical fruit samples was made by comparison of their mass fragmentation patterns and Kovats retention indices to those of authentic compounds.

C. Gas-Liquid Chromatography (GLC)

A Hewlett-Packard Model 5710-A gas chromatograph equipped with a flame ionization detector was used for routine gas chromatographic analysis. Two types of wall coated open tubular (WCOT) fused silica capillary columns were used: 50 m x 0.22 mm i.d. coated with Carbowax 20M and 60 m x 0.22 mm i.d. coated with OV-101. The oven temperature was programmed to increase 2° C/min. The injector and detector temperature were 250° C. The injector split ratio was 1:100. A Hewlett-Packard Model 3385-A reporting integrator was used to determine the peak area.

D. Gas Chromatography/Mass Spectrometry (GC/MS)

A Hitachi Model M-80 combination mass spectrometer-gas chromatograph (Hewlett-Packard Model 5710A) equipped with Hitachi Model M-6010 and 10 II/A data system was used under the following conditions: ionization voltage, 70 eV; emission current, 80 μA; ion accel voltage, 3100V; ion source temperature, 200° C. The gas chromatographic column and oven conditions were as described for the Hewlett-Packard gas chromatograph.

TABLE I. Sources, amounts, and yields of volatiles of tropical fruits used.

Fruits	Sources	Amounts used (Kg)	Yields of volatiles(%)
Mango (*Mangifera indica L.*)			
fruits	havested in Philippines	1.5	0.003
puree	canned in Philippines	1.5	0.007
Papaya (*Carica papaya L.*)			
fruits	Harvested in Hawaii	1.3	0.008
puree	canned in Taiwan	1.5	0.009
Passion fruit (*hybrid F₁, Passiflora edulis Sims X P. edulis f. flavicapa*)			
fruits	harvested in Taiwan	1.0	0.011
juice	canned in Taiwan	1.5	0.019
Avocado (*Persea americana Mill*)			
fruits	harvested in Mexico	1.0	0.002

III. RESULTS AND DISCUSSION

A. Mango

Mango grows on an evergreen tree native to Southeastern Asia. It has been cultivated in India for more than 4000 years. It gradually spread from Southeastern Asia to tropical

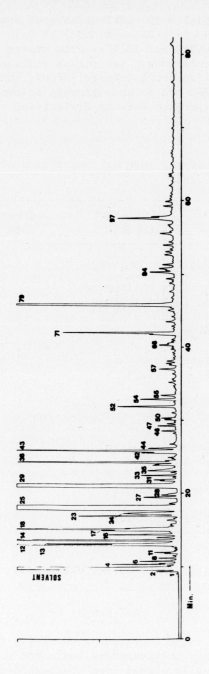

FIGURE 1. Gas chromatogram of mango fruit extract. See experimental section for gas chromatographic conditions.

and subtropical countries. The genus contains forty-one valid
varieties distributed throughout South Asia and the Malaya
Archipelago with innumerable varieties existing in other tropical areas of the world. Among those, the genus *Alphonso* is
most commonly found in the world market.

The mango fruits have a unique flavor. Angelini et al.
(1973) performed preliminary analysis of mango flavor components. Later, Hunter et al. (1974) reported forty compounds
in Alphonso mango extract.

In this study, twenty-one constituents were identified in
the Mango fruit extract. The compounds identified are shown
in Table II and a typical gas chromatogram of the mango fruit
extract is shown in Figure 1. The main components are monoterpenes of which the total gas chromatographic peak area% is
42.32. Terpinolene (peak no. 25 in Figure 1) is the major
constituent of the mango fruit extract (peak area% = 37.25).
p-Cymen-8-ol (79) is an oxidation product of monoterpenes.
p-Cymen -8-ol found in mango fruit comprises over 5% of the
total extract.

Various researchers have reported the occurrence of monoterpenes in a mango fruit. Hunter et al. (1974) reported that
cis-ocimene was a major component. Gholop and Bandyopadhyay
(1977) found cis-ocimene and myrcene as main constituents of
mango fruit extract. Craveiro et al. (1980) reported 3-carene
as a major constituent, whereas El-Baki et al. (1981) found α-
pinene, myrcene and limonene as main components.

TABLE II. Volatile components identified in the Mango
fruit extract.

Peak No. in Figure 1.	Compounds	Peak area %	Kovats index
1.	acetaldehyde	0.15	
2.	acetone	0.22	
4.	unknown	0.39	960
6.	2-methyl-3-buten-3-ol	0.22	1000
8.	ethyl n-butyrate	0.17	1024
11.	unknown	0.17	1073
12.	cis-3-hexenal	3.32	1122
13.	unknown	1.13	1130
14.	cis-3-carene	2.54	1148
15.	unknown	0.23	1162
16.	α-terpinene	0.62	1177
17.	limonene	0.84	1194
18.	trans-2-hexenal	2.56	1200
23.	acetoine	0.80	1257

24.	p-cymene	1.07	1262
25.	terpinolene	37.25	1300
27.	trichloropropane	0.42	1317
28.	unknown (M^+ = 134)	0.24	1326
29.	cis-3-hexen-1-ol	19.72	1354
31.	unknown (M^+ = 134)	0.22	1370
33.	unknown (M^+ = 134)	0.29	1381
35.	acetic acid	0.52	1400
36.	unknown (M^+ = 152)	0.22	1410
38.	unknown (M^+ = 132)	1.61	1417
42.	unknown	0.31	1442
43.	unknonw (M^+ = 152)	1.93	1449
44.	unknonw	0.31	1455
46.	unknown	0.16	1497
47.	linalool	0.17	1510
50.	unknown	0.15	1531
52.	2,5-dimethyl-4-methoxy-2H-3-furanone	0.70	1560
54.	diethyleneglycol monoethyl ether	0.40	1577
55.	unknown	0.20	1587
57.	nerol	0.23	1650
66.	unknown	0.21	1700
71.	4-methyl acetophenone	1.33	1735
79.	p-cymen-8-ol	5.07	1800
84.	unknown	0.31	1883
97.	unknown	0.82	2018

The compounds identified in mango puree are shown in Table III.   Figure 2 shows a gas chromatogram of mango puree extract
The composition of mango puree volatiles is somewhat more complex than that of mango fruit volatiles. Hunter et al. (1974) reported forty volatiles in mango puree extract. The compounds identified were monoterpenes, furans, furanones, and lactones. The main constituents found in this study were fatty acids; n-butyric acid (peak no. 78, 22.42%), n-caproic acid (105, 5.70%). The organic acids do not go through gas chromatographic columns easily. Free fatty acids have not been isolated in large quantities from fruit extracts prior to this study in which we used an inert fused silica capillary column.

Some formation of heat-induced products was observed. Furfural (55), 5-methylfurfural(74), 2-acetylfuran (61), and furfuryl alcohol (82) may be formed during heat processing. Monoterpene hydrocarbons, which were major volatiles in the mango fruit extract, were minor component in the mango puree extract. The quantities of C-6 aldehydes and alcohols which give the green-note were much less in the puree extract than in the fruit extract.

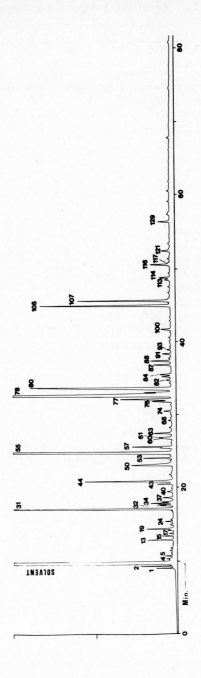

FIGURE 2. Gas chromatogram of mango puree extract. See experimental section for gas chromatographic conditions.

TABLE III. Volatile components identified in the mango puree extract.

Peak No. in Figure 2.	Compounds	Peak area %	Kovats index
1.	acetaldehyde	a	
2.	acetone	a	
5.	2-methyl-3-buten-2-ol	0.36	1000
13.	n-butanal	0.89	1107
15.	3-penten-2-ol	0.37	1133
17.	β-pinene	0.17	1149
18.	pyridine	0.20	1165
19.	isoamyl alcohol	1.06	1173
22.	myrcene	a	1202
24.	limonene	0.29	1209
31.	acetoin	9.27	1262
31.	p-cymene	a	1262
32.	unknown (M^+ = 134)	0.37	1273
33.	acetol	0.41	1276
34.	ocimene	0.66	1281
35.	unknown (M^+ = 134)	0.48	1285
37.	cis-3-hexenyl acetate	0.40	1300
37.	dimethyl formamide	a	1300
40.	n-hexanol	0.27	1320
43.	acethyl butyryl	0.52	1345
44.	cis-3-hexen-1-ol	3.18	1354
50.	acetic acid	2.53	1400
53.	unknown (M^+ = 132)	a	1419
53.	linalool oxide (I)	0.97	1419
55.	furfural	18.48	1434
55.	cis-3-hexenyl n-butyrate	a	1434
56.	linalool oxide (II)	0.30	1446
57.	unknown (M^+ = 152)	1.26	1448
60.	2,5-dimethyl-2H-furan-3-one	0.40	1470
61.	2-acetylfuran	0.45	1472
63.	ethyl 3-hydroxy butanoate	0.68	1482
68.	linalool	0.17	1514
74.	5-methylfurfural	0.34	1538
76.	2,5-dimethyl-4-methoxy-2H-furan-3-on	0.84	1562
77.	isophorone	2.12	1567
78.	n-butyric acid	22.42	1575
79.	dimethyleneglycol monoethyl ether	0.93	1582
80.	γ-butyrolactone	8.82	1593
81.	phenylacetaldehyde	a	1608
82.	furfuryl alcohol	0.26	1612
83.	2-methylbutyric acid	0.42	1621

84.	N-methylpyrolidone	0.47	1627
87.	nerol	0.67	1650
88.	α-terpineol	0.76	1659
91.	pentanoic acid	0.33	1676
93.	benzyl acetate	0.17	1686
100.	4-methyl acetophenone	0.44	1734
105.	n-caproic acid	5.70	1790
113.	unknown	0.31	1855
114.	2-phenylethyl alcohol	0.50	1860
116.	unknown	0.91	1891
117.	β-ionone	a	1894
121.	unknown	0.42	1925
129.	n-octanoic acid	0.49	2000

B. Papaya

Papaya belongs to the family *Caricacea*, which includes four genera and more than twenty species of *Carica* native to tropical and subtropical areas of North and South America. Since papaya was introduced to Hawaii in 1919, has become the most important commercial fruit produced in Hawaii. Most of the crop is exported in the fresh form. Some papaya is commercially available as a frozen puree.

The development of off-flavors has been a major problem in the production of papaya puree. In this study, we attempted to investigate the difference between volatiles in fruit and puree in order to determine the nature and origin of off-flavors in papaya puree.

Relatively little information about the composition of papaya fruit appeared in the literature until the report by Flath and Forrey (1977) who identified 106 compounds in volatile components of fresh papaya fruit. We identified twenty components in papaya fruit extract (TABLE IV). Figure 3 shows a gas chromatogram of the papaya fruit extract. The major components were fatty acids which were not reported in Flath and Forrey (1977) as free acids but rather as methyl esters derivatives. Butyric acid accounted for 37.34% of total area% of the extract. Flath and Forrey (1977) identified a series of fatty acid methyl esters in the diazomethane treated extract. Prior to the invention of fused silica capillary columns, it was difficult to measure the acid content using gas chromatography. An inert fused silica capillary column allowed us to determine the exact content of free fatty acids in the fruit extracts. The total area% of free fatty acids was 54.4% in this study.

The presence of linalool (24) in large quantity gives characteristic papaya flavor. A unique constituent, benzyl

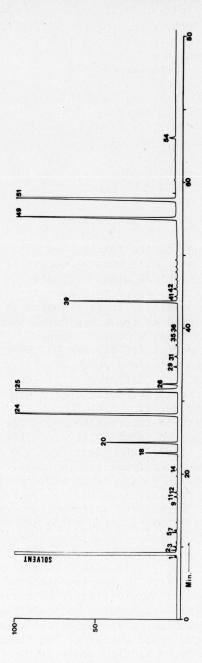

FIGURE 3. Gas chromatogram of papaya fruit extract. See experimental section for gas chromatographic conditions.

isothiocyanate (51) was identified. On the other hand, benzyl isothiocyanate was not found in the puree extract. The presence of benzyl isothiocyanate and it's glucosinolate precusor have been previously reported (Ettlinher and Hodgkins, 1956; Tang and Syed, 1972).

TABLE IV. Volatile components identified in the papaya fruit extract.

Peak No. in FIGURE 3	Compounds	Peak area %	Kovats index
1.	acetone	a	500
2.	unknown	a	940
3.	unknown	a	985
5.	n-butyl alcohol	a	1100
7.	o-xylene	a	1125
9.	acetoin	a	1248
12.	N,N-dimethylformamide	a	1309
14.	unknown	a	1365
18.	linalool oxide (furanoid)	1.50	1417
20.	linalool oxide (furanoid)	3.81	1443
24.	linalool	14.59	1504
25.	n-butyric acid	37.34	1539
26.	n-hexyl n-hexanoate	0.74	1585
29.	γ-hexalactone	a	1630
31.	n-valeric acid	a	1655
35.	linalool oxide (pyranoid)	a	1700
36.	linalool oxide (pyranoid)	a	1713
39.	n-hexanoic acid	6.19	1788
41.	geraniol	a	1800
42.	benzyl alcohol	0.15	1818
49.	n-octanoic acid	10.90	2000
51.	benzyl isothiocyanate	22.59	2059
54.	n-decanoic acid	0.45	2207

[a] peak area% less than 0.01.

The volatiles identified in the papaya puree extract are listed in TABLE V. FIGURE 4 shows a typical gas chromatogram of papaya puree extract. The major component of papaya puree extract in this study was n-butyric acid (25) which comprised 74.36% of total area% of the extract. The concentrated extract gives a strong acidic flavor due to the large quantity of fatty acids content (total area% = 83.4). The formation

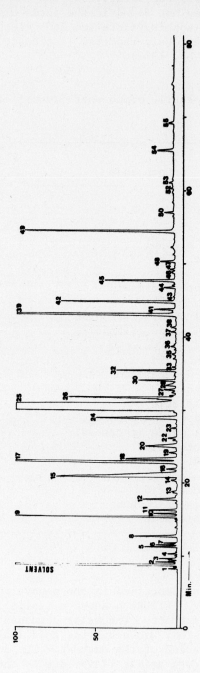

FIGURE 4. Gas chromatogram of papaya puree extract. See experimental section for gas chromatographic conditions.

of a series of fatty acids were observed in the puree extract.
Acetic acid, n-butyric acid, isovaleric acid, n-hexanoic acid,
n-heptanoic acid, n-octanoic acid, and n-decanoic acid are
probably the major source of off-flavor in the puree products.

Obvious heat-induced products furfural and furfuryl alcohol were recovered in fairly large quantities. As Chan et
al. (1973) described, it is hard to avoid the development of
off-flavors during papaya puree production.

TABLE V. Volatile components identified in the papaya
puree extract.

Peak No. in FIGURE 4	Compounds	Peak area %	Kovats index
1.	acetone	a	500
2.	unknown	a	940
3.	unknown	0.10	985
4.	unknown	a	1043
5.	n-butyl alcohol	0.23	1100
6.	p-xylene	0.12	1112
7.	o-xylene	a	1125
8.	n-amyl alcohol	0.45	1165
9.	acetoin	1.35	1248
10.	acetol	0.15	1260
11.	unknown	0.22	1271
12.	N,N-dimethylformamide	0.38	1315
13.	unknown	a	1332
14.	unknown	a	1365
15.	acetic acid	3.00	1370
16.	unknown	0.11	1395
17.	furfural	4.10	1410
18.	linalool oxide (furanoid)	0.77	1421
19.	unknown	0.10	1435
20.	linalool oxide (furanoid)	0.55	1454
21.	unknown	a	1460
22.	unknown	0.12	1465
23.	unknown	a	1484
24.	linalool	1.17	1515
25.	n-butyric acid	74.36	1539
26.	n-hexyl n-hexanoate	1.20	1585
27.	furfuryl alcohol	0.15	1609
28.	isovaleric acid	0.21	1620
30.	cis-3-hexenyl n-valerate	0.31	1634
32.	α-terpineol	0.52	1658
33.	unknown	a	1665

34.	benzyl acetate	a	1684
35.	linalool oxide (pyranoid)	a	1700
36.	linalool oxide (pyranoid)	a	1713
37.	unknown	a	1748
38.	nerol	a	1760
39.	n-hexanoic acid	3.94	1788
40.	dihydro-β-ionone	a	1795
41.	geraniol	0.23	1800
42.	benzyl alcohol	1.18	1818
43.	benzyl n-butyrate	a	1834
44.	β-phenyl ethyl alcohol	0.12	1858
45.	phenylacetonitrile	0.77	1876
46.	β-ionone	a	1900
47.	n-heptanoic acid	a	1908
48.	unknown	0.16	1923
49.	n-octanoic acid	1.68	2000
50.	unknown	0.10	2050
52.	eugenol	a	2112
53.	o-,tert-butylphenol	a	2127
54.	n-decanoic acid	0.22	2207
55.	unknown	a	2265

[a]peak area% less than 0.01.

C. Passion Fruit

There are about four hundred known species of passion fruit, genus *Passiflora*. About fifty species bear edible fruits but only a few have commercial value. Passion fruit is consumed fresh or as a juice. It has a mixed fruity flavor and is rich in nutrients.

We identified forty-four compounds in passion fruit extract. The compounds identified are shown in TABLE VI. FIGURE 5 shows a gas chromatogram of the passion fruit extract The main components were C-6 alcohol derevatives which give a strong green-note: n-hexyl-n-butyrate (33), cis-3-hexenyl-n-butyrate (38), n-hexyl-n-hexanoate (55), and cis-3-hexenyl-n-hexanoate (61). The peak area% of those esters is over 50 of the extract. Many esters possesing fruity flavors were also isolated. Ethyl butyrate (6) is an important flavor ingredient in soft drinks and confectionaries. The level of β-ionone is high in the passion fruit compared with other tropical fruits. β-Ionone must contributes characteristic flavor to passion fruit because of its low odor threshold.

Parliment (1972) isolated several free fatty acids in large quantities from the passion fruit. We isolated, however only two fatty acids; n-hexanoic acid and n-heptanoic acid.

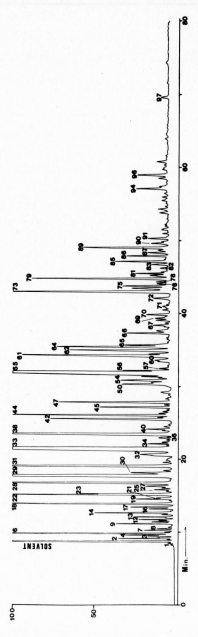

FIGURE 5. Gas chromatogram of passion fruit extract. See experimental section for gas chromatographic conditions.

Acetol and α-humulene have not been reported in passion fruit
volatiles prior to this study.

TABLE VI. Volatile components identified in the passion
fruit extract.

Peak No. in FIGURE 5	Compounds	Peak area %	Kovats index
1.	acetone	a	500
2.	unknown	0.43	957
3.	unknown	0.11	990
4.	unknown	0.83	1010
6.	ethyl n-butyrate	9.11	1031
7.	n-butyl acetate	0.39	1052
8.	isoamyl acetate	0.26	1075
9.	n-butyl alcohol	0.89	1112
10.	p-xylene	0.25	1116
11.	m-xylene	0.17	1121
12.	o-xylene	0.38	1125
13.	cyclopentanone	0.45	1157
14.	n-amyl alcohol	1.22	1175
16.	unknown	0.25	1200
17.	n-butyl n-butyrate	0.49	1207
18.	ethyl n-hexanoate	2.54	1223
19.	unknown	0.38	1243
21.	unknown	0.12	1247
22.	acetoin	1.60	1259
23.	n-hexyl acetate	1.23	1260
25.	unknown	0.27	1275
27.	2-heptanol	0.45	1289
28.	N,N-dimethylformamide	0.49	1305
28.	cis-3-hexenyl acetate	3.97	1305
29.	n-hexanol	10.60	1322
30.	trans-3-hexenol	0.62	1333
31.	cis-3-hexenol	5.50	1352
32.	unknown	0.81	1388
33.	n-hexyl n-butyrate	14.34	1394
34.	unknown	0.17	1411
36.	unknown	0.18	1429
38.	cis-3-hexenyl n-butyrate	4.87	1440
40.	isoamyl n-hexanoate	0.28	1452
42.	ethyl 3-hydroxybutyrate	1.60	1477
44.	benzaldehyde	2.44	1485
45.	linalool	0.81	1506
47.	n-octanol	1.45	1519

50.	4-terpineol	1.07	1567
52.	unknown	0.86	1575
54.	unknown	0.43	1586
55.	n-hexyl n-hexanoate	8.32	1590
56.	unknown	0.48	1600
57.	unknown	0.28	1613
60.	unknown	0.31	1624
61.	cis-3-hexenyl n-hexanoate	2.50	1631
62.	unknown (M^+ = 204)	1.61	1645
64.	α-terpineol	1.69	1659
65.	α-humulene	0.64	1669
66.	benzyl acetate	0.61	1689
67.	unknown	0.28	1706
68.	unknown	0.14	1720
69.	unknown	0.16	1724
70.	unknown (M^+ = 204)	0.33	1735
71.	methyl salycilate	a	1754
72.	unknown (M^+ = 204)	0.35	1774
73.	n-hexanoic acid	2.68	1788
74.	dihydro-β-ionone	0.76	1791
75.	geraniol	0.73	1800
76.	unknown	0.13	1806
78.	unknown	0.16	1812
79.	benzyl alcohol	2.05	1818
80.	unknown	0.34	1821
81.	benzyl isobutyrate	0.73	1834
82.	unknown	0.16	1844
83.	β-phenylethyl alcohol	2.74	1852
85.	unknown	0.69	1865
86.	n-heptanoic acid	0.58	1882
87.	butyrated hydroxytoluene	0.31	1888
88.	β-ionone	1.34	1893
89.	unknown	0.26	1923
90.	unknown	0.29	1935
93.	unknown	0.35	2045
95.	-decalactone	0.49	2084
96.	dihydroactinidiolide	0.25	2260

a peak area% less than 0.01.

Forty-six volatile contituents were isolated and identified in the extract of passion fruit juice in this study (
TABLE VII). FIGURE 6 shows a gas chromatogram of the passion
fruit juice extract. The major component was furfural (37),
a sugar degradation product. It is somewhat surprising that
furfural was present in a large quantity (21.09%) since we
could not detect any furlfural in the extract of the fresh

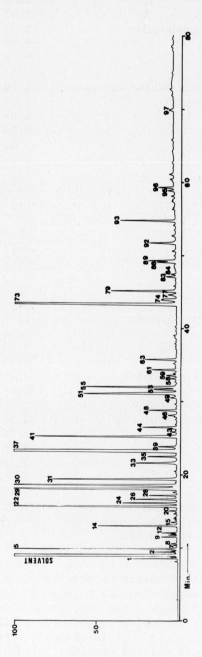

FIGURE 6. Gas chromatogram of passion fruit juice extract. See experimental section for gas chromatographic conditions.

fruit. This may be due to the juice canning process. Trace
amounts of furfural were reported in passion fruit extract by
Winter and Kloti (1972). The free fatty acids contents were
also higher in the fruit juice extract than in the fresh
fruit extract. Chan et al. (1973) reported that formation of
furan derivatives and fatty acids caused off-flavor of passion fruit products.

The composition of C-6 alcoholderivatives (total area% =
22.1), which play an important role in fruit flavors, is
similar to that of the fresh fruit. However, the concentrated extract did not give a fresh fruity-note due to the presence of furfural and fatty acids.

TABLE VII. Volatile components identified in the passion
fruit juice extract.

Peak No. in FIGURE 6	Compounds	Peak area %	Kovats index
1.	acetone	0.57	500
2.	unknown	0.31	957
5.	2-methyl-3-butene-2-ol	2.98	1015
7.	n-butyl acetate	0.12	1052
8.	isoamyl acetate	0.24	1075
9.	n-butyl alcohol	0.37	1112
10.	p-xylene	0.10	1116
11.	m-xylene	a	1121
12.	o-xylene	0.26	1124
14.	n-amyl alcohol	1.76	1175
15.	unknown	0.15	1182
20.	2-methyl tetrahydrofuran-3-one	0.19	1245
22.	acetoin	5.97	1259
24.	acetol	1.51	1270
26.	3-methyl-2-butene-1-ol	1.07	1281
28.	N,N-dimethylformamide	0.68	1305
29.	n-hexanol	8.39	1322
30.	trans-3-hexenol	5.86	1333
31.	cis-3-hexenol	3.64	1352
33.	n-hexyl n-butyrate	0.25	1394
33.	acetic acid	1.02	1394
35.	linalool oxide (furanoid)	0.91	1421
37.	furfural	21.09	1434
39.	linalool oxide (furanoid)	0.62	1442
41.	acetylfuran	1.93	1466
41.	2,5-dimethyl-2H-furan-3-one	3.07	1466
43.	unknown	0.10	1480

44.	benzaldehyde	0.89	1485
46.	isobutyric acid	0.32	1511
48.	5-methylfurfural	0.79	1524
49.	2,2,6-trimethyl-2-hydroxycyclo-		
	hexanone	0.26	1558
51.	n-butyric acid	4.01	1566
53.	3,5,5-trimethyl-2-cyclopentenone	0.81	1578
55.	n-hexyl n-hexanoate	3.21	1590
58.	unknown	0.23	1616
59.	2-methylbutyric acid	0.33	1619
61.	cis-3-hexenyl n-hexanoate	0.75	1631
63.	γ-hexalactone	0.99	1648
73.	n-hexanoic acid	13.34	1788
74.	dihydro-β-ionone	1.22	1791
77.	unknown	0.36	1810
79.	benzyl alcohol	2.23	1818
83.	β-phenyl ethyl alcohol	0.38	1852
84.	γ-octalactone	0.10	1856
86.	n-heptanoic acid	0.61	1882
88.	β-ionone	1.10	1893
91.	methyl furoate	1.19	1937
92.	n-octanoic acid	1.18	1989
94.	ethyl cinnamate	0.27	2068
95.	γ-decalactone	0.77	2084
96.	dihydroactinidiolide	0.25	2260

[a]peak area% less than 0.01.

D. Avocado

Avocados, the genus *Persea*, are grown in tropical and subtropical climates, generally from 40° S. Lat. to 40° N. Lat. (Gustafson, 1976). The major producers of avocado in the world are the U.S., Israel, and South Africa.

There are virtually no reports on the volatile constituents of avocado. Avocado contains lipids in large quantities (Kikuta and Erickson, 1968), which makes analysis difficult.

We identified fifty-two components in the avocado fruit extract (TABLE VIII). FIGURE 7 shows their gas chromatogram. The major constituent of the extract was trans-2-hexenol which gives a strong green-note. The characteristic green-fruity flavor is due to the presence of C-6 alcohols (n-hexanol, cis-3-hexenol, trans-2-hexenol) which comprise over 80% of the total area% of the extract. The ester content was significantly low compare with the other fruit extracts. A characteristic feature of the avocado volatiles was the presence of a large numbers of alcohol derivatives. Isoamyl

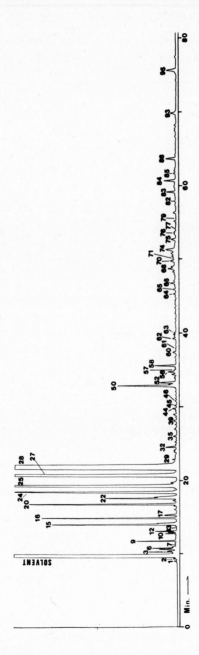

FIGURE 7. Gas chromatogram of avocado fruit extract. See experimental section for gas chromatographic conditions.

alcohol (15) and amyl alcohol (17) have not been found in
other tropical fruits.

    TABLE VIII. Volatile components identified in the avoca-
do fruit extract.

Peak No. in FIGURE 7	Compounds	Peak area %	Kovats index
1.	acetaldehyde	a	
2.	acetone	a	836
3.	2-methyl-2-butanal	0.18	968
5.	diisopropyl ketone	0.19	1000
6.	unknown	0.22	1013
7.	2-methyl-1-propanol	a	1038
8.	isoamyl formate	a	1053
9.	n-hexanal	0.39	1078
10.	n-butanol	0.12	1130
11.	p-xylene	0.10	1137
12.	m-xylene	0.21	1143
13.	2-hexanone	a	1159
14.	pyridine	a	1172
15.	isoamyl alcohol	1.88	1178
15.	o-xylene	0.30	1185
16.	trans-2-hexenal	1.61	1208
17.	n-amyl alcohol	0.24	1220
20.	acetoin	2.11	1265
22.	cyclohexanone	1.04	1285
23.	3-methyl-2-buten-1-ol	0.49	1287
24.	N,N-dimethyl formamide	3.58	1304
25.	n-hexanol	9.82	1324
26.	unknown	0.15	1334
27.	cis-3-hexenol	13.50	1356
28.	trans-2-hexenol	58.95	1380
29.	tetradecane	a	1400
32.	furfural	0.16	1432
35.	2-ethyl hexylalcohol	a	1456
39.	benzaldehyde	a	1500
44.	linalool	0.10	1543
45.	n-octanol	a	1543
46.	linalyl acetate	a	1556
50.	diethyleneglycol monoethyl ether	1.03	1581
51.	unknown	0.10	1585
52.	unknown	0.35	1589
56.	menthole	0.17	1613
57.	methyl benzyl ketone	a	1617

58.	N-methylpyrolidone	0.36	1630
60.	α-terpineol	a	1659
61.	styrallyl acetate	a	1670
62.	benzyl acetate	a	1693
63.	naphthalene	a	1707
64.	octadecane	a	1800
65.	o-methoxy phenol	a	1810
66.	benzyl alcohol	0.10	1824
68.	phenyl ethyl alcohol	0.14	1860
70.	benzyl cyanide	a	1879
71.	butyrated hydroxy toluene	a	1885
74.	benzothiazole	0.23	1907
75.	unknown	0.11	1932
76.	phenol	a	1948
77.	biphenyl	a	1960
79.	γ-nonalactone	a	1986
82.	methyl cinnamate	a	2030
83.	butyrated hydroxy anisole	0.15	2053
84.	ethyleneglycol monophenyl ether	0.20	2080
85.	ethylphenol	a	2100
88.	tetradecanol	0.21	2130
93.	unknown	0.14	2231
95.	hexadecanol	0.31	2314

[a] peak area% less than 0.01.

IV. SUMMARY

Table IX shows the area% of each chemical group in the extracts. Mango fruit contains monoterpenes in large quantities. The alcohol compositions of mango fruit and passion
fruit are quite similar. The major difference between mango
fruit and passion fruit is that passion fruit contains many
more esters than does mango fruit. Monoterpenes give a characteristic flavor to mango fruit and esters give passion
fruit a sweet fruity flavor. High levels of esters in the
papaya fruit extract is due to the presence of benzyl isothiocyanate (22.59%). It is surprising that avocado contain
a high level of alcohols. The characteristic flavor of avocados comes from C-6 alcohols. It is obvious that processed
products such as puree and juice contain some heat-induced
compounds. Sugar degradation products such as furans were
isolated in large quantities from the processed products,
mango puree, papaya puree, and passion fruit juice. These
products also contained free fatty acids in large quantities.
Among the fatty acids identified, n-butyric acid seems to be

a characteristic constituents of tropical fruits. Passion
fruit volatiles included six n-butyrates (TABLE VI) of which
the total area% was 31.14. n-Butyric acid constituted 37.34%
of the papaya fruit extract. No butyrates were found, however
in the same extract. The unique constituents for each fruit
are benzyl isothiocyanate in papaya, terpinolene in mango,
n-hexyl-n-butyrate in passion fruit, and trans-n-hexenol in
avocado. There are many important edible tropical fruits
other than the ones reported in this study. It will be interesting to investigate their volatile constituents to dis-
cover their flavor characteristics.

TABLE IX.   Compositions of chemical groups in the extracts
obtained from tropical fruits.

Fruits	Esters	Acids	Alcohols	Mono-terpenes	Furans	Ketons aldehyde
Mango fruit	0.17	0.52	25.01	42.32	0.70	2.50
Mango puree	9.90	31.56	6.63	7.85	20.32	9.26
Papaya fruit	23.33	54.40	0.15	19.90	a	a
Papaya puree	1.51	83.41	1.98	3.01	4.25	a
Passion fruit	51.24	3.26	25.07	7.53	a	3.38
Passion juice	5.66	20.81	28.19	3.85	25.35	6.73
Avocado fruit	a	a	85.76	0.10	0.16	4.27

[a] peak area% less than 0.01.

REFERENCES

Angelini, P., Bandyopadhyay, C., Rao, B. Y. K., Gholap, A. S.,
    and Bazinel, M. L. (1973). *Prec. 33rd Ann. Mtg. Inst. Food
    Technol.* Miami Beach, Florida. Paper No. 366.
Chan, Jr., H. T., Flath, R. A., Forrey, R. R., Cavaletto, C.
    G., Nakayama, T. O. M., and Brekke, J. E. (1973). "Deve-
    lopment of off-odors and off-flavors in papaya puree." *J.
    Agric. Food Chem. 21,* 566.
Craveira, A. A., Andrade, C. H. S., Matos, F. J. A., Alencar,
    J. W., and Machado, M. L. L. (1980). "Volatile constitu-
    ents of *Mangifera indica Linn.*" *Rev. Latinoamer. Qufm.
    11,* 129.
El-Baki, M. M. Abd, Askar, A., El-Samahy, S. K., and El-Fadeel

M. G. Adb. (1981). "Studies on mango flavor." *Deutsche Lebensm.-Rund. 77,* 139.

Ettlinger, M. G., and Hodgkins, J. E. (1956). "The synthesis of isothiocyanates from amines." *J. Org. Chem. 21,* 404.

Flath, R. A., and Forrey, R. R. (1977). "Volatilecomponents of papaya *(Carica papaya L. Solc Variety). J. Aric. Food Chem. 25,*103.

Gholop, A. S., and Bandyopadhyay, C. (1977). "Characterisation of green aroma of raw mango *(Mangifera indica L.). J. Sci. Fd. Agric. 28,* 885.

Gustafson, D. (1976). "World avocado production." *Proc. First Int. Tropical Fruit Short Course, The Avocado.* Gainseville Fla.

Hunter, G. L. K., Bucek, W. A., and Radford, T. (1974). "Volatile components of canned Alphonso mango." *J. Food Sci. 39,* 901.

Kikuta, Y., and Erickson, L. C. (1968). "Seasonal changes of avocado lipids during fruit development and storage." *Calif. Avocado Soc. Year bk. 52,* 102.

Parliment, T. H. (1972). "Some volatile constituents of passion fruit." *J. Agric. Food Chem. 20,* 1043.

Tang, C. S., and Syed, M. M. (1972). "Localization of benzyl glucosinolate and thioglucosidase in Carica papaya fruit." *Phytochemistry, 11,* 2531.

Winter, von M., and Klot, R. (1972). "Uber das aroma der gelben Passions frucht. *(Passiflora edulis f. flavicarpa)." Helv. Chim. Acta, 55,* 181.

VOLATILE COMPONENTS MODIFICATIONS DURING HEAT TREATMENT OF FRUIT JUICES

Jean Crouzet [1]
Griansak Chairote [2]
Freddy Rodriguez

Laboratoire de Biochimie Appliquée
Université des Sciences et Techniques du Languedoc
Montpellier, FRANCE

Souleymane Seck

Ecole Nationale Supérieure Universitaire de Technologie
Dakar, SENEGAL

Fruit juices, tomato and apricot juices, were heated at several temperatures during different times using a special device working in continuous. With this device the temperature of treatment was known with less than 2°C and the residence time with less than 1%.

Volatile components modifications occuring during these treatments were studied by GLC after extraction and concentration. In the case of tomato juices we observed an increase of the quantities of benzaldehyde, phenylacetaldehyde, benzyl alcohol and principaly of furanic compounds. These last compounds were not detected in apricot juices heated in similar conditions. In these juices the main changes observed were increases of the quantities of terpernic alcohols, of linalol oxides and of γ-butyrolactone.

The pathways implied in the formation of these different components are discussed.

The increase of the quantities of components such as furanic derivatives or terpenic alcohols, according to the nature of the product, may be uséd for the detection of fruit juices prepared from products submitted to a heat treatment such as concentrates.

[1] *Present address : University of Chiang Mai, Thailand.*

[2] *Present address : University of Caracas, Venezuela.*

I. INTRODUCTION

During the last feeften years the study of volatile components of fruits and vegetables have been led to numerous works,
several reviews concerning this subject are available (1,2).
In contrary data concerning volatile components modifications
during heat treatment occuring in juice fruits processing are
more scarse and dispersed.

Kirchner and Miller (3) shows that in pasteurized grapefruit juice 3-hexenol and geraniol lacked and that the treatment induces a decrease of limonene content. On the other hand
an increase of linalol oxides, α-terpineol and furfural was
observed. Such an increase of α-terpineol was also noticed
during flash pasteurization or pasteurization of orange juice
(4). Heat treatment leads also in this juice to the increase
of 4-terpinenol, carveol and 2,8-menthadienol and to the formation of 3-hydroxy-2-butanone and 3-methyl-2-butene-1-ol.

Chan et al (5) stated that linalol oxides, benzaldehyde
and α-terpineol content increases during heat treatment at
95°C during 2 minutes of acidified papaya puree.

Studying the influence of some processing paramaters on
the aroma of black curants and using correlations between
instrumental and sensory data Karlsson-Ekstrom and Von Sydow
(6,7) have shown that during heating of this fruit the sensory
changes involved are linked to a decrease of terpenic hydrocarbons and to an increase of methyl sulphide and aldehydes.
The effect of the heat treatment is more important when the
temperature was more elevated.

In the case of tomato juices (8-11) the compounds formed
during heat processing of juice and concentrates or during
heat treatment at 100°C in a closed system occurs from sugar
and ascorbic acid degradation : furanic and aromatic compounds
or from carotenoïd pigments degradation : 6-methyl-5-heptene-
2-one, β-ionone, geranyl and farnesyl acetone. Furanic derivatives are also the main volatile components formed during
heating of plum, strawberry(12) and apple juices or concentrates (13). In this last case pyrolic and phenolic compounds
were also detected. In most case laboratory studies concerning
heat induced compounds formation were performed using a batch
reactor. In these conditions, according to the heat transfert
resistance, the temperature of treatment and the residence
time of the product were not suffcently controled.

In these conditions a special device allowing a continuous
heat treatment with a good control of temperature and residence time was conceived. Using this device, the heat treatment of two fruit juices formely studied in our laboratory
(14,15,16) was undertaken.

II. EXPERIMENTAL

A. *Materials*

Tomatoes (Vee ROMA variety) were grown in an open field at the experimental station of Puyricard (13 France).Juice was prepared by cold break : the fruits were crushed at room temperature and after refining the juice obtained was poured without any heat treatment in enamel lined jars and stored at -20°C during 6-8 months before heat treatment and analysis.

An industrial puree of the variety Rouge du Roussillon was picked from the juice processing line of a factory (Société Coopérative Agricole Roussillonnaise des Aspres des Albères et du Littoral) in Elne (Pyrénées Orientales, France). After washing, fruits were blanched at 85°C for 3 min, crushed and refined, the puree was then flash pasteurized at 125°C for a few seconds. Picking was undertaken after quick cooling at 20°C. The puree was frozen at -20°C and stored at this temperature for 5 months.

B. *Heat treatment*

In order to control temperature and residence time, the device described fig. 1 was used, the fruit juice contained in a 3 l feeding tank was pumped by means of a peristaltic pump through a steam-heated exchanger ; under these conditions the sample quickly reaches the temperature chosen for heat treatment. The liquid was maintained at this temperature through a 3.4 m x 8 mm i.d. coiled stainless steel column set in a thermostated furnace ; at the exit of the column the liquid was cooled rapidly by a second exchanger and collected in a cold trap ; temperature was controlled by five thermocouples distributed along the column.

With this device it was shown (17) that the accuracy obtained from temperature was 2°C and that we had a laminar flow with a vessel dispersion number (18) less than 0,05, so that the residence time was known with a precision of 1 %.

C. *Isolation of volatile components*

The volatile components were obtained by stripping in a cyclone apparatus working under vacuum (19). The extraction of 1 litre of juice was carried out over a 2 h period at 30°C under 80 mm pressure. The traps were cooled with ice water and liquid nitrogen successively. The condensed fractions in the two traps were extracted with methylene chloride, the organic extract was dried over a small quantity of anhydrous sodium

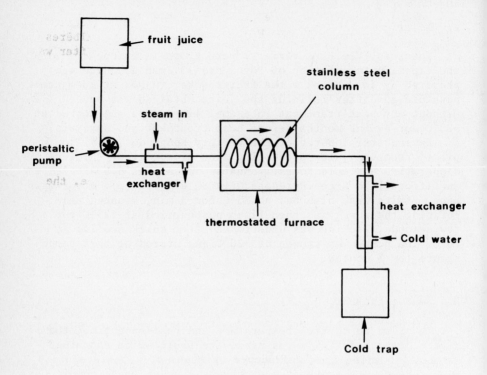

FIGURE 1. Device used for heat treatment of juices.

sulphate and filtered. Most of the excess methylene chloride
was then removed under a stream of nitrogen to yield a concentrate.

For quantitative determination 15 µg of 3-methyl pentanol
and 15 µg of n-decanol used as internal standards were added
to the aqueous extract obtained from apricot juice before the
extraction step.

D. *Chromatography*

Gas liquid chromatography was performed on a Varian-
Aerograph gas chromatograph fitted with a flame ionization
detector.

For qualitative works two columns were used :
       - for tomato juice study a glass SCOT capillary
column 50 m x 0,5 mm i.d. coated with Carbowax 20 M, the

temperature was programmed from 50 to 170°C at 20C per min. and held.

    - for apricot juice study a glass WCOT capillary column 54 m x 0,5 mm coated with Carbowax 20 M, the temperature was programmed from 50 to 170°C at 4°C per min. and held.

    For quantitative determinations a 1,5 % Carbowax 20 M on Chromosorb W 80/100 mesh DMCS 3 m x 3,18 mm operating from 50 to 190°C at 2°C per min. was used. The output signal was fed through a Spectra Physics SP 4 000 central processor and plotted on a SP 4050 printer plotter.

E. Gas chromatography-Mass spectrometry

    An LKB 2091 mass spectrometer was coupled with a WCOT glass capillary column 25 m x 0,3 mm Carbowax 20 M column operating from 30-190°C at 2°C per min used. The flow rates of the carrier gas (He) was 0.8 ml per min.

    The ionizing energy was 70 eV and source temperature 230°C

F. Ascorbic acid determination

    The method of Roe and Kuetner (20) was used.

III.   RESULTS

A. Tomato juice

    Tomato juice obtained by "cold break" was treated using the apparatus described in the experimental part at 85°C for 14 minutes and at 98°C during 2,5 minutes.

    Chromatogram obtained after extraction of volatile components for the non treated juice is given fig. 2.

    The relative proportions of the most characteristic components identified using previously reported works (8,9,11, 14) are listed table 1.

    Three phenomenons acting in inverse order may be advanced in order to explain the fluctuations observed in the amount of volatile aldehydes and alcohols formed by oxidation of poly-insaturated fatty acids : hexanal, cis-3-hexenal, trans-2-hexenal, hexanol, cis-3-hexenol ands trans-2-hexenol :

    - we can suppose that enzymatic systems are always in activated form during the stripping of volatile compounds from the non treated juice and during the first steps of heating,

    - under the condition used for heat treatment an increase

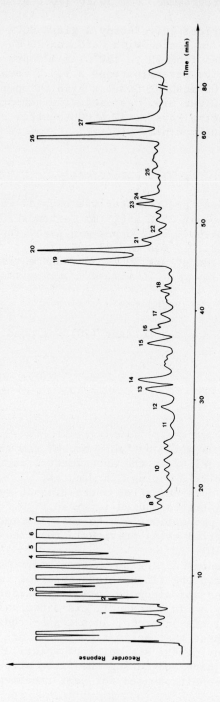

FIGURE 2. Gas chromatogram of tomato volatile components (*variety Vee Roma*) Carbowax 20 M on Chromosorb 80/100 mesh. DMCS. Temperature programmed from 50 to 150°C at 2°C per minute.

of chemical oxidations of insaturated fatty acid occurs certainly (10). The increase of relative per cent of 2-octenal
and decadienal agrees with these findings,
   - the speed of isomerisation of trans-2-hexenal in cis-3
hexenal which is considered as a chemical isomerisation (21)
is increased as the temperature rises.
   Our results show a two fold increase of the relative per
cent of isobutylthiazole in the juice treated at 98°C for
2,5 minutes relatively to the non treated juice. The mechanism involved in the formation of these compounds is generaly
considered as being biosynthétic, Stevens (22) have shown
that isobutylthiazole concentration was geneticaly determined.
In contrary according to the data obtained in the present
work a degradative pathway can also be envisaged. Pittet and
Hruza (23) think that if some thiazoles are biogeneticaly
produced a great number of these products have been isolated
from heat treated food - stuffs and Mulders (24) gives a
mechanism for the formation of acyl thiazoles.
   As stated previously (8-11,25) heat treatment of tomato
juice inducesan increase of the relative percent of furanic
compounds : furfural, 2-acetyl furan, furfuryl alcohol. Tatum
et al (26) have shown that these compounds were formed during
heating of ascorbic acid aqueous solutions and our results
(11) on model systems : ose-amino-acids or ascorbic acid -
amino acids indicates that the main pathway involved in furfural formation occurs from ascorbic acid degradation.
However it was postulated that other furanic compounds are
produced during the thermal degradation of polyosides : cellulose and pectic substances.
   It must be pointed out that an increase of the amount of
5-methyl furfural was observed only when the temperature reaches 98°C and a decrease of the relative percent of 5-methyl-2
acetyl furane during heat treatment. So only furfural, 2-ace-
tyl furan and furfuryl alcohol can be considered as markers
of heat treatment of tomato juice.
   The increase of aromatic aldehydes ; phenyl acetaldehyde
and benzaldehyde can be explained by a Strecker degradation
of phenylalanine in phenyl acetaldehyde (11,27) followed by a
degradation of this product conducting to the production of
benzaldehyde (28).
   The formation of 6-methyl-5-heptene-2-one and geranyl
acetone from carotenoïds is well documented (10,28).
   However the decrease of the relative amount of these
compounds after heating at 98°C in the closed system used may
be attributed to further degradative reactions.
   The modifications of relative percent of terpenic compounds are more surprising. The observed increase of the
content of α-terpineol and geraniol agrees with the results

TABLE 1. Relative percent of main volatile components in tomato juice (variety Vee ROMA) before and after heat treatment.

Peak number (fig. 2)	Compounds	Non treated juice	Heated juice 85°C, 14 min	Heated juice 98°C 2,5 min
1	hexanal	0,8	1,0	0,8
2	cis-3-hexanal	1,0	1,2	3,4
3	trans-2-hexanal	2,9	4,2	2,0
4	6-Me-5-heptene-2-one	2,1	4,1	2,6
5	hexanol	15,1	10,9	22,6
6	cis-3-hexenol	19,7	14,7	22,7
7	trans-2-hexenol	6,9	6,1	1,7
8	isobutylthiazole	0,08	0,13	0,2
9	2-octenal	0,14	0,4	0,6
10	furfural	0,12	0,2	0,5
11	2-acetyl furan	0,08	0,3	0,2
12	benzaldehyde	0,6	0,6	1,3
13	linalol	0,5	0,4	0,3
14	5-Me-furfural	0,7	0,7	0,9
15	2-Me-5-acetyl furan	0,5	0,1	0,3
16	δ-butyrolactone	0,3	0,3	0,3
17	phenylacetaldehyde	0,3	0,2	0,5

126

18	furfuryl alcohol	0,15	0,3	0,7
19	α-terpineol	2,4	7,0	4,3
20	carvone	2,0	1,4	2,7
21	geraniol	0,4	0,9	0,8
22	geranylacetone	0,1	0,2	0,1
23	geranial	0,2	0,6	2,7
24	decadienal	0,2	0,3	0,8
25	benzyl alcohol	0,1	0,4	0,8
26	2,6-Di Me- undeca-10-one-2	4,4	2,1	4,1
27	2-phenyl ethanol	2,4	0,8	2,6

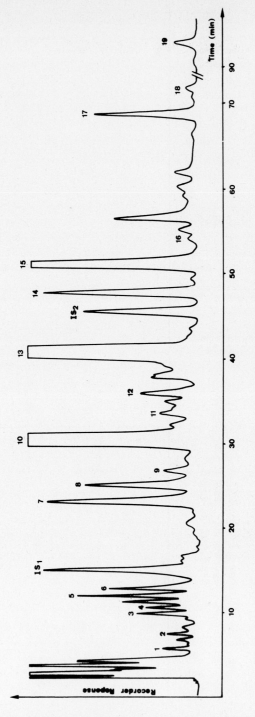

FIGURE 3. Chromatogram of volatile compounds of apricot variety "Rouge du Roussillon" 1,5% Carbowax 20 M on Chromosorb W 80/100 mesh DMCS. Temperature programmed from 50 to 190°C at 2°C per minute. IS_1 = internal standard 1 (3-methyl pentanol/IS_2 = internal standard 2 (nonanol).

of several authors, such an increase of terpenic alcohols was observed during heating of blackcurrants (30), orange (31) and tomato juices (21) or during tea processing (32).

Particularly BUTTERY et al (25) have shown that during tomato volatile components extraction at 100°C under atomospheric pressure a considerable increase of the amount of α-terpineol and linalol was observed. According to these authors the increase of linalol in heated tomato allows this product to play an important part in the aroma of cooked tomato.

So the decrease of linalol observed in our work is very surprising and cannot be explained.

B. Apricot juice

The data obtained for concentration (µg/1) of the main volatile compounds isolated from apricot juice heated at different temperatures (75°C, 90°C, 98°C) for 5 minutes are listed table 2.

The compounds so analyzed were identified (fig. 3) using data obtained, for the variety Rouge du Roussillon used in this work, in our laboratory (15,16).

Differences observed are only quantitative, none new compound, thermally induced was observed. It must be pointed out that during heating of apricot juice, the production of furanic compounds was not observed whereas these compounds are characteristic of heat treated tomato juice, as well as plum, strawberry and apple juices (12,13). However as in tomato juice an increase of benzaldehyde content was found.

Concerning terpenic hydrocarbons, we observed as the temperature go from 75 C to 98°C a decrease of the amount of myrcene, limonene, and γ-terpinene whereas an increase for p-cimene was noticed. These variations can be connected as follow : the decrease of myrcene and limonene concentrations can be explained by rearrangement of these compounds in γ-terpinene (33), then the deshydrogenation of γ-terpinene conducts to p-cymene (fig. 4).

One other way for explain the increase of limonene consist in an oxidative reaction conducting to the formation of α-terpienol (33-35), the increase of relative amount of α-terpineol was of about 50 per cent between 75°C and 90°C. When the temperature reaches 98°C a decrease, probably resulting of degradative reactions, occurs. Oxidation of monoterpenes is involved in the formation, during heating, of the other terpenics alcohols linalol, 4-terpinenol, nerol and geraniol.

One other way is the degradative oxidation of carotenoïds principally β-carotene (36) which is the main carotenoïd pigment identified in apricot.

TABLE 2. Concentrations of main volatile components in apricot juice (variety Rouge du Roussillon) before and after heat treatment.

Peak number (fig. 3)	Compounds	Non treated juice	Concentration µg/l		
			Heated juice 75°C, 5 min	Heated juice 90°C, 5 min	Heated juice 98°C/5 min
1	camphene	2,0	2,2	3,0	3,0
2	β-pinene	2,3	2,8	4,1	3,5
3	myrcene	6,3	15,0	9,0	12,8
4	limonene	4,7	3,2	2,2	2,0
5	γ-terpinene	11,9	13,4	11,1	8,1
6	p-cymene	9,0	9,6	11,5	8,1
7	trans linalol oxide (A)	25,7	39,7	65,0	86,5
8	cis linalol oxide (B)	18,7	28,2	40,0	35,9
9	benzaldehyde	2	2,2	7,9	8,2
10	linalol	258	152,7	295,4	240,4
11	4-terpinenol	7,9	9,8	13,9	20,7
12	γ-butyrolactone	8,9	20,7	43,0	35,4
13	α-terpineol	221	225,0	346,0	257,6
14	nerol	24,4	29,7	50,0	34,4
15	geraniol	70,7	103,2	187,2	112,6
16	2-phenylethanol	1,5	1,5	6,1	2,3
17	γ-decalactone	17,5	24,8	46,7	34,7
18	ethyl pentanoate	2,4	0,7	2,5	0,9
19	farnesol	5,0	19,1	25,2	18,0

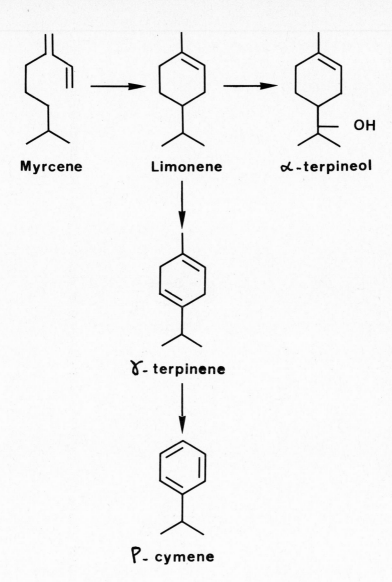

Fig. 4 - Rearrangment of terpenic hydrocarbons occuring
during heat treatment of apricot juice.

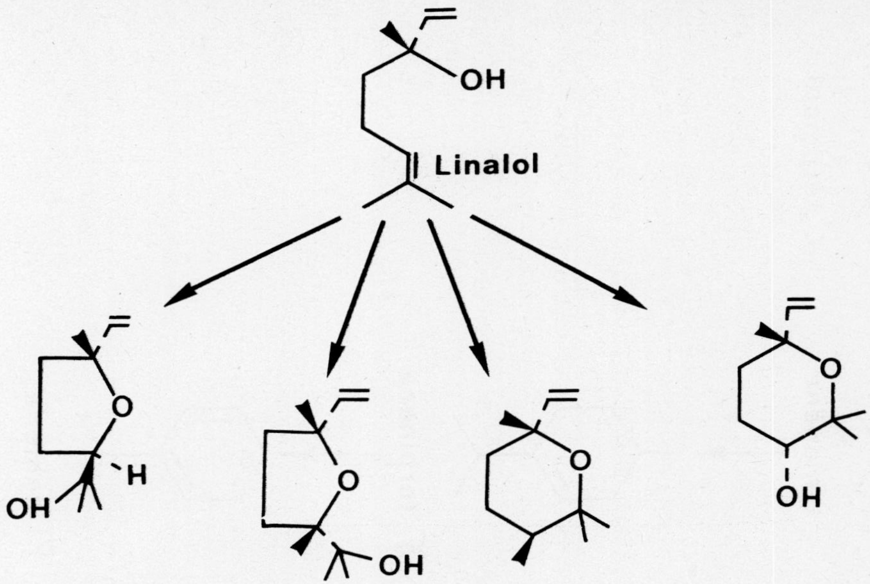

Linalol oxide (A) Linalol oxide(B) Linalol oxide(C) Linalol oxide(D)

FIGURE 5. Formation of linalol oxides from linalol.

Preliminary work conducts in your laboratory have shown a decrease of this compound during heat treatment of apricot juice.

In the case of linalol, part of this alcohol was ultimatly oxidized into linalol oxides (fig. 5), the amount of the furanic oxides A and B was found increase during heating of apricot juice.

A decrease of linalol during heat processing of peaches was observed by Souty et Reich (37).

Increase of the amount of γ-butyrolactone and γ-decalactone are also indicative of heat treatment of this juice. γ-butyrolactone was found to be formed during ascorbic acid degradation (26) or during fructose heating in basic medium (38), the first pathway being the more probable during heating of apricot juice. The ascorbic acid content of apricot juice was determined along the different treatments studied (table 3) and we observed a decrease of the content according to a first order law.

TABLE 3. Ascorbic acid retention during heat treatment

treatment	ascorbic acid mg/100 ml	ascorbic acid retention %
no treatment	9,4	100
75°C 5 min.	6,8	72
90°C 5 min.	5,7	60
98°C 5 min.	5,2	55

As no formation of furanic compounds was observed during the heating of apricot juice we can assumed that the losses observed for ascorbic acid are related to γ-butyrolactone formation.

In these conditions it is possible to detect the use of concentrates for the preparation of juice fruits by the study of the increase of the quantities of components such as furanic compounds for tomato juice or terpenic alcohols or γ-butyrolactone for apricot nectars.

REFERENCES

1. Johnson, A.E., Nursten H.E. and Williams, A.A., *Chim.
    and Indust.*, 556 and 1212 (1971).
2. Salunkhe, D.K. and Do, J.Y., *Crit. Rev. in Food Sci. and
    Nutr.* 161 (1976).
3. Kirchner, J.G., and Miller, J.M., *J. Agric. Food Chem.*
    1, 510 (1953).
4. Schreier, P., Drawert, F., Jukker, A. and Mickw,
    Z. Lebensm. Unters. Forsch. 164, 188 (1977).
5. Chan, H.T.Jr., Flath, R.A., Forrey, R.R., Cavaletto, C.G.
    Nakayama T.O.M. and Brekke J.E., *J. Agric. Food Chem.*
    21, 556 (1973).
6. Karlsson-Ekstrom G., and Von Sydow, E., *Lebensm. Wiss. u.
    Technol.* 6, 86 (1973).
7. Larlsson-Ekstrom, G., and Von Sydow, E., *Lebensm. Wiss.
    u. Technol.* 6, 165 (1973).
8. Seck, S., Crouzet, J., and Piva, M.T., *Ann. Technol. Agric.*
    25, 85 (1976).
9. Sieso, V., and Crouzet, J., *Food Chem.* 2, 241 (1977).
10. Schreier, P., Drawert, F., and Junker, A., *Z. Lebensm.
    Unters. Forsch.* 165, 23 (1977).
11. Seck, S., and Crouzet, J., *J. Food Sci.* 46, 790 (1981).
12. Sloan, J.L., Bills, D.D., and Libbey, L.M., *J. Agric.
    Food Chem.* 17, 1370 (1969).
13. Brule, G., *Ann. Technol. Agric.* 22, 45 (1973).
14. Seck, S., and Crouzet, J., *Phytochemistry* 12, 2925 (1973).
15. Rodriguez, F., Seck, S., and Crouzet, J., *Lebensm. Wissen
    u. Technol.* 13, 152 (1980).
16. Chairote, G., Rodriguez, F., and Crouzet, J. *J. Food Sci.*
    46, 1898 (1981).
17. Seck, S.,Etude des constituants volatils de la tomate.
    Influence des traitements thermiques. These de Docteur
    d'Etat Montpellier 1979.
18. Levenspied, O., *in* "Chemical reaction engineering"
    p. 253. John Wiley and sons Inc. New-York (1972).
19. Bartholomew, W.H., *Anal. Chem.* 21, 527 (1949).
20. Roe, J.H. and Kuether, C.A., *J. Biol. Chem.*, 147, 399
    (1943).
21. Kazeniac, S.J., and Hall, R.M., *J. Food Sci.* 35, 519
    (1970).
22. Stevens, M.A., *J. Am. Soc. Hort. Sci.*, 95, 9, (1970).
23. Pittet, O., and Hruza, D.H., *J. Agric. Food Chem.*, 22,
    264 (1974).
24. Mulders, E.J., *Z. Lebensm. Unters. Forsch.*, 152, 193
    (1973).

25   Buttery, R.G., Seiffert, B.H., Guadagni, D.G. and Ling, L.C., *J. Agric. Food Chem.*, 19, 524 (1971).

26.   Tatum, J.H., Shaw, P.E. and BERRY, R.E., *J. Agric. Food Chem.*, 17, 38 (1969).

27.   Seck, S., and Crouzet, J., *Sciences des Aliments*, 2, 187 (1982).

28.   Nursten, H.E. and Woolfe, M.L., *J. Sci. Food Agric.* 23, 803 (1972).

29.   Cole, E.R. and Kapur, N.S., *J. Sci. Food Agric.*, 8, 360 (1957).

30.   Von Sydow, E. and Karlsson-Ekstrom, G., *Lebensm. Wiss. u. Technol.*, 4, 54 (1971).

31.   Schreier, P., Drawert, F., Junker, A. and Mick, W., *Z. Lebensm. Unters. Forsch.*, 164, 188 (1977).

32.   Saijo, R., and Takeo, T., *Agric. Biol. Chem.*, 37, 1367 (1973).

33.   Dieckman, R.H. and Palamand, S.R., *J. Agric. Food Chem.*, 22, 498 (1974).

34.   Proctor, B.E., and Kenyon, E.M., *Food Technol.*, 3, 387 (1949).

35.   Buckholz, L. and Daun, H., *J. Food Sci.*, 43, 535 (1978).

36.   Sanderson, G.W., Co, H. and Gonzalez, J.G., *J. Food Sci.* 36, 231 (1971).

37.   Souty, M. and Reich, M., *Ann. Technol. Agric.*, 27, 837 (1978).

38.   Shaw, P.E., Tatum, J.H. and Berry, R.E., *J. Agric. Food Chem.*, 16, 979 (1968).

ANALYTICAL PROCEDURES FOR EVALUATING
AQUEOUS CITRUS ESSENCES

Manuel G. Moshonas
Philip E. Shaw

U. S. Citrus and Subtropical Products Laboratory[1]
Winter Haven, Florida

I. INTRODUCTION

Citrus essences are aqueous solutions collected as the
first distillate from evaporators during the production of
juice concentrates. Volatile components of citrus juice are
concentrated in this fraction which has the quantitative and
qualitative composition essential to the flavor and aroma
characteristics of the fresh juice from which it is produced,
making it a desirable natural flavoring fraction. An oily
phase, known as essence oil, is also condensed as part of the
first distillate. This layer is separated from the aqueous
essence and is also used as a flavoring agent. Aqueous orange
essence is the most important and most widely used of the
citrus essences. It is added back to orange concentrate to
restore the "fresh" flavor and aroma and is also used as a
natural flavoring fraction to impart an orange flavor and
aroma to synthetic drinks and other products. Currently,
10 million pounds of aqueous orange essence are being utilized
annually in the United States; however, the potential annual
production is approximately 20 million pounds. The full
commercial potential of orange essence has not been realized
primarily because of the difficulty of evaluating its
strength and quality and thus, an inability to produce a

[1]Southern Region, Agricultural Research Service,
U. S. Department of Agriculture.

137

consistent, standard product. The strength and quality
characteristics of essence vary from lot to lot because of
variations in processing methods, fruit cultivar, maturity
and/or season of harvest. Although many methods have been
developed for evaluating citrus essences none have been
satisfactory to the citrus industry. Subjective organoleptic
evaluations are still necessary to adequately evaluate aqueous
essences for use in flavoring citrus products. However, a
new, simple direct gas chromatographic (GC) method for
determining strength and quality of aqueous fruit essences
developed by Moshonas and Shaw (1983) shows real promise in
solving this problem and being accepted as an objective method
for essence evaluation by the citrus industry.

II.   DEVELOPMENT OF ANALYTICAL METHODS

A.   Chemical Oxygen Demand

In 1968, Dougherty reported a method for determining the
strength of citrus essences in which he subjected aqueous
essence to potassium dichromate-sulfuric acid oxidation. He
then measured the percent light transmission which gave a
chemical oxygen demand value for the essence constituents.
This value was then compared with a standard curve drawn from
values of known oxidized solutions of sucrose to determine the
strength of essence.

B.   Concentration by Liquid-Liquid Extraction

Attaway et al. (1962) described a procedure for the
detection and identification of carbonyl components in trace
amounts from orange essence. Volatile organic components were
separated by saturation of essence with anhydrous Na_2SO_4
extraction with diethyl ether, concentration by steam bath,
and reextraction with isopentane to remove the water. Most of
the solvent was removed by evaporation leaving an organic
fraction to be further analyzed by gas chromatography.

C.   Concentration by Methylene Chloride Extraction

Wolford et al. (1962) developed a method of analysing
orange essence in which aqueous essence was saturated with
Na_2SO_4 and extracted three times with distilled methylene
chloride. Most of the methylene chloride was then removed by

distillation at atmospheric pressure. Most of the ethanol and methanol which might interfere with further analysis were not extracted. However, about 95% of the remaining essence compounds could be recovered for further analysis by gas chromatography.

D.   Concentration by Ether Extraction

Schultz et al. (1964) used distilled ethyl ether to extract the organic constituents from aqueous essence. Essence was extracted with ether in a single pass through a 2-in. by 28-in. glass column extractor equipped with three vibrating perforated stainless steel disks. Solvent was continuously removed with a 40-plate 1-in. diameter Oldershaw column. Final solvent removal was done with a small helice packed distillation column. Gas chromatography was used for further separation and mass spectrometry for identification of constituents.

E.   Concentration by Ethyl Chloride Extraction

Moshonas and Shaw (1972) extracted orange essence with ethyl chloride (b.p. 12°C) in an effort to retain more of the most volatile constituents in the extract than had been retained with methylene chloride (b.p. 40°C). Although several new orange constituents were identified in this study, use of the low boiling solvent was not a significant improvement over methylene chloride in retaining low boiling components.

F.   Thin-Layer Chromatography

Attaway et al. (1964B) developed a gas-liquid and thin-layer chromatographic technique for orange essence analysis. Samples of non-aqueous essence were obtained using the method of Wolford et al. (1962). These samples were injected into a preparative gas chromatographic (GC) column and samples of essence components transferred from the GC effluent to thin-layer chromatographic plates. For aldehydes, ketones and esters, either benzene or trifluorotrichloroethane-methylene chloride was used to develop the plates. For alcohols, methylene chloride was used for development of plates, and for terpene hydrocarbons, Skelly-solve B was the developing solvent. Plate spots were detected by either $KMnO_4/H_2SO_4$,

5% vanillin in 96% H_2SO_4 or dinitrophenylhydrazine. This
method was useful for identification of esters and for the
partial elucidation of molecular structure of several types
of compounds.

G. Subtractive Analysis Using Girard-T Reagent or Sodium Bisulfite

Wolford et al. (1962) and Attaway et al. (1964A) analysed
orange essence by employing two methods which removed
carbonyl components prior to analysis of the remaining
organic constituents. In the first method, aqueous orange
essence was extracted with methylene chloride. After the
removal of solvent, carbonyl components were separated from
the essence extract by Girard-T reagent according to the
method of Teitelbaum (1958) and Stanley et al. (1961),
yielding a carbonyl-free fraction. The second method involved
the use of sodium bisulfite to form soluble α-hydroxysulfonate
derivatives followed by methylene chloride extraction to yield
the noncarbonyl fraction. Carbonyl-free fractions were
further separated by gas, column or paper chromatography and
individual compounds were identified.

H. Colorimetric Techniques

Colorimetric techniques for evaluating oxygenated terpenes
as $C_{10}H_{18}O$, saturated aliphatic aldehydes as octanal and
α,β-unsaturated aldehydes as citral were developed by
Attaway et al. (1967). The procedure for the oxygenated
terpenes was based on their reaction with acidic vanillin
solution to produce colored products. The saturated aliphatic
aldehyde procedure was based on the ability of aldehydes to
catalyze the oxidation of p-phenylenediamine by hydrogen
peroxide to yield a dark product known as Bandrawski's base
(Feigl, 1946). The unsaturated aldehyde method was based on
the ability of aldehydes to combine with aromatic amines to
form colored Schiff bases (Feigl, 1946). The colored
solutions which resulted from these reactions were analysed
with a colorimeter and the various functional groups of orange
essence were evaluated.

Braddock and Petrus (1971A,B) identified malonaldehyde in
orange essence and developed a procedure to quantify it. They
added 2-thiobarbituric acid reagent in citrate buffer to
aqueous essence to form a pink-orange complex; cellulose
powder was added and the mixture was filtered. The
remaining "film pad" was washed with 0.1N HCl, then with

0.1N NaOH, and the filtrate was collected in a flask
containing 1.2N HCl. Finally, the adsorbance of the pink
solution was determined with a spectrophotometer.

In 1970, Peleg and Mannheim reported a modified
Critchfield method (1963) for determining the carbonyl
concentration in citrus essences. This method was based on a
colorimetric determination of total carbonyls as their
dinitrophenylhydrozones. Ismail and Wolford (1970) and
Petrus et al. (1970) described colorimetric procedures for
estimation of total aldehydes in aqueous orange essence using
N-hydroxybenzenesulfonamide based on a spot test described by
Feigl (1966).

I.   Concentration by Adsorption on Organic Polymer Powders

Adsorption of orange essence components on an organic
polymer was first reported by Moshonas and Lund (1971) in an
analytical procedure wherein a Porapak Q precolumn was placed
in the GC oven and coupled to a packed Carbowax 20M column by
means of a three-way valve. The Porapak column trapped the
essence components other than water, methanol, acetaldehyde
and ethanol. The essence components were then stripped off
the pre-column into a cold trap followed by flash heating to
inject the trapped sample onto the Carbowax column for
completion of the analysis.

Schultz et al. (1971) and Dravnieks and O'Donnell (1971)
developed analytical methods in which headspace aroma
volatiles from orange essence or other materials were adsorbed
on organic polymers, including Porapak Q and Chromosorbs 101
and 102. Then by rapid heating, these volatiles were quickly
eluted from the polymer and injected onto Carbowax 20M columns
for further separation and evaluation.

J.   Concentration by Microextraction

In 1973, Dinsmore and Nagy reported a procedure in which a
10 ml sample of orange essence, 3.6 g of NaCl and 0.20 ml of
diisopropyl ether were combined in a microextraction
apparatus. The mixture was shaken until the organic
components were partitioned between the ether and water phases.
A saturated NaCl solution was added forcing the upper ether
layer through an open stopcock. The ether layer was then
analysed by gas chromatography.

K. Ultraviolet Absorption

Randall et al. (1973) measured ultraviolet (UV) absorption of essence in the range between 200-300 nm. They noted that UV absorption would be useful for industrial control work, since monitoring the absorption at a particular wavelength would show whether a process was in steady state or the organic content of the stream was changing.

L. Bromate Titration

Moshonas and Shaw (1975) developed a method of determining the strength of essence based on bromate titration of the unsaturated components in orange essence. To 50 ml of aqueous essence, 10 µl of HCl containing methyl orange indicator was added. The solution was titrated with .025 N bromide-bromate solution until color disappeared. A linear relationship between bromine uptake and essence strength was demonstrated.

M. Gas Chromatography

In 1975, Moshonas and Shaw reported a GC method for orange essence strength analysis. A 1/4 in. O.D. column with only the first 3 in. packed with Chromosorb 101 and stoppered with glass wool was placed in the GC oven. A 10 ml sample of aqueous essence was directly injected into the column which was held at 80°C for 8 min to allow water, ethanol, methanol and acetaldehyde to elute. Temperature was immediately raised to 210°C forcing the remaining constituents to elute. The strength of the essence was estimated by calculating the combined area under the two peaks which contained all the remaining constituents. Lund and Bryan (1977) quantitatively analysed orange essence for methanol, ethanol and acetaldehyde by injecting 2 µl of aqueous essence onto a 1/8 in. x 5 ft column packed with Porapak Q with the temperature held at 120°C. Concentration of these three compounds were determined from GC peak areas. Analysis of the other essence components was made by injecting 2 µl of essence or concentrated essence onto a 1/8 in. x 15 ft column packed with either 5% Carbowax 20M or 5% diethylene glycol succinate (DEGS).

III. LIMITATIONS

Although there has been limited success in estimating aqueous citrus essence strength and quality with one or more of these analytical methods, results have been variable, inconsistent and incomplete. None of these methods provided the kind of quantitative, qualitative or constituent relationship data necessary for acceptance by the citrus industry as an accurate, objective means for essence evaluation.

The chemical oxygen demand method primarily measured only the large quantity of ethanol and methanol present. The various extraction methods were useful for identifying individual constituents. However, incomplete extraction and loss of volatiles during concentration were sources of error. The aldehyde measuring methods lacked consistency, primarily because they measured the relatively large quantity of acetaldehyde which can vary widely from sample to sample because of its extreme volatility that makes it difficult to capture. The colorimetric, UV and other methods based on measuring the quantity of particular groups within the essence do not accurately reflect the strength or quality of the whole essence. Finally, the polymer adsorption methods required frequent calibration with known standards to produce reliable results.

Most of the above methods have been used by various citrus processors to try to obtain a standardized product, but without success. Large-scale blending of batches of essence to achieve a uniform product followed by taste evaluation of the blended product is currently the most effective way of standardizing aqueous orange essence.

IV. MOST RECENT DEVELOPMENT

A simple analytical gas chromatographic method developed by Moshonas and Shaw (1983) shows considerable promise in solving the problem of developing an objective method for evaluating strength and quality of aqueous citrus essences. This method makes it possible, for the first time, to obtain detailed quantitative and qualitative analysis of the volatile flavor and aroma constituents of aqueous essences from a direct injection of 1 µl or less of the whole essence. To employ this method, the GC must be equipped with a flame ionization (FID) detector and with a 50 meter cross-linked SE-54 wide bore capillary (Hewlett Packard Company, Avondale, PA), or similar non-polar fused silica column. The injection and detector temperatures are set at 275°C and the oven

temperature is initially set and held for 3 min at 40°C. The
oven temperature is then increased at 6°C/min to 175°C and
held there for 5 min. All aqueous essence analyses can be
made thereafter by simply injecting 1 μl samples of whole
essence into the injection system and pushing the "start"
button. The instrument then goes through the program and
prints a complete report which includes a GC curve of the
separated constituents, calculated area percentages and
retention times of each component. Identities of each
constituent can also be printed once their identity has been
programmed into the instrument. Gas chromatograms of
commercial aqueous orange essence at three different strengths
are shown in Figure 1 and of grapefruit and lemon aqueous
essences in Figure 2.

One important area of information that had eluded
researchers studying aqueous fruit essences has been a means
of determining the quantitative relationship of individual
constituents to each other. The importance of such
information stems from the common knowledge that a change in
the quantitative relationship of constituents in a flavor
fraction can have a dramatic effect on flavor and aroma.
This GC analytical method now makes that information
available. An example of this kind of change and its effect
on quality on citrus essences was demonstrated when this
method was used to analyse and compare two orange essences
obtained from oranges harvested from the same grove and
processed in the same plant with the same equipment. The
only difference was that one sample of essence was obtained
from oranges harvested before the January 1982 freeze, while
the second sample came from oranges harvested after the
freeze. The GC compositional profile of each sample is shown
in Figure 3. The profile of the before-freeze sample of
essence compared very well to the profile of a normal, good
quality commercial essence sample. The profile of the after-
freeze sample showed quantitative and qualitative differences
from that for normal essence, indicating an adverse effect on
quality. To confirm this conclusion, taste tests were
conducted using an expert taste panel. The panel determined
that there was a significant difference (99% confidence level)
in the flavor of the two samples and, through a second taste
test, determined a significant preference (95% confidence
level) for the essence obtained from oranges before the
freeze.

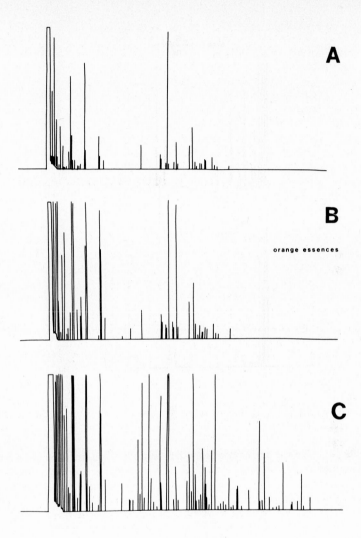

FIGURE 1. Gas chromatograms of aqueous orange essence at three different strengths: A, low; B, average; C, high.

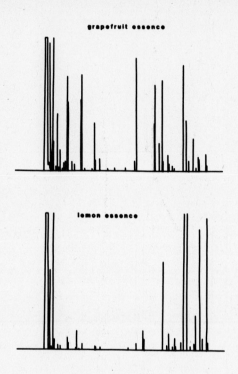

FIGURE 2. Gas chromatograms of aqueous grapefruit and lemon essences.

An example of strength evaluation of aqueous essence by this GC method was demonstrated by the analysis of two series of aqueous orange essences produced from the same lot of oranges, on the same day and in the same plant. The only difference was that each series of samples was produced by a different essence recovery unit. Analyses by this method showed that both samples were qualitatively indentical and that constituent relationship was the same. However, quantitatively, the flavor and aroma constituents from one recovery unit were consistently 35% less than the constituents from the other unit. This information led to equipment modifications of the unit producing the weaker essence which increased the essence strength to the level of the stronger essence.

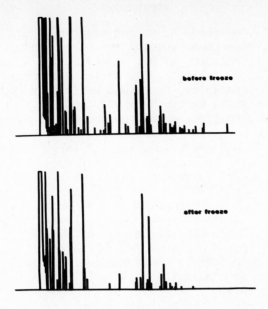

FIGURE 3. Gas chromatogram of aqueous orange essence from oranges harvested before and after freeze.

A GC analytical method has been developed which not only provides an objective method for evaluating aqueous fruit essence strength and quality, but which can be utilized to determine differences in flavor characteristics due to production methods, cultivar, horticultural practices and unusual weather conditions. It can also be used to monitor efficiency of recovery units and individual constituents, and, to detect adulteration. Although this method was developed for evaluating aqueous citrus essences, it has also been demonstrated that the method can be used to evaluate other aqueous fruit essences, including apple, grape, pineapple, strawberry and banana.

REFERENCES

Attaway, J. A., Wolford, R. W. and Edwards, G. J. (1962).
J. Agric. Food Chem. 10, 102.

Attaway, J. A., Wolford, R. W., Alberding, G. E. and
Edwards, G. J. (1964A). J. Agric. Food Chem. 12, 118.

Attaway, J. A., Hendrick, D. V. and Wolford, R. W. (1964B).
Proc. Fla. State Hortic. Soc. 77, 305.

Attaway, J. A., Wolford, R. W., Dougherty, M. H. and
Edwards, G. J. (1967). J. Agric. Food Chem. 15, 688.

Braddock, R. J. and Petrus, D. R. (1971A). J. Food Sci. 36,
1095.

Braddock. R. J. and Petrus, D. R. (1971B). Proc. Fla. State
Hortic. Soc. 84, 223.

Critchfield, F. E., "Organic Functional Group Analysis",
pp. 78-80, Pergamon Press, Oxford, 1963.

Dinsmore, H. L. and Nagy, S. Report of the Citrus Chemistry
and Technology Conference, Winter Haven, FL, Oct. 10, 1973.

Dougherty, M. H. (1968). Food Technol. 22, 1455.

Dravnieks, A. and O'Donnell, A. (1971). J. Agric. Food
Chem. 19, 1049.

Feigl, F., "Quantitative Analysis by Spot Tests", 3rd ed.,
pp. 340, 345, Elsevier Publishing Co., New York, 1946.

Feigl, F., "Spot Tests in Organic Analysis", p. 196, Elsevier
Publishing Co., New York, 1966.

Ismail, M. A. and Wolford, R. W. (1970). J. Food Sci. 35, 300.

Lund, E. D. and Bryan, W. L. (1977). J. Food Sci. 42, 385.

Moshonas, M. G. and Lund, E. D. (1971). J. Food Sci. 36, 105.

Moshonas, M. G. and Shaw, P. E. (1973). J. Food Sci. 38, 360.

Moshonas, M. G. and Shaw, P. E. (1975). Int. Flavours 6, 133.

Moshonas, M. G. and Shaw, P. E. (1983). J. Agric. Food
Chem., in press.

Peleg, Y. and Mannheim, C. H. (1970). J. Agric. Food
Chem. 18, 176.

Petrus, D. R., Dougherty, M. H. and Wolford, R. W. (1970).
J. Agric. Food Chem. 18, 908.

Randall, J. M., Bryan, W. L., Bissett, O. W. and Berry, R. E.
(1973). J. Food Sci. 38, 1047.

Schultz, T. H., Teranishi, R., McFadden, W. H., Kilpatrick,
P. W. and Corse, J. (1964). J. Food Sci. 29, 790.

Schultz, T. H., Flath, R. A. and Mon, T. R. (1971). J. Agric.
Food Chem. 19, 1060.

Stanley, W. L., Ikeda, R. M., Vannier, S. and Rolle, L. A.
(1961). J. Food Sci. 26, 43.

Teitelbaum, C. L. (1958). J. Org. Chem. 23, 646.

Wolford, R. W., Alberding, G. E. and Attaway, J. A. (1962).
J. Agric. Food Chem. 10, 297.

Wolford, R. W., Dougherty, M. H. and Petrus, D. R. (1968).
Int. Fruchtsaftunion, Ber. Wiss. Tech. Komm. 9, 151.

SPECTROPHOTOMETRIC AND SPECTROFLUOROMETRIC CHARACTERIZATION OF ORANGE JUICES AND RELATED PRODUCTS

Donald R. Petrus
Steven Nagy

Florida Department of Citrus
Lake Alfred, Florida

I. INTRODUCTION

During the past 25 years spectral characteristics, including both visible and ultraviolet absorbances at specific wavelengths, have been studied to monitor both desirable and undesirable properties of citrus juice. Investigations have included the effect of processing variables on the ultraviolet absorption of grapefruit juice (1); determination of hesperidin content in orange juices and peel extracts by ultraviolet absorption (2); spectral analysis of carotenoids and carotenes extracted from commercial samples of Florida frozen orange juice concentrates (3); determination of the polyphenolic content of lemon juices by ultraviolet absorption (4) and ultraviolet spectral properties of lemon juice under adverse storage conditions (5). More recent investigations (6, 7) have utilized the unique spectral characteristics of citrus juices, including room temperature fluorescence, to assess probable adulteration.

The many variables associated with the growing and processing of citrus fruit make it difficult to document the exact chemical and physical nature of a citrus product. However, spectrophotometric and spectrofluorometric analyses have proved to be a powerful tool in defining many of the physico-chemical characteristics of these nutritious products.

In this review, we report on research conducted by

Instrumental Analysis of Foods
Volume 2

our laboratory over the past 15 years on the
spectrophotometric and spectrofluorometric analyses of
orange juice products. Methodology employed in our work
has been extensively reported elsewhere (6-9).

II. ORANGE JUICE

        Petrus and Dougherty (8) recognized that much useful
information could be gained by obtaining both the visible
and ultraviolet absorption spectra of an alcoholic
extract of an orange juice sample. To this end, these
workers used this technique to show changes in Florida
Hamlin, Pineapple and Valencia orange juices during fruit
maturation and processing. A typical ultraviolet-visible
absorption spectrum of Florida Valencia orange juice is
shown in Figure 1. Spectral curves for Pineapple and
Hamlin orange juices were similar except for differences
in absorption intensities (8). Figure 1 shows
inflections or peaks at 465, 443 and 425 nm in the
visible spectrum and at 325, 280 and 245 nm in the
ultraviolet spectrum. Peaks in the visible spectrum are
mainly due to the presence of carotenoids (3, 9, 10).
Petrus and coworkers (10) showed that the sum of
carotenoid absorbances at 465, 443 and 425 nm were highly
correlated (r = .973) with color scores obtained with the
Hunter Model D-45 Citrus Colorimeter. The Hunter
Colorimeter is a reflectance instrument that measures
citrus redness and yellowness and is used by the citrus
industry and, exclusively in Florida, to determine color
scores of orange juices (11).
        Typical visible absorption spectra of alcoholic
solutions of Hamlin, Pineapple and Valencia orange juices
are shown in Figure 2. Valencia juice shows the most
intense visible absorption spectrum, whereas Hamlin shows
the weakest. In an extensive study on the color
characteristics of orange juice, Petrus and Dougherty (8)
noted that absorbance at 443 nm increased with fruit
maturity.
        The ultraviolet spectrum shown in Figure 1 is due
mainly to ascorbic acid (245 nm), total flavonoids (280
nm) and total polyphenols (325 mm) (2, 4, 7). Absorbance
at 280 nm is indicative of flavonoid content; however, it
is important to note that the majority of flavonoids do
not have their peak absorbances at this wavelength.
Horowitz and Gentili (12) compiled an extensive list of
ultraviolet wavelength maxima for many flavanone

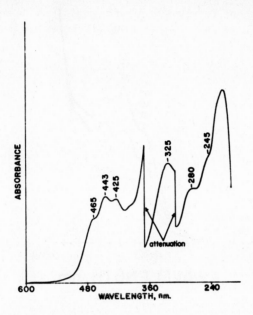

FIGURE 1.   Absorption spectrum of Florida Valencia orange juice.

glycosides, flavone glycosides, C-glycosylflavones and flavone aglycones found in citrus fruit. The main flavonoids found in orange juice and their λ max (12) are: hesperidin (285 nm), naringenin 7B-rutinoside (285 nm), isosakuranetin 7B-rutinoside (283 nm), naringenin 7B-rutinoside 4'B-D-glycoside (283 nm) and eriocitrin (285 nm).

Absorbance in the region 325-335 nm has been used by Vandercook and Rolle (4) to measure total polyphenolics in lemon juice and by Petrus and Dougherty (8) in orange juice and orange pulpwash (detailed explanation of this product is given in a following section). Vandercook and Rolle (4) showed a high correlation between absorbance at 325 nm (total polyphenolics) and citric acid content of lemon juice.

Oranges for processing are classified conveniently according to the interval between blossoming and harvesting. In Florida, oranges that mature between October and January are classed as early season (Hamlin is typical); those that mature between January and March are designated mid-season (Pineapple is typical), and

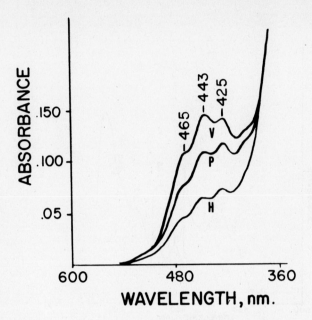

FIGURE 2. Visible absorption spectra of solutions
of Hamlin (H), Pineapple (P) and Valencia (V) orange
juices.

fruit harvested between mid March and July are
characterized as late season (Valencia is typical).
Absorption characteristics of Hamlin and Valencia oranges
at various harvest dates and at two extraction pressures
(8) are shown in Tables I and II. The absorbance at 443
nm (Table I) is an indicator of juice color and shows an
increase with fruit maturity (Hamlin orange; November 3,
1971 to February 2, 1972) as mentioned eariler, but not
with extraction pressure (light versus hard squeezes).
In contrast, there is a decrease in absorbance at 443 nm
for the Valencia variety (Table II) with maturity but,
again, no noticeable change caused by extraction pressure
is evident.

Absorption at 325 nm (indicator of total polyphenolics) and 280 nm (indicator of total flavonoids)
increased with maturity and extraction pressure for both
the Hamlin and Valencia oranges (Tables I and II). High
extraction pressures tend to cause more incorporation of
rag (carpellary membranes) and albedo (inner peel layers)
into the juice product (8). Since rag and peel contain

TABLE I. Absorption Characteristics for Hamlin
Orange Juice Obtained at Two Extraction Pressures

Date	443 nm	325 nm	280 nm	245 nm
Light Squeeze				
November 3, 1971	.038	.444	.842	1.528
November 17, 1971	.043	.475	.932	1.740
December 1, 1971	.057	.523	1.090	1.980
December 29, 1971	.051	.503	1.078	1.960
January 12, 1972	.070	.529	1.270	1.920
February 2, 1972	.071	.539	1.300	1.950
Hard Squeeze				
November 3, 1971	.039	.687	1.272	1.692
November 17, 1971	.048	.740	1.382	1.960
December 2, 1971	.053	.709	1.382	2.100
December 29, 1971	.057	.727	1.498	2.140
January 13, 1972	.073	.724	1.600	2.040
February 3, 1972	.078	.715	1.580	2.120

From reference (8).

TABLE II. Absorption Characteristics for Valencia Orange
Juice Obtained at Two Extraction Pressures

Date	443 nm	325 nm	280 nm	245 nm
Light Squeeze				
March 15, 1972	.139	.640	1.090	1.859
April 13, 1972	.140	.639	1.130	1.755
April 26, 1972	.140	.627	1.137	1.678
May 10, 1972	.137	.679	1.170	1.739
May 24, 1972	.130	.689	1.113	1.580
June 8, 1972	.132	.653	1.115	1.470
June 21, 1972	.129	.682	1.170	1.518
Hard Squeeze				
March 16, 1972	.148	.855	1.338	2.040
April 13, 1972	.146	.873	1.550	1.965
April 27, 1972	.143	.858	1.488	1.840
May 10, 1972	.143	.921	1.490	1.975
May 24, 1972	.137	.922	1.413	1.820
June 8, 1972	.130	.930	1.603	1.760
June 22, 1972	.130	.931	1.605	1.708

From reference (8).

high concentrations of flavonoids and bitter principles,
juice quality is lowered when excessive amounts of these
constituents are processed into the product (13).
Spectra of alcoholic solutions of rag and albedo show
strong absorbances at 325 nm and 280 nm (8).

The absorption attributed to vitamin C (245 nm)
increased slightly during maturation of Hamlin oranges;
also, a slight increase in absorbance at 245 nm resulted
from higher extraction pressures (Table I). Since the
highest concentration of vitamin C is found in the peel
(14), high extraction pressures would tend to express
more vitamin C. In Table II a decrease in absorbance
values at 245 nm was noted during maturation of Valencia
oranges, whereas at higher extraction pressures an
increase was observed. The vitamin C content of Valencia
oranges decreases during maturation (14).

Fluorescence spectral characteristics of alcoholic
solutions of citrus juices have been used by Petrus and
Attaway (6) to define product quality. Table III lists
the excitation and emission wavelengths utilized by these
workers for sample characterization. A typical
fluorescence excitation and emission spectra obtained
from an alcoholic solution of Florida Valencia orange
juice is shown in Figure 3. Strong excitation was noted
at 283 nm with a shoulder at 290 nm and slight
inflections at 270 nm and 302 nm with emission maximum at
333 nm.

TABLE III. Fluorescence Excitation and Emission
         Characterization of Orange Juice and Pulpwash

Excitation (nm)	Emission (nm)	Range setting	Excitation spectra	
			Orange juice	Pulpwash
340	423	0.1	Strong	Much stronger
230	310	0.03	Weak	Weaker
283	333	0.1	Strong maximum	Minimum
290	343	0.1	Strong maximum	Minimum or inflection
302	353	0.1	Inflection	Strong maximum
270	333	0.1	Inflection	Shoulder or maximum

From reference (6).

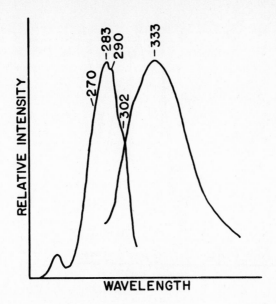

FIGURE 3. Fluorescence excitation (283 nm maximum) and emission (333 nm maximum) spectra obtained from Valencia orange juice.

III. WATER EXTRACTION OF SOLUBLE ORANGE SOLIDS

Water extraction of soluble orange solids (WESOS) is a by-product of the citrus processing industry. This byproduct is also known as pulpwash (PW), orange washed pulp (OWP) or water extraction of orange solids (WEOS). WESOS is made in the following manner: after extraction of juice from oranges, the juice is separated from the rag, excess pulp, seeds and other components by a finisher. The residual components contain adsorbed orange juice, and are further processed by washing with water (to extract the adhered juice) using one or a number of washing stages and by concentrating the washings to 45 to 65°Brix (15, 16). The purpose for development of this byproduct was to increase juice yield. This extra juice was processed, along with the original expressed juice, into frozen concentrated orange juice or used as a beverage base in drinks. Unfortunately, the original intent was abused, and this beneficial byproduct became the major adulterant of

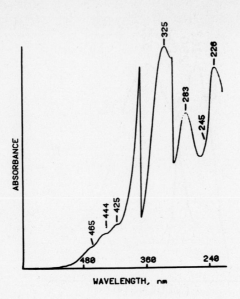

FIGURE 4. Spectral absorption curve of commercial
Florida-produced orange pulpwash.

single strength and frozen concentrated orange juice (6).
Department of Citrus Rule 20-64.07 (17) now prohibits the
addition of WESOS to frozen concentrated orange juice and
concentrated juice for manufacturing packed in Florida.
Because of this byproduct's importance, some representative
WESOS spectra are included in this report.

A typical ultraviolet-visible absorption curve for a
WESOS sample is shown in Figure 4. As noted, the visible
portion of the curve is characterized by weak, unresolved
absorption at 465, 443 and 425 nm, whereas absorption in
the ultraviolet region at 325 and 283 nm are noticeably
strong and well resolved. Water washing of orange pulp
would tend to incorporate more rag, pulp and water-
extractable soluble solids (8, 20). Washing frees adhered
orange juice from the pulp particles but it also causes
extraction of large quantities of flavonoids (9, 18).

The absorption characteristics of late-season orange
juices and their corresponding pulpwashes are shown in
Table IV. As noted, visible absorption at 443 nm
(indicative of carotenoids) decreased, whereas
ultraviolet absorption at 325 nm (polyphenolics) and 280
nm (flavonoids) increased from that of the parent orange
juices. In pulpwash samples, the absorbance at 443 nm

TABLE IV. Absorption Characteristics of Late Season
Orange Juices and Corresponding Pulpwashes

Plant	Juice[a]	443 nm	325 nm	280 nm
1	OJ	0.118	0.835	1.282
	PW	0.050	1.735	2.404
1	OJ	0.122	0.792	1.285
	PW	0.059	1.611	2.404
2	OJ	0.161	0.892	1.318
	PW	0.075	1.422	2.010
2	OJ	0.158	0.951	1.475
	PW	0.062	1.388	2.055
3	OJ	0.158	0.842	1.270
	PW	0.079	1.288	1.835
3	OJ	0.139	0.880	1.278
	PW	0.072	1.427	1.950
4	OJ	0.156	0.898	1.360
	PW	0.065	1.235	1.718

[a]OJ = Parent orange; PW = Pulpwash.
From reference (9).

was about half that of the parent orange juice. However,
absorbances at 325 nm were about 1.4 to 2-fold higher and
at 280 nm about 1.3 to 2-fold higher for the pulpwash
samples. The weak visible absorption of pulpwash
solutions is due to the concentration of residual parent
orange juice, whereas the very strong ultraviolet
absorption is the result of the parent orange juice and,
also, the presence of rag, pulp and their water-
extractable soluble solids. As noted in Table IV,
absorbance values vary, and this is due to different
pulpwashing practices at different processing plants and
to different varieties of fruit used in processing (8,
19, 20).

A comparison of fluorescent excitation and emission
curves obtained from alcoholic solutions of orange juice
and WESOS is shown in Figure 5. Maxima for orange juice
appeared at 290 nm excitation and 343 nm emission with
excitation shoulders or inflections at 270, 283 and 302
nm. Pulpwash revealed maxima excitation at 270 and 302
nm with shoulders at 283 and 290 nm, and maximum emission
at 353-355 nm (see also Table III). Alcoholic solutions
of rag and albedo components produced fluorescence
spectra (7) similar to the WESOS spectrum of Figure 5.

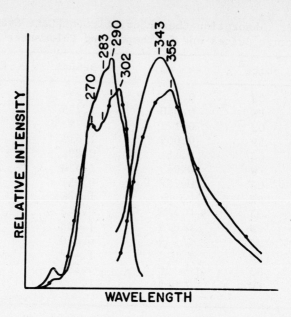

FIGURE 5. Comparison of fluorescence excitation and emission spectra of orange juice (—) and orange pulpwash (●—●).

IV. ORANGE JUICE ADDITIVES

The United States Code of Federal Regulations (21) identifies orange juice and the addition of preservatives to orange juice products. Currently, two preservatives are approved, namely, sodium benzoate and sorbic acid (either may be used in orange juice products but at an amount not to exceed 0.2 percent by weight). Any added preservative must be declared on the product label. Unfortunately, an added preservative is not always declared, and when this occurs, the product is considered illegally branded. An example of an undeclared preservative in single-strength orange juice is shown in Figure 6. Visible absorption at 465, 443 and 425 nm appeared normal; however, the ultraviolet spectrum revealed a noticeable aberration. Ultraviolet absorbance at 325 nm appeared weak and absorbance at 280 and 245 nm were obscured by a strong absorbance at 254 nm. Further investigation of the 254 nm peak by high performance liquid chromatography and spiking studies identified the peak as the preservative, sorbate (22).

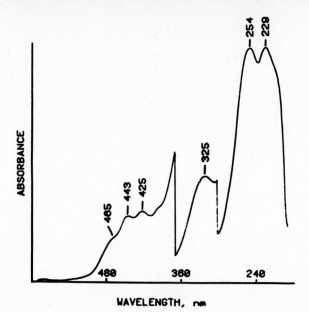

FIGURE 6. Visible and ultraviolet absorption curve obtained from retail single-strength orange juice from concentrate showing sorbate preservative absorption at 254 nm.

Florida processors import large quantities of bulk concentrated orange juice for manufacturing which may be blended during the production of Florida frozen concentrated orange juice and other products. Regrettably, some of these imported products also contain undeclared adulterants; one of which is a non-citrus derived colorant. In grading processed orange juice using USDA standards, score points are given within specified ranges for 3 factors: color, defects and flavor. The grading system allows up to 40 quality points for color, out of a total of 100 (23). Thus, juices with a deeper orange color are generally accorded a higher grade than juices having a more yellow-orange color (higher grades often translate into higher retail prices). An example of an adulterated product containing an added colorant is shown in Figure 7. The curve labelled "A" is the normal fluorescence excitation and emission background curve obtained for pure orange juice, whereas the curve labelled "B" is that from an imported product. Research of the fluorescence characteristics of the colorant, turmeric, either as

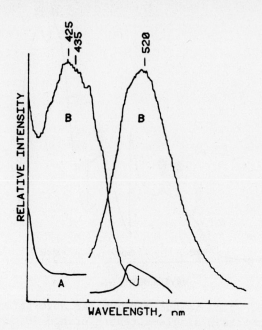

FIGURE 7. Turmeric fluorescence excitation (435 nm)
and emission (520 nm) curves obtained from imported bulk
concentrated orange juice for manufacturing by Florida
processors; curve A is normal fluorescence background for
pure orange juice, whereas curve B represents the imported
product showing added colorant.

obtained or in spiked orange juice, produced fluorescence
curves almost identical to the "B" curves of Figure 7. The
oleoresin, turmeric, is obtained by solvent extraction of
the dried rhizomes of Curcuma longa L. and is a yellow
orange to red brown viscous liquid. Turmeric has been
detected, of late, in many adulterated citrus products
(18, 24).

V. ORANGE DRINKS

The principal juice products derived from expression
of fruits are juices, drinks and nectars. A fruit drink is
a diluted fruit juice which may contain added water, sugar,
citric acid, gum arabic, artificial colors and flavors
(25). The spectrum shown in Figure 8 was obtained from an

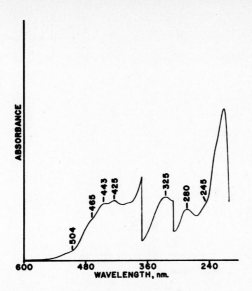

FIGURE 8.   Absorption curve obtained from a solution
of commercial orange drink fortified with ascorbic acid.

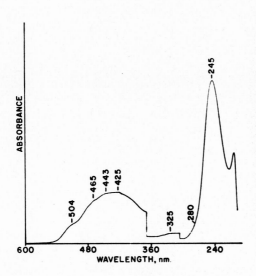

FIGURE 9.   Absorption curve of synthetic powdered
orange drink fortified with ascorbic acid.

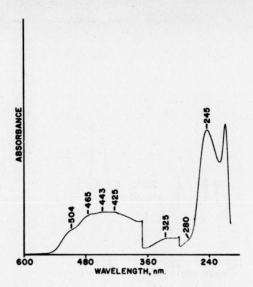

FIGURE 10. Visible-ultraviolet absorption of synthetic powdered orange drink.

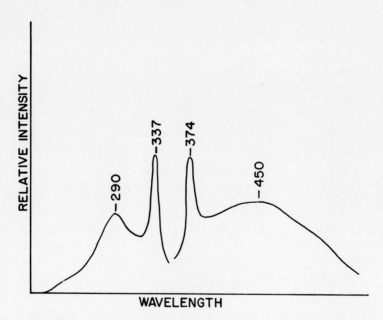

FIGURE 11. Fluorescence excitation and emission curves for a commercial synthetic orange drink.

alcoholic solution of a commercial orange drink with a label declaration stating 'fortified with ascorbic acid'. The spectral shape and absorption maxima indicated the presence of orange juice, and the absorbance at 504 nm indicated the addition of an artificial color. Ultraviolet absorption was much lower than observed for pure juices, and the absorbance at 245 nm indicated little if any addition of ascorbic acid. Two orange drinks (made by addition of water to the synthetic orange drink powder) are shown in Figures 9 and 10. The absorption spectra of alcoholic solutions of these two reconstituted drinks revealed the addition of ascorbic acid (absorbance at 245 nm) but the virtual absence of any orange juice. The fluorescent characteristics of a representative powdered orange drink is shown in Figure 11. As noted, this fluorometric pattern is quite different from that of an orange juice (see Figure 3). For the powdered orange drink (Figure 11), maximum excitation was at 337 nm with a peak at 290 nm; maximum emission was at 374 nm with a broad peak at 450 nm. The excitation and emission spectrum are similar to those obtained with approved FD & C dyes (24). The broad emission at 450 nm is probably due to one or more additives.

REFERENCES

1. Hendrickson, R., Kesterson, J. W., and Edwards, G. J., Proc. Fla. State Hortic. Soc. 71:190 (1958).
2. Hendrickson, R., Kesterson, J. W., and Edwards, G. J., Proc. Fla. State Hortic. Soc. 72:258 (1959).
3. Ting, S. V., Proc. Fla. State Hortic. Soc. 74:262 (1961).
4. Vandercook, C. E., and Rolle, L. A., J. Am. Assoc. Off. Anal. Chem. 46:359 (1963).
5. Vandercook, C. E., J. Food Sci. 35:517 (1970).
6. Petrus, D. R., and Attaway, J. A., J. Assoc. Off. Anal. Chem. 63:1317 (1980).
7. Petrus, D. R., and Dunham, N. A., in "Citrus Nutrition and Quality" (S. Nagy and J. A. Attaway, eds.), p. 423. American Chemical Society, Washington, D. C., 1980.
8. Petrus, D. R., and Dougherty, M. H., J. Food Sci. 38:659 (1973).
9. Petrus, D. R., and Dougherty, M. H., J. Food Sci. 38:913 (1973).
10. Petrus, D. R., Huggart, R. L., and Dougherty, M. H.,

J. Food Sci. 40:922 (1975).

11. Steward, I., in "Citrus Nutrition and Quality" (S. Nagy and J. A. Attaway, eds.) p. 129. American Chemical Society, Washington, D. C., 1980.

12. Horowitz, R. M., and Gentili, B., in "Citrus Science and Technology" (S. Nagy, P. E. Shaw and M. K. Veldhuis, eds.) Vol. 1, p. 397. Avi Publishing Co., Westport, CT, 1977.

13. Nagy, S., and Rouseff, R. L., in "The Analysis and Control of Less Desirable Flavors in Foods and Beverages" (G. Charalambous, ed.) p. 171. Academic Press, New York, 1980.

14. Nagy, S., J. Agric. Food Chem. 28:8 (1980).

15. McKinnis, R. B., Andrews, R. A., and Jones, H. L., Trans. Citrus Conf. (Fla. Sect. ASME) 10:1 (1964).

16. Belk, W. C., Trans. Citrus Conf. (Fla. Sect. ASME) 10:19 (1964).

17. Official Rules Affecting the Florida Citrus Industry Pursuant to Chapter 601, Florida Statutes 1975, Amended 1982, State of Florida, Department of Citrus, Lakeland, FL.

18. Rouseff, R. L., unpublished data, Florida Department of Citrus, 1982.

19. Rouse, A. H., Atkins, C. D., and Moore, E. L., Proc. Fla. State Hortic. Soc. 72:227 (1959).

20. Huggart, R. L., Olsen, R. W., Wenzel, F. W., Barron, K. W., and Ezell, G. H., Proc. Fla. State Hortic. Soc. 72:221 (1959).

21. Code of Federal Regulations 21, Food and Drugs, Parts 100 to 169, Revised as of April 1, 1980.

22. Bennett, M. C., and Petrus, D. R., J. Food Sci. 42:1220 (1977).

23. Nordby, H. E., and Nagy, S., in "Fruit and Vegetable Juice Processing Technology", 3rd Edition (P. E. Nelson and D. K. Tressler, eds.), p. 35. Avi Publishing Co., Westport, CT, 1980.

24. Petrus, D. R., unpublished results, Florida Department of Citrus, 1982.

25. Nagy, S., in "Encyclopedia of Food Science" (M. S. Peterson and A. H. Johnson, eds.), p. 334. Avi Publishing Co., Westport, CT (1978).

Florida Agricultural Experiment Stations Journal Series No. 4444.

FLAVOR CHARACTERISTICS OF THE COMPONENTS OF ORANGE BLOSSOM CITRUS AURANTIUM

Felix Buccellato

Custom Essence Inc.
Somerset, New Jersey

Orange blossoms are the flowers from the bitter orange tree, Citrus Aurantium Linnaeus. Originating in China, the bitter orange tree was probably introduced to southern France somewhere between the tenth and eleventh centuries A.D. by the conquering Arabs. It remained the only orange known to Europeans for about five centuries. Since the fruit of this tree is so unpalatable, though more sour than bitter from this author's first hand experience, it is no wonder that products for fragrances were developed. Distilled oil of orange flower appeared as early as the sixteenth century, but only when it was used by the Duchess Flavio Orsini, a member of a noble and powerful Roman family, did it emerge in 1680 as a fashionable fragrance article. Since she was also known as the Princess of Neroli, or Nerola, a town situated roughly thirty kilometers northwest of Rome, the name Oil of Neroli was adopted and remains in current use. While this is probably the most glamorous tale of orange blossom use, it is by no means the only reason for the success of products from Citrus Aurantium. The peel of the fruit is used in marmalade, a bitter but popular jam. In Italy, for instance, where bitter-tasting products are more accepted than in the United States, the juice from the fruit is used in beverage flavors such as Chinotto and Aranciata, both popular bitter orange sodas. Orange blossom and related products have been used in beverages, ice cream, ices, candy, baked goods, chewing gum, gelatins and puddings. It may well be that orange flower water absolute became commercially available as a result of an effort to reduce the exported volume of orange flower water. This was done to decrease the cost of freight

Instrumental Analysis of Foods
Volume 2

FIGURE I

PRODUCTS DERIVED FROM CITRUS AURANTIUM

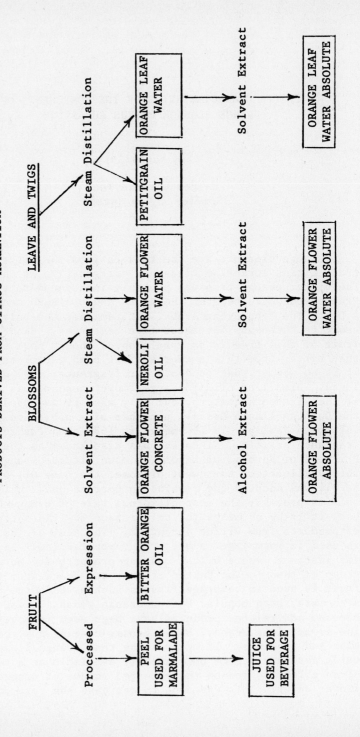

and custom duties levied on orange flower water.

Along with the esthetic uses come the pharmacological uses of Neroli oil. It has a weak antiseptic effect, belongs to the group of cramp inducing oils and has a strong bactericidal effect on staphylococcus aureus. Neroli Water or Eau de Neroli is classifed as a hypnotic, sedative, antispasmodic and capable of calming heart rhythm. For all those flavorists and perfumers who have sleepless nights, it is a well-known remedy for insomnia.

Overall, the citrus industry in Europe is literally built on the bitter orange tree. The roots of the tree are quite resistant to disease. As a result, lemon, sweet orange, mandarin, grapefruit and bergamot are commonly propagated on the root stock.

Figure 1 shows a scheme of products from the bitter orange tree for general reference.

Through the years, orange flower has been the subject of some study. There is some information in the literature on the composition of Neroli. As early as 1928, Professor G. V. Villavecchia identified d-pinene, l-camphene, dipentene, decanal, jasmone, l-linalool, linalyl acetate, geraniol, nerol, d-nerolidol, phenyl ethyl alcohol, α-terpineol, indole, methyl anthranilate along with esters of phenyl acetic acid and benzoic acid. Information on orange flower absolute is more obscure. For a flavorist or perfumer, the composition of the natural flower is of paramount interest, and it is in the analysis of orange flower absolute that one might best look for that information. In Table I, analyses of orange flower absolute, neroli and orange flower water absolute can easily be compared. In theory, if one were to combine neroli and orange flower water absolute, one should arrive at the total fragrance composition that exists in the flower. Of course, this is unrealistic and does not take into account the reality of changes introduced by processing.

TABLE I

Components	Orange Flower Absolute	Neroli	Orange Flower Water Absolute
Alpha Pinene	trace	0.8	--
Beta Pinene	0.4	15.0	1.1
Myrene	0.1	1.6	--
Limonene & Cis Ocimene	5.1	16.1	0.5
Trans Ocimene	0.6	6.0	0.2
Linalool	32.0	30.6	44.2
Benzyl Nitrile	0.6	--	2.5
Phenyl Ethyl Alcohol	4.5	--	1.9
Alpha Terpineol	2.4	3.0	18.5
Citronellol	0.5	0.2	0.2
Nerol	0.9	0.5	2.8
Linalyl Acetate	16.8	9.1	--
Geraniol	1.5	2.0	6.4
Indole	1.0	0.1	0.1
Eugenol	0.3	--	0.5
Neryl Acetate	0.8	1.7	0.5
Citronellyl Acetate	0.1	--	--
Geranyl Acetate	0.6	2.9	0.5
Nerolidol	7.6	4.0	1.7
Farnesol	7.7	4.0	0.5

As can be seen in Table I, there are some components
which vary in concentration due to their water solubility
such as phenyl ethyl alcohol and methyl anthranilate.
Further, if we focus on linalyl acetate, we notice a reduction of this chemical in neroli and practically a complete
loss in orange flower water absolute. This is probably
a result of the harsh distillation process which hydrolizes
the ester. Overall, this may be viewed as an indicator
of what may happen to other oils that contain linalyl
acetate and undergo a steam distillation. products such
as lavender and lavandin may in fact have higher concentrations of linalyl acetate in the plant than the analysis
of their essential oils would lead us to believe.

With problems of processing aside, we can begin to examine the constituents for their flavor-frangrance contribution. As a guideline, I have listed the main characteristics of orange blossom in Table II. Under each heading, I have placed the chemicals that contribute most to that character. You will note that methyl anthranilate is under two categories, floral and fruity. I feel that this particular chemical plays an important dual role in the overall flavor and fragrance of orange blossom.

TABLE II

CITRUS	FLORAL	FRUITY
alpha pinene	phenyl ethyl alcohol	neryl acetate
beta pinene	alpha terpineol	citronellyl acetate
myrcene	citronellol	geranyl acetate
limonene	nerol	methyl anthranilate
ocimene	geraniol	
linalool	indole	
linalyl acetate	methyl anthranilate	
	nerolidol	
	farnesol	

SPICY	GREEN/EARTHY
eugenol	benzyl nitrile
	pyrazines

After categorization by odor, I have often found it helpful to examine the constituents with regard to their chemical class. While each component is different in its character and performance, examination by class can give an overall appreciation for the construction of the odor of orange blossom.

HYDROCARBONS - alpha pinene, beta pinene, myrcene, limonene, and ocimene. This class provides the bright, fresh citrus character that is so characteristic of neroli. Collectively, the hydrocarbons represent about forty percent of neroli oil and they are the reason neroli is so bright and diffusive, adding a fresh note to any composition. The role of hydrocarbons in the absolute is less important than in the distilled oil. Representing only about six percent, this group boosts the floralcy of the other components, giving the overall composition brightness and diffusion. The most important component in this class is ocimene, which provides a very fresh citrus note along with a subtle green nuance.

ALCOHOLS - linalool, phenyl ethyl alcohol, alpha terpineol,
citronellol, nerol, geraniol, nerolidol and farnesol. This
group provides the floral body of orange blossom. As can
be seen in Table I, this group is much more important to
the character of the absolute than the distilled oil.
Linalool is the major contributor to odor impact and character in this group. This chemical can be viewed as the link
that bridges the citrus character with the floral character.
Displaying characteristics of each, linalool, well-known
as it may be to all in the industry, is a very important
chemical to the flavor and fragrance of orange blossom.

ESTERS - linalyl acetate, methyl anthranilate, neryl acetate,
citronellyl acetate, geranyl acetate. While these esters
can generally be characterized as fruity, the part they play
in orange blossom can best be described as supporting and
modifying the citrus character. The single most important
member of this class is methyl anthranilate. It is both
fruity and floral and is a major characterizing component
in both the absolute and the distilled oil. Even though
there is only one tenth the amount in the distilled oil,
the presence of methyl anthranilate can be easily detected
by odor and taste.

PHELOLS - The only phenol reported here is eugenol. While
it is a minor constituent, the effect of this chemical should
not be overlooked or under-estimated. The spicy clove like
flavor adds a particular brightness and natural character
to orange blossom. Phenols as a class are extremely important odor contributors. They often occur at trace levels,
and it is certain that more are waiting to be discovered
in orange blossom.

NITROGENOUS COMPOUNDS - Indole and methyl anthranilate are
very important character contributors to orange flower.
Indole is a nitrogen-bearing molecule and acts as a powerful
floral chemical tha can also be perceived as quite animalic
when not in extreme dilution. While methyl anthranilate
is listed along with the esters, it differs in one important aspect that it is also an amino ester and as previously
stated, is a major flavor characterizer for orange blossom. Benzyl Nitrile is an interesting part of the flavor
of orange blossom. It adds to the peculiar green somewhat
leafy vegetable character.

     After all the previously discussed components are com-
bined at their proper concentration, it becomes quite clear

that the odor and flavor of neroli has been vaguely imitated
but the unique character has not been captured. It is at
this crucial point that we come to grips with the nature
of orange blossom.

The occurrence of pyrazines in nature is well-known
and the odor importance of this class of chemicals is just
beginning to be understood and appreciated. In 1971, Duprey
and James identified 2-methoxy-3-isopropyl, 2-methoxy-3-
isobutyl and 2-methoxy-3-isopropyl 5-methyl pyrazines in
oil of petitgrain. It would come as no surprise to find
them in the blossoms of the flower. It is, in my opinion,
precisely that peculiar pyrazine-like note with the green
yet dry and earthy character than is so characteristic of
orange blossom. Granted, there may be many other unidentified constituents and many of them important to the aroma of
orange blossom but I would suspect that pyrazines, along
with thiazoles and other nitrogenous compounds, are the most
important contributors to the unique somewhat bread-like
character of orange flower. It is well-known to perfumers
and flavorists alike, that orange blossom has the ability
to add a unique freshness to the top note while exhibiting
a long lasting character with surprising tenacity. This
is a common characteristic of the performance of pyrazines.
The pyrazines do for orange flower what damascenone and damas-
cone do for rose.

If the rose can be dubbed "Queen of the Florals," I
suppose we could call the orange blossom the princess. It
is an expensive product and finds its way into many creations.
It has been used as a main theme in frangrances like L'Origan
and Fidgi and appears extensively as top notes of modifiers
in many colognes like Eau Sauvage and 4711.

ENHANCING NATURE

There are many chemicals commercially available that
have become successful because they impart orange flower
notes quite economically. Nerone = 1-(Para menthene-6yl)-
1-propanone is a green and somewhat earthy note reminiscent
of petitgrain. This chemical is rather unusual in the sense
that there are not many of this type or class exhibiting
a character of orange flower or petitgrain.

An interesting group of chemicals are the various substituted naphthalenes.

        alpha methyl naphthyl ketone
        beta methyl naphthyl ketone
        nerolin
        yara-yara

They all bear a resemblance to each other and to methyl

anthranilate which is probably the reason they have come
to be associated with the orange flower character initially.
There are a few other chemicals reported in the literature, for example, gamma terpinene, para cymene, alpha
terpinyl acetate, sabinene, alpha phellandrene, terpinolene,
terpinene-4-ol, cis-3-hexenol, decanal, neral, geranial and
thymol. Guenther makes mention of benzaldehyde; an unidentified basic compound with a nicotinic odor; nitrile of phenyl-
acetic acid; jasmone; and an unidentified nitrogenous compound.
Additionally, Corbier and Teisseire identified cis-8-hepta-
decene and 2,5-dimethyl 2, vinyl-4-hexenal. Unfortunately,
they did not characterize the odor or flavor of these components.

Many unidentified or discovered and not yet publicized
chemicals exist in orange blossom, and in order to gain a
proper understanding of the nature of orange blossom much
additional work is needed.

REFERENCES

Corbier, B., and Teisseire, P., (1974). Contribution to
    the Knowledge of Neroli Oil Recherches, No. 19.
Fenaroli Handbook of Flavor Ingredients, (1971). 713.
Gildemeister & Hoffman. (1900). 480.
Gildemeister, Huetig Verlag, Die Physiologischen & Pharmacolo-
    gischen Wirkungen der ae/oele. 91.
Guenther. (1949). The Essential Oils. Vol. III., 254.
Lawrence, B. (1980). Progress in Essential Oils, Perfumer
    & Flavorist. Vol 5, No. 6, 28.
Takken. (1975). Olfactive Properties of Polysubstituted
    Pyrazines, Journal of Agriculture & Food Chem. Vol. 23,
    4.
Villavecchia, G.V. (1928). Dizionario Di Merceologia e di
    Chimica Applicata. 407.

APPLICATION OF MULTIVARIATE ANALYSIS TO CAPILLARY GC PROFILES: COMPARISON OF THE VOLATILE FRACTION IN PROCESSED ORANGE JUICES

R. S. Carpenter
D. R. Burgard
D. R. Patton
S. S. Zwerdling

The Procter and Gamble Company
Miami Valley Laboratories
Cincinnati, Ohio

I. INTRODUCTION

High resolution capillary GC (CAP-GC) separations play an increasingly important role in the analysis of food and beverage flavor. One of the objectives in using this technique is to develop a better understanding of the chemical changes that occur in flavor and aroma as a result of processing. An obvious use for such information is in defining optimum process conditions for maximizing product quality and eliminating undesirable effects such as the formation of off-flavor. However, even in simple cases detailed evaluation and interpretation of CAP-GC results are often complicated due to the complexity of these separations.

With the application of multivariate analysis (MVA) effective techniques can be developed for evaluating CAP-GC data and correlating GC results with objective (e.g., process conditions, stability, aging) and sensory (e.g., flavor, aroma) aspects of food and beverage quality (1,2). The success of this approach has been demonstrated in the characterization of a variety of flavor materials (3-6).

Instrumental Analysis of Foods
Volume 2

173

Although commercial orange juice has also been
investigated regarding the effects of storage, stability,
and volatile contribution to flavor quality (7-10), few
reports have dealt with a multivariate approach (11). We
report here the results of CAP-GC/MVA analyses comparing
the volatile fraction in both Valencia and blended single
strength orange juice from three types of processing (high
temperature pasteurization, freeze, and evaporative
concentration). Compositional variations in juice starting
material due to varietal and blending differences were
minimized by using single lots of either pure Valencia or
blended juice to generate all processed samples.

II. EXPERIMENTAL

A. Orange Juice Samples

Samples of freeze (44.9° Brix) and evaporative (50.5°
Brix) concentrate were prepared from a single stock of
fresh Valencia orange juice collected during the 1982
season. Evaporative concentrate, produced by thermally
accelerated short time evaporation (TASTE) was formulated
with added Valencia peel oil (0.02%) and Valencia cutback
juice (approximately 15%) in accordance with standard
industry practices. Samples of blended single strength
juice (feedstock, 15° Brix) and high temperature
pasteurized juice (processed, 12° Brix) were also obtained.
Processed single strength juice was prepared from feedstock
by heat treatment through high temperature pasteurization
at 275°F (135°C) with a 10.4 second residence time. All
samples were stored frozen at -60°C prior to analysis.

B. Sample Preparation

Juice samples were prepared for analysis using a
combined vacuum distillation-extraction procedure. Juice
concentrates were diluted to single strength levels (12°
Brix) with distilled water prior to analysis. Single
strength and Valencia juice were prepared directly.

Hydrocarbon standards (C9 and C13 n-alkanes) were spiked
into 300 ml of sample at a level of 5 ppm and the sample
distilled under vacuum (1-5 torr) at 40°C until 250 ml of
distillate was collected. Distillate collection was

accomplished using a series of three cold traps (ice/water, dry ice/methanol, and liquid N_2). After distillation the contents of the traps were combined, made up to 10% (w/v) with KCl, and extracted three times with methylene chloride (5 ml/100 ml distillate). The combined extracts were dried over anhydrous Na_2SO_4 and then concentrated by evaporation in an ice bath under a stream of N_2 to a final volume of about 0.5-1.0 ml.

C. Chromatographic Analysis

High resolution capillary GC (Hewlett Packard HP5880A) was used to assay juice extracts according to the conditions in Table I. Both peak retention times and integrated peak areas were retained for further analysis.

TABLE I.   CAP-GC Assay Conditions

Carrier Gas: Hydrogen
Linear Velocity: μCH_4 @ $0^\circ C$ = 66 cm/sec;
                  μCH_4 @ $250^\circ C$ = 39 cm/sec
Column: Fused Silica Methyl Silicon
         (50 m x 0.32 mm ID)
Detector: FID (Air/H_2/N_2) (300:30:20)
Sample Volume: 1.5 μL (Split 1:50)
Temperature Program: Level I 0°-$85^\circ C$ at 2°/min
                      Level II 85°-$250^\circ C$ at 5°/min

D. Multivariate Data Analysis

Replicate analyses of each juice were used to establish the statistical validity of chromatographic profile data. Four samples from each Valencia juice (fresh, freeze and TASTE concentrates), and six samples of each blended single strength juice were used in the analysis. Duplicate injections were also made for each sample.

123 peaks in Valencia juice chromatograms and 84 peaks in the blended single strength juice chromatograms were selected for multivariate analysis. Unretained peaks in

chromatograms not selected for data analysis were either
small (< 10 ppb as determined by internal standard
response) or they did not occur in all chromatographic
replicates.

The nearest neighbor algorithm was used to compare GC
profiles for the different juices (12). This consisted of
treating each chromatogram, in turn, as an unknown and
determining which of the other profiles was most similar to
it based on a comparison of peak areas. This was
accomplished by a difference calculation (subtraction) for
one variable (1-dimensional analysis) or a Euclidean
distance calculation, $[(X_{11}-X_{21})^2 + (X_{12}-X_{22})^2 + ...]^{1/2}$,
for several variables simultaneously. Peaks, or
combinations of peaks that clustered chromatograms of the
same type together and separated the different types of
juices were retained as useful discriminators. From these
results both normalized peak areas and peak area ratios
were used for further calculations. All multivariate
analyses were performed on a Hewlett Packard (HP-1000)
computer.

III.   RESULTS AND DISCUSSION

A.   Effects of High Temperature Processing
on Single Strength Juice

A comparison of representative profiles for both the
feedstock and high temperature processed juices is shown in
Figure 1. Of the 84 peaks selected for MVA, 14 components
were identified which showed reproducible differences
between the feedstock and processed juice. These are shown
in Figure 2 where their relative position in chromatograms
and standard deviation of measurement are indicated. This
dispersion of components across the chromatographic profile
indicates that process induced changes occurred over a wide
range of component volatilities.

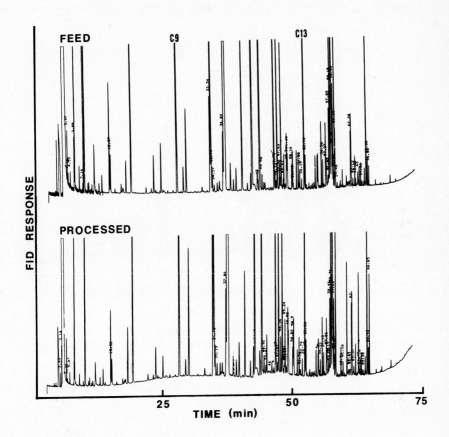

FIGURE 1. Representative, CAP-GC profiles comparing feedstock and thermally processed single strength orange juices.

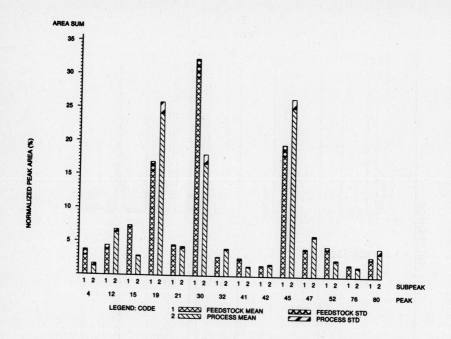

FIGURE 2. Histogram profile of components
discriminating feedstock from thermally processed single
strength juice. Subpeak axis indicates treatment code: 1)
feedstock, 2) processed juice. Peak numbers are listed on
bottom x-axis. Component levels are plotted as normalized
peak areas + standard deviation.

Identification of peak components based on retention
time comparison with standards demonstrated that a majority
of the 14 components were saturated alcohols and aldehydes
(C6-C12) (Table II). Although this product/precursor
relationship was not demonstrated for other classes of
compounds in the juice (e.g., terpenes), it does suggest
general oxidation of alcohols to their corresponding
aldehydes during processing. Similar observations have
been reported in other studies (8,9).

TABLE II.   Identification of Saturated (C6–C12) Alcohols
and Aldehydes in Feedstock and Processed Single
Strength Juice from Figure 2.

Peak Number	Identity
12	Hexanal
15	Hexanol
19	Octanal
30	Octanol
32	Nonanal
41	Nonanol
45	Decanal
52	Decanol

Ratio data for aldehydes and alcohols in table II yielded
an estimate of the magnitudes of these changes which
occurred during processing (Figure 3). The mean value for
each of these ratios is listed in Table III. Overall
aldehyde/alcohol ratios were about three times higher in
the processed juice with the largest increases occurring in
the shorter chain compounds. Comparison of peak area
ratios between the two juices for other identified
components (e.g., limonene, myrcene, and α-pinene) showed
no change as a result of processing. This is in contrast
to some previous studies on the effects of heat treatment
in processed juice (8,9). α-terpineol (peak #42) was found
to increase slightly as a result of processing. These
components were noted because of their reported implication
in juice flavor quality (8,13).

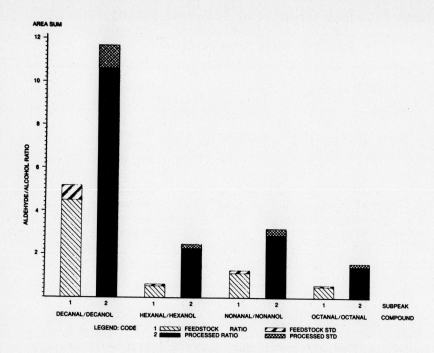

FIGURE 3. Aldehyde/Alcohol Ratios in 1) Feedstock and 2) Processed Juice. Ratios are plotted as mean values ± standard deviation.

TABLE III. Relative Increase in the Aldehyde/Alcohol Ratio for Processed Single Strength Juice

Compound	Mean Ratio[1]		Increase
	Feedstock	Processed	
decanal/decanol	4.6	10.7	2.3 x
hexanal/hexanol	0.54	2.3	4.3 x
nonanal/nonanol	1.2	2.9	2.4 x
octanal/octanol	0.53	1.5	2.8 x
		Average Increase	3.0 x

[1]Peak area ratios to C9 and C13 internal standards.

B.   Effect of Freeze and Evaporative Concentration
on Fresh Valencia Juice

Representative chromatographic profiles for fresh,
freeze, and evaporative concentrate are shown in Figure 4.
Of the 123 components selected for MVA in Valencia juice,
24 components were identified that showed statistically
significant differences as a result of either freeze or
evaporative processing relative to fresh juice. These
differences resulted from peak level changes in both
processed juices with the same changes not always being
observed in both juice types.

Ratio analysis of these 24 components indicated
several different patterns or variations in peak level
changes as a function of processing (Table IV). These were
classified as individual or groups of key components that
either increased, decreased, or showed no change as a result
of processing when compared to fresh juice. Table IV
represents the average of 8 replicate analyses for each
juice type. A value of 1.0 implies that the component level
in the processed juice and the fresh juice are the same.
Changes (or lack of changes) were determined to be
significant at the 95% confidence level (% relative standard
deviation about 25%). Ratio data were calculated from
individual peak area/internal standard ratios using an
average of both C9 and C13. This average was used since the
distillation-extraction efficiency was not known for each
component.

In comparing peak levels, the most significant feature
observed is the decrease or loss of volatile components in
the evaporative concentrate relative to both fresh and
freeze concentrate juice (variation #2; Table IV). This is
interesting, although not unexpected, since evaporative
concentrate is rectified with added peel oil and cutback
juices to restore some of the flavor volatiles that are lost
as a result of the TASTE process. The relative amount of
these 17 components (peak area ratio to C9 and C13 internal
standards) in each juice is shown graphically in Figure 5.

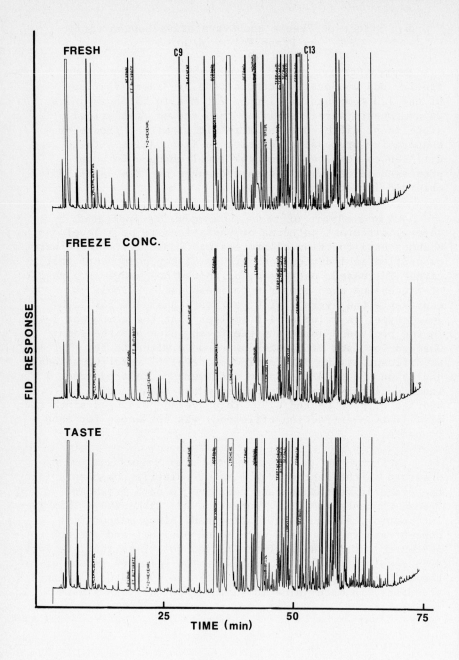

FIGURE 4. Representative CAP-GC profiles for fresh, freeze, and evaporative (TASTE) concentrates from Valencia orange juice.

TABLE IV. Variation in Peak Component Levels Between Freeze and Evaporative (TASTE) Concentrate Relative to Fresh Juice.

Variation	Direction of Peak Changes[1]		Peak Number	Mean Concentration Ratio[2]	
				Freeze	TASTE
1	F	Decrease	16	0.37	0.36
	T	Decrease			
2	F	No change	9	0.80	N.D.[3]
	T	Decrease	10	1.01	N.D.
			19	1.01	N.D.
			29	0.91	0.26
			34	1.15	0.10
			35	1.50	0.08
			43	1.01	0.28
			45	0.82	0.28
			46	0.88	0.07
			49	1.18	0.47
			50	0.93	0.41
			58	0.94	0.35
			60	0.91	0.38
			61	1.15	0.34
			62	1.03	0.02
			70	0.70	N.D.
			99	1.55	N.D.
3	F	Increase	66	1.43	N.D.
	T	Decrease	102	1.70	0.29
			103	1.63	0.18
			114	1.81	0.44
4	F	Increase	42	1.99	1.58
	T	Increase	116	21.13	1.68

(1) Freeze (F) and Evaporative Concentrate(T)
(2) Peak area ratios to C9 and C13 internal standards relative to fresh juice.
(3) Peak was below detectable limit in evaporative concentrate.

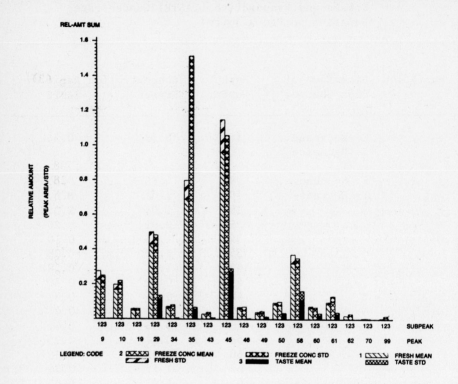

FIGURE 5. Comparative profile of peak levels in
Valencia juice showing no change in fresh and freeze
concentrate but decreases in evaporative (TASTE) juice.
Values are reported at the 95% confidence interval.

Although most of the comparative changes observed in this study were the result of volatile losses in evaporative juice, several other types of changes in both evaporated and freeze concentrate were observed (Peaks #16, 42, 66, 102, 103, 114, 116; Table IV). Since a number of factors affect juice volatile composition (e.g., seasonal and varietal differences, process fluctuation) the significance of these components has yet to be determined.

IV. CONCLUSIONS

Multivariate analysis plays a very useful role in evaluating complex chromatographic data resulting from CAP-GC separations of flavor volatiles. Its illustration in this study, comparing the volatile profiles from several kinds of processed orange juices, resulted in identifying a number of component differences which accurately discriminated among process juice types. Knowledge of the structural identity for many of these components also yielded information into the chemical nature of process effects (e.g., component oxidation, volatile losses). Although limitations still exist in the standard application of multivariate techniques in CAP-GC flavor analysis (primarily GC data manipulation and computer accessibility), the potential exists for these methods to be used in the routine evaluation of food and beverage flavor quality.

V. ACKNOWLEDGMENTS

Special thanks go to Messrs. Pedro Rodriguez, Ev Kitchen, Richard Hafner, John Powers, and Ms. Kathrine Moore for their help in collecting orange juice samples. Also, special thanks go to Ms. Michaelle Jones for technical assistance in graphic reproductions and to Mrs. Carol Dalton for preparing this manuscript.

VI.   REFERENCES

1. Powers, J. J., in "Food Flavours" (I. D. Morton and A. J. Macleod, eds.), p. 121. Elsevier Publishing, New York (1982).
2. Spencer, M. D., Pangborn, R. M., and Jennings, W. G., J. Agric. Food Chem. 26, 725 (1978).
3. Powers, J. J., and Keith, E. S., J. Food Sci., 33, 207 (1968).
4. Young, L. L., Bargmann, R. E. and Powers, J. J., J. Food Sci., 35, 219 (1970).
5. Qvist, I. H., von Sydow, E. C. F., and Akesson, C. A., Lebensm. Wiss. Technol., 9, 311 (1976).
6. Aishima, T., Nagasawa, M., and Fukushima, D., J. Food Sci., 44, 1723 (1979).
7. Kealey, K. S., and Kinsella, J. E., CRC Critical Rev. Food Sci. Nutr., 11, 1 (1978).
8. Durr, P., Analytik, 19, 35 (1980).
9. El-Samahy, S. K., El-Baki, M. M. A., Askar, A., Taha, R. A., and El-Fadeel, M. G. A., Mitt. Klosterneuburg, 30, 250 (1980).
10. Shaw, P. E., in "Citrus Science and Technology," (Nagy, S., Shaw, P. E., Veldhuis, M. K. eds.), p. 427, Vol. 1 (1977)
11. Bayer, S., McHard, J. A., and Winefordner, J. D., J. Agric. Food Chem. 28, 1306 (1980).
12. Kowalski, B. R., and Bender, C. F., J. Am. Chem. Soc., 94, 5632 (1972).
13. Ahmed, E. M., Dennison, R. A., and Shaw, P. E., J. Agric. Food Chem., 26, 368 (1978).

QUANTITATIVE HEADSPACE ANALYSIS
OF SELECTED COMPOUNDS IN EQUILIBRIUM
WITH ORANGE JUICE

Pedro A. Rodriguez
Cynthia R. Culbertson

The Procter and Gamble Co.
Miami Valley Laboratories
Cincinnati, Ohio

I. INTRODUCTION

Knowledge of the concentrations of chemicals present in air is necessary to assess the safety as well as the organoleptic properties of the compounds we breathe. These assessments often require the measurement of compounds present at the ppb or ppt levels. These levels are low enough to belong in the category of ultra-trace analysis.

The ultra-trace analysis of organic compounds by gas chromatography is difficult because of sensitivity and selectivity limitations inherent in the state-of-the-art detectors and columns.

To overcome the sensitivity limitations of the flame ionization detector (FID), it is necessary to concentrate the sample. The need to concentrate poses problems of its own. Variable trapping efficiencies and low recoveries from the trap are frequent occurrences. The occurrence of these problems and the presence of interferences often preclude accurate ultra-trace analysis of gaseous samples. The need for selectivity is particularly acute in ultra-trace analyses because as the concentration of sought-for materials decreases, the higher is the probability of encountering interferences.

Aside from sensitivity and selectivity needs, another related problem is the need to prepare accurate standards for equipment calibration and method validation. Accurate gaseous standards are difficult to prepare because of the surface adsorption of compounds dissolved in gases and because of the purity requirements of the gas itself.

To overcome the concentration problem we developed an inert cryogenic concentrator for use with fused-silica capillary columns (1). To prepare accurate gaseous standards and to validate the methods we used ^{14}C-labeled compounds. This use necessitated the development of a radioactivity detector (RAD) for capillary columns (2). The RAD output was used to establish radiopurity of the standards and to perform mass-balance calculations. These data, along with specific activity data, were used to obtain FID response factors. These factors were used to calculate the concentration of non-labeled materials in an unknown sample.

In this communication we report the use of GC-FID-RAD instrumentation for the analysis of selected components of orange juice headspace.

II. EXPERIMENTAL

A. Instrumentation

A Perkin-Elmer 3920 gas chromatograph equipped with a quartz injector/trap (1) and a Hewlett-Packard OV-101 50m X 0.32mm id fused-silica column was used to analyze the samples. This system handles volumes up to 20 mL of a gas at S.T.P.

The effluent from the column was split so that 54% of the sample reached the FID while the remaining 46% reached the RAD. The instrumentation is shown schematically in Figure 1. The RAD consists of a combustion furnace, a flow-through proportional counter tube (G-M) and associated electronics. Make-up gases and a trickle of oxygen were added to minimize band broadening and to maximize transfer of the radiolabel. The RAD was operated at 70% counting efficiency and the resulting background was 3-5 cpm.

B. Radiolabeled Standards

^{14}C-labeled n-decane (7.5 mCi/mM) and ethanol (4.5mCi/mM) were purchased from ICN (Irvine, CA.).

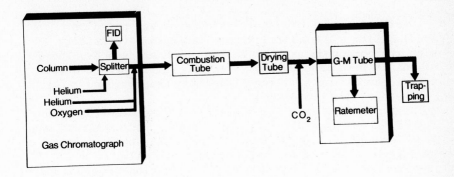

FIGURE 1. GC-FID-RAD instrumentation for quantitation of ^{14}C-labeled and unlabeled compounds.

C. Preparation and Sampling of Juice and Standards

Juice samples were obtained from Valencia oranges processed within a 48-hour period. These samples were frozen for transportation and storage. Juice samples designated (I) in Table 1 were analyzed immediately after thawing the stored juice. Samples designated (II) in Table 1 indicate the thawed juice was refrigerated for 24 hours prior to analysis.

A 5-mL sample of juice or an aliquot of a radiolabeled standard was placed in a 150-mL Hypo-vial [R] along with a magnetic stirring bar. After capping with a mininert valve [R] which had the back-up septum removed, the materials were allowed to equilibrate at room temperature (22 ± 3°C). Sampling of the headspace was started after a minimum equilibration time of 30 minutes. During the last 5 minutes of a given equilibration period, the sample was gently stirred via a magnetic stirrer. Headspace samples were withdrawn with Hamilton 1- or 5-mL gas-tight syringes.

III. RESULTS AND DISCUSSION

Selection of radiolabeled standards was based on availability and similarity in polarity and volatility with the compounds present in the juice headspace. The standards were used to study reproducibility and transfer efficiency of the sampling, injection-trapping and chromatographic processes.

The relative standard deviation of gas sampling via gas-tight syringes was 10%. No residual radioactivity was found on the syringe walls or barrel after the injection, indicating quantitative transfer of materials.

The transfer efficiency of the injection-trapping and chromatographic processes, measured by the recovery of radiolabeled ethanol and n-decane, is better than 90%. This recovery was derived after correction for radiopurity of the two standards. A radiochromatogram of ^{14}C-n-decane is shown in Figure 2. The sample of ^{14}C-n-decane has three minor radiolabeled impurities that account for 5% of the sample radioactivity.

The specific activity (corrected for radiopurity) and the split ratio were used to calculate the mass of compound reaching the FID. The mass and the corresponding integrator area counts were used to calculate response factors for n-decane and ethanol. The response factors were 1.05 and 0.50 X 10^{13} counts-per-gram of compound, respectively. The ethanol value is 47% of that measured for n-decane and is in good agreement with the fraction reported by Dietz (3). Furthermore, the n-decane response factor corresponds to a FID sensitivity of 0.010 coulomb-per-gram of compound. This agrees with the value expected from the specifications of our unit.

Calculation of an unknown concentration using the hydrocarbon response factor based on weight of compound may produce large errors. For example, calculation of ethanol concentration with the decane response factor results in a 100% concentration error.

Higher accuracy can be achieved by the use of response factors based on the carbon content of the compound. These factors are 1.25 and 1.0 X 10^{13} counts-per-gram of carbon for n-decane and ethanol, respectively. Therefore, a smaller (25% vs. 100%) concentration error will result from calculating the ethanol concentration with the hydrocarbon response factor based on carbon content. Because of the lower error expected by the use of the carbon-based response factors, the concentrations of oxygen-containing materials in orange juice headspace were

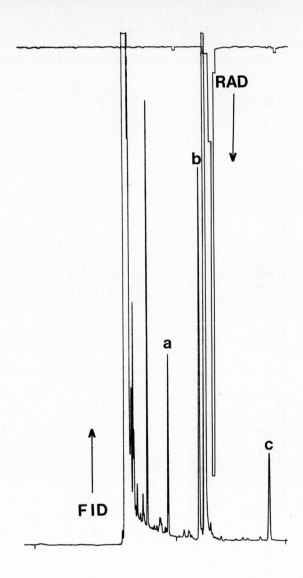

FIGURE 2. Chromatogram of ^{14}C-n-decane. Peaks labeled a,b and c are radiolabeled impurities. Total mass of n-decane injected: 500 ng. Isothermal run at 120°C.

calculated using the carbon-based ethanol response factor. The carbon-based response factor of n-decane was used to compute the concentration of α-pinene, myrcene and limonene.

A chromatogram of the headspace of fresh Valencia juice is shown in Figure 3.

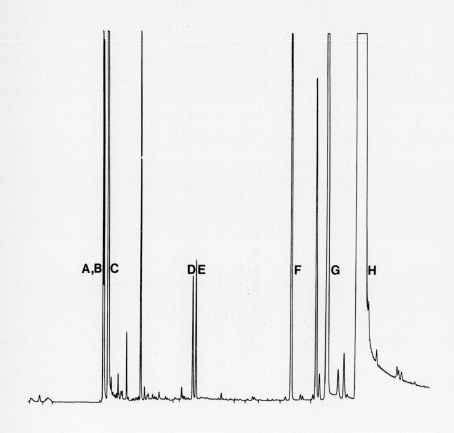

FIGURE 3. Chromatogram of fresh Valencia juice headspace. A: acetaldehyde, B: methanol, C: ethanol, D: hexanal, E: ethyl butyrate, F: α-pinene, G: myrcene H: limonene. Temperature programmed run: 50° to 130°C at 4°C/minute. Sample size: 5 mL.

TABLE 1

Headspace concentration* of selected volatiles in equilibrium with fresh (FRESH), freeze concentrate (F.C.) and conventional concentrate (T.A.S.T.E.)

SAMPLE	ACETALDEHYDE	METHANOL	ETHANOL	HEXANAL
FRESH (I)	7.4 ± 3.0	11.0 ± 4.0	61 ± 2	0.60 ± 0.06
FRESH (II)	8.5 ± 0.5	8.8 ± 0.5	66 ± 2	0.70 ± 0.01
F.C. (I)	6.2 ± 0.7	7.5 ± 0.7	78 ± 7	0.36 ± 0.06
F.C. (II)	6.0 ± 0.2	7.3 ± 0.3	77 ± 4	0.37 ± 0.02
T.A.S.T.E. (I)	0.74 ± 0.11	1.4 ± 0.3	3.2 ± 0.16	0.032 ± 0.004
T.A.S.T.E. (II)	0.64 ± 0.07	1.3 ± 0.3	3.4 ± 0.21	0.043 ± 0.005

SAMPLE	ETHYL BUTYRATE	α-PINENE	MYRCENE	LIMONENE
FRESH (I)	0.51 ± 0.005	1.2 ± 0.4	4.7 ± 1.6	208 ± 52
FRESH (II)	0.53 ± 0.03	1.6 ± 0.7	6.2 ± 2.2	278 ± 99
F.C. (I)	0.65 ± 0.06	1.6 ± 0.5	6.0 ± 1.3	256 ± 53
F.C. (II)	0.62 ± 0.03	1.3 ± 0.3	5.0 ± 0.8	217 ± 33
T.A.S.T.E. (I)	0.036 ± 0.003	1.3 ± 0.4	4.8 ± 1.0	208 ± 52
T.A.S.T.E. (II)	0.036 ± 0.002	1.5 ± 0.5	5.2 ± 1.5	218 ± 57

*Concentrations in ppm (v/v) calculated from carbon-based response factors (see text).

The gas-phase concentrations of 8 major components of this and other juices are shown in Table 1.

The concentrations shown in Table 1 range from 30 ppb to several hundred ppm. The limit of detection is around 1 ppb. The relative standard deviation of the more volatile materials (acetaldehyde through ethyl butyrate) is about 10%. This figure reflects the rapid equilibration between liquid and headspace.

Higher relative standard deviations are sometimes observed for the peaks corresponding to acetaldehyde and methanol. The higher deviations may be due to integration errors caused by a large number of small, unresolved peaks in the region where acetaldehyde and methanol elute. The small peaks would make it difficult for the integrator to decide where to split the acetaldehyde/methanol pair. As the samples are processed, or aged, it is conceivable that these peaks are lost or decrease so that the integrator has less difficulty splitting the pair -- compare FRESH (I) with FRESH (II).

The highest relative standard deviations (30-40%) correspond to peaks identified as α-pinene, myrcene, and limonene. These high deviations are a consequence of reporting a time-independent value for these concentrations. Agreement between two samples is typically within 15% relative when comparing results obtained over similar equilibration periods. The agreement observed under these conditions suggests a slow gas-liquid equilibrium for the three compounds.

To minimize cultivar differences, representative samples of fresh and processed Valencia juice were obtained within a 24-to 48-hour span after the arrival of the fruit. Therefore, the concentrations shown in Table 1 are a reflection of the processing methods used.

ACKNOWLEDGMENTS

We thank G. O. Kinnett for specific activity measurements and R. S. Carpenter for valuable discussions.

REFERENCES

1. P. A. Rodriguez, C. L. Eddy, G. M. Ridder and C. R.
    Culbertson, J. Chromatogr., 236 (1982) 39-49.
2. P. A. Rodriguez, C. R. Culbertson, C. L. Eddy.
    Submitted to J. Chromatogr.
3. W. A. Dietz, J. Gas Chromatogr., (Feb.) 68, 1967.

EXTRUSION COOKING OF HIGH VISCOSITY
THIN BOILING AND THICK BOILING STARCHES
IN A NEW DOUBLE EXTRUSION PROCESS*

Itamar Ben-Gera and *Oak B. Smith*

Wenger International, Inc.

and

Galen Rokey

Wenger Manufacturing, Inc.

INTRODUCTION

Starches of a wide range of different properties are
available today to the food, feed, paper, textile, oil, pharmaceutical and other industries. Most of these products are
produced by conventional starch gelatinization and modification technology. This conventional technology, with drum
dryers and reaction vessels is being challenged now by modern
extrusion cooking systems. Applying the technology of extrusion cooking to the needs of the different industries which
use gelatinized and modified starches has not been an easy
task. Although certainly an area of great interest, motivated
by several reasons, one of which is the desire for lower production costs, considerable research and development work had

Prepared for presentation at the Third International
Flavor Conference, Corfu, Greece, July, 1983.

Instrumental Analysis of Foods
Volume 2

to be done, and was done, over a period of some 30 years now.
This is not surprising, in view of the fundamental differences
between extrusion cooking and the conventional and older technology, and in view of the need if not always than at least in
many cases, to duplicate existing products, with which the end
users are already familiar, and do accept or even require their
exact functional properties.

In the past, success of extrusion cooked starches has been
achieved consistently mainly in the field of low viscosity
thin boiling starches. This paper reports on a new process,
developed by Wenger, which permits the production of pregelatinized starches of pasting characteristics of anywhere from
thin to thick boiling.

When starch and water are worked together in the extruder
to form starch doughs, these doughs are extensible and expand
easily. The extrudate of such cooked dough sets up quickly
after extrusion and shows a characteristic cellular structure.

Starch or starch containing flours can be gelatinized
during an extrusion cooking process. They can be extruded as
thick boiling starches cooked at high moisture contents and
at low extrusion temperatures. Conversely, starch can be extrusion cooked as a thin boiling starch at low moisture contents and at relatively high extrusion temperatures.

THIN AND THICK BOILING STARCHES

Raw native starch does not dissolve in cold water and takes
up water very slowly. Pregelatinized starch does dissolve in
cold water. It has attained its water absorption, water holding capacity, viscosity characteristics in cold water due to
the fact that it had been pregelatinized.

As a practical illustration I would say that the use of
adequate quantity of thin boiling starch with lukewarm water
will result in a thin slurry, with a consistency similar to
that of buttermilk, while mixing adequate quantity of thick
boiling starch with warm or hot water will produce a viscous
paste, comparable to that of a breakfast porridge.

It is possible to produce pregelatinized starches through
extrusion cooking. These products can be tailor made and once
rehydrated, to exhibit their maximum viscosity in different
temperatures as required. The difference between the processing conditions of these different starches relate mainly to
their moisture content during extrusion cooking.

It is important to note that:
1. From an economical point of view, it would be not only
advantageous, but probably crucial to obtain higher levels of
viscosity throughout the temperature range shown in this slide.

2. It is important to bring moisture levels down from 50% for the thick boiling starch. Such reduction will make economics of the extrusion process more attractive.

EXTRUSION COOKING OF HIGH VISCOSITY, THICK AND THIN BOILING STARCHES

The new process for the gelatinization of starches or cereal and tuber flours which is discussed here, has a unique combination of advantages over other methods of pregelatinizing starchy raw materials. This new process is a double extrusion process, in which the starch is moistened and gelatinized in an extrusion cooker, after which it is conveyed into a secondary extruder. This second extruder is a cooling and forming extruder.

The extrudate is cut at the die of the second unit into thin sheets. Thereafter the extrudate is dried and cooled. After drying and cooling, the product is put through suitable size reduction equipment to reduce particle size as desired. By the proper choice of processing temperatures, moisture contents, pressures, feed rates, barrel assembly length and choice of components like flow constrictors, screws and sleeves, die configurations and other factors, pregelatinized starches can be produced in which viscosity characteristics, solubility rate of rehydration, thickening power, gel strength, and gel setting times are tailored.

The new process has a unique combination of advantages over other methods of pregelatinizing thick boiling starches:

1. Pregelatinized thick boiling starches can be made at high production rates of up to 2,200 kilograms per hour using one cooking extruder and one forming extruder together with a dryer/cooler.

2. Viscosity of pastes prepared from the precooked materials can be controlled over a wide range. At one extreme, a starch can be precooked by the new process which when added to water and heated or cooled, gives the temperature/viscosity profile equivalent to that of a thick boiling drum dried starch. At the other extreme, pregelatinized thin boiling starches can be produced by extrusion cooking which gives high solids/low viscosity solutions that do not gel upon cooling.

3. Rate of rehydration can be controlled. Slow hydration with no lumps when added to water, or quick rehydration and increased viscosity in cold water which could give a pulpy or a smooth texture, as desired, can be achieved.

Low viscosity, thin boiling starches can be produced in a single cooking extruder, by bypassing the cooling and forming extruder, as mentioned earlier. This is shown in Fig. 6, for

Fig. 1. Photograph of Wenger F-25/20 starch cooker including cooking extruder and cooling and forming extruder.

Fig. 2. Photograph of X-255 extrusion cooker/cooling and forming extruder (capacity to 2,200 kilograms per hour.)

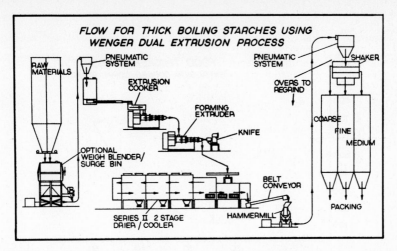

Fig. 3. Flow sheet for the extrusion cooking of thick boiling starches.

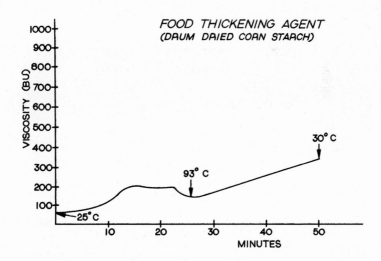

Fig. 4. Viscosities of typical drum dried food thickening agent.

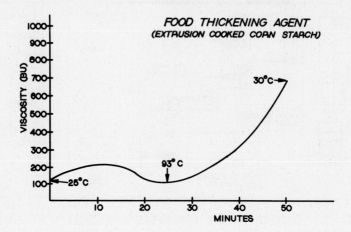

Fig. 5.   Viscosities of extrusion cooked food thickening
agent - Wenger process.

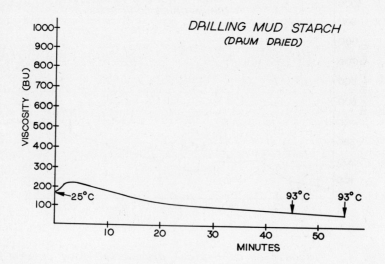

Fig. 6.   Plots viscosities of drum dried drilling mud
starch as a thin boiling starch.

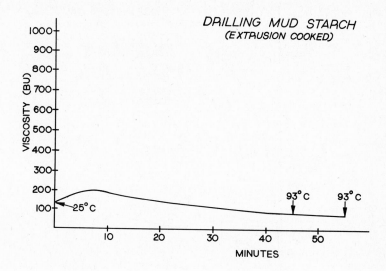

Fig. 7. Illustrates viscosities of an extrusion cooked thin boiling, drilling mud starch.

a drum dried product and in Fig. 7, for an extrusion cooked product. These are drilling muds.

It should be noted that drilling muds are thin boiling pregelatinized starch products. Since both drum dried and extrusion cooked processes produce almost identically the same end drilling mud viscosities, the advantage of the extrusion cooked drilling mud over the drum dried drilling mud will be the great reduction in processing cost per ton and the savings in investment in drum dryers to produce the same capacity as could be produced on a single extrusion system as the one shown before.

Figure 8 compares a rate of amylose leaching which is an indicator of shear level during processing, in drum dried drilling mud starch, drum dried food thickening agent, and extrusion cooked food thickening agent.

DRUM DRIED PREGELATINIZED STARCHES AS COMPARED TO EXTRUSION COOKED PREGELATINIZED STARCHES

A drum dryer is the most widely used equipment to manu-facture pregelatinized starches and flours, especially those

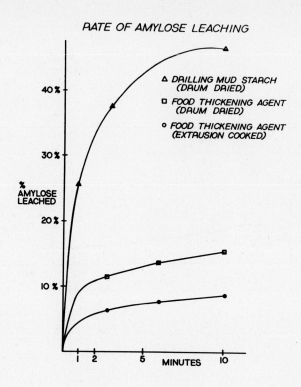

Fig. 8. Rate of Amylose leaching, drum dried and extrusion cooked starches.

processed for high paste viscosity. A drum dryer operates at low speeds and imparts low shearing forces to a starch slurry during cooking. Upon rehydration, the starch granules imbibe water and swell and produce high viscosities. However, cooking starch on a drum dryer has many limitations. The production rate is low, maintenance costs and operating costs are high, and the process is difficult to control. The drum dryer is inefficient in transferring heat to cook and dry the starch. Drum dryers require much higher moisture contents (60-70% moisture) than do extrusion cookers (15-30%). Energy costs will continue to be of a major concern and an important factor in the overall processing costs of pregelatinized starch produced on a drum dryer.

Extrusion processing is advantageous over drum drying because extruders require less energy and lower equipment and lower labor costs per unit of dry product; the process is

TABLE I. Extrusion System

	Energy consumption per hour		
Capacity	*Electricity*	*Gas*	*Steam*
Wenger model F-25/20		Dryer/cooler	Extruder
at 12% MCWB kg/h	kW/h	m3/h	kg/h
	Extruder Dryer/Cooler		
180	28.7 40.0	3.6	9.0
230	30.0 40.0	4.6	11.5

easier to control; it produces at much higher production rates and is much less costly to maintain than are drum dryers.

Total energy used per kilo of product in the process which is presented here is much lower than the energy used per kilogram of same product when produced by a drum dryer.

Tables I, II, III and IV show the energy requirements of drum dried starch as contrasted to energy requirements for the new Wenger process.

As one can see from Table II, increasing the throughput or utilizing extrusion systems of higher capacity will result in lower energy requirements per kilogram of throughput.

The energy consumption values for extruder throughputs of 180, 230, 350, 2200 kilograms per hour are based on Wenger model numbers as shown in Table II.

The 350 kilograms per hour rate makes a practical direct comparison of utility requirements for Wenger systems and drum dryers of 350 kilograms per hour capacity.

Table III converts energy requirements to actual costs of electricity, gas, and steam. Utility costs will, of course, vary from one location to another but since the utility costs are shown at the bottom of the table, it will be a simple matter to calculate utility costs, based on local costs for these utilities.

Table IV shows the total energy consumption per kilogram of finished product on extrusion cooking systems of various capacities as well as for the drum dryer at capacity of 350 kilograms per hour.

TABLE II. Extrusion System and Drum Dryer

Process	Capacity at 12% MCWB kg/h	Electricity kWh/kg	Gas m3/kg	Steam kg/kg	Wenger model
		Energy consumption per kg of finished product			
Extrusion	180	0.378	0.02	0.05	F-24/20
Extrusion	230	0.304	0.02	0.05	F-25/20
Extrusion	350	0.267	0.02	0.05	X-25/25
Drum Dryer	350	0.051	--	1.33	
Extrusion	2200	0.088	0.02	0.05	X-155/200

TABLE III. Extrusion System and Drum Dryer

		Energy cost per kg of finished product				
					Total	
Process	Capacity kg/h	Electricity $/kg	Gas $/kg	Steam[a] $/kg	without steam $/kg	with steam $/kg
Extrusion	180	0.026	0.003	0.001	0.029	0.030
Extrusion	230	0.021	0.003	0.001	0.024	0.025
Extrusion	350	0.019	0.003	0.001	0.022	0.023
Drum dryer	350	0.004	----	0.027	0.031	0.031
Extrusion	2200	0.006	0.003	0.001	0.009	0.010

[a]Many extrusion cooked starch products do not require steam during processing. The addition of steam depends upon the type of starch being processed and the desired functional properties of the end products.

Assumption: Electricity = US$ 0.07/kWh
            Gas = US$ 0.15/m3 (9000 kcal/m3)
            Steam = US$ 0.02/kg (550 kcal/kg)

TABLE IV. Extrusion System and Drum Dryer

Total energy consumption per kg of finished product[a]

Process	Capacity kg/h	Total energy without steam kcal/kg	with steam kcal/kg
Extrusion	180	500	531
Extrusion	350	406	435
Drum dryer	350	---	781
Extrusion	2200	254	282

[a]*Accumulated total energy additions made from Table II.*

CONCLUSION

The double extrusion method developed by Wenger and described here, is versatile in that it allows the production of a whole spectrum of pregelatinized starches with manageable, preselected functional properties.

The equipment used in the process to produce pregelatinized high viscosity, thin and thick boiling starches and flours are: 1) a cooking extruder, 2) a forming extruder, 3) a dryer.

By the proper choice of processing temperatures and moisture contents, as well as machines parameters and configuration, pregelatinized starches can be produced which vary greatly in viscosity characteristics, solubility, rate of rehydration, thickening power, gel strength and other important functional properties.

Processing costs per ton are much lower for extrusion cooking systems than for drums, be they thick boiling or thin boiling starches. We feel that the versatility of the system is greater than other systems used or promoted today.

MICROSTRUCTURE OF PROTEIN GELS
IN RELATION TO THEIR RHEOLOGICAL PROPERTIES

Toshimaro Sone
Shun'ichi Dosako
Toshiaki Kimura

Technical Research Institute
Snow Brand Milk Products Co., Ltd.
Kawagoe, Saitama
Japan

I. INTRODUCTION

The microstructure of protein gels is of great importance in understanding their rheological properties, since the microstructure is determined by their chemical composition and physical forces (1). To date, in light of this significant suggestion, many studies have been carried out to elucidate the relationship between the rheological properties of protein gels and their microstructure. Some microstructures compatible with the rheological properties have been revealed: typical network structure is responsible for high hardness and equilibrium modulus in a heat-induced soybean protein gel (2); and dense microstructure of milk gels is related to high stress relaxation (1) or firmness (3). Harwalkar and Kalab (4) discussed the relationship between firmness or water holding capacity of milk gels induced by acidulunts as well as heating and ultrastructure of casein micelles in terms of their size and shape. A heat-induced soybean curd varies in texture depending on the procedure employed for its coagulation; microstructure of the curds coagulated at pH 4.6 or with calcium chloride (2) and with calcium sulfate or glucono-δ-lactone (GDL) (6) were appreciably different.

In most cases, microstructure is observed though a scanning electron microscope (SEM), by preparing specimens for SEM by freeze-drying or critical-point drying. However, in some cases, artifacts due to specimen preparation are

Instrumental Analysis of Foods
Volume 2

encountered when observing the microstructure, and it is difficult to determine whether the microstructure observed is in
its native state. A cryo-SEM, on the other hand, can basically remove such artifacts (7).

　　This paper describes the microstructure of protein gels
in relation to their rheological properties, based on molecular behavior, although the mechanism for gelation is still
under debate.

II. MECHANICAL PROPERTIES OF SOYBEAN PROTEIN CURDS AND THEIR MICROSTRUCTURE

Figure 1 shows SEM-images of soybean protein curds

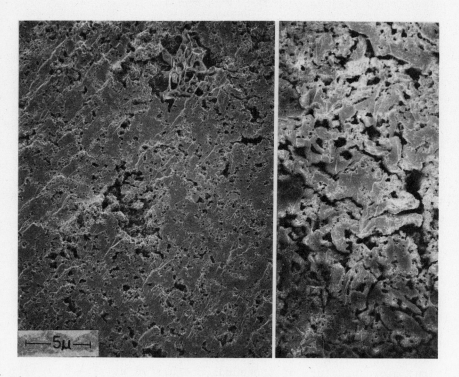

Figure 1. Microstructures of soybean protein curds aggre-
gated at pH 4.6 (left) or by addition of 20mM CaCl$_2$ (right),
observed through a scanning electron microscope equipped with
a cryounit (x 400). Specimens were immersed in a 10% solution of ethanol without any chemical fixation.

obtained with different procedures. Soybean protein isolated at pH 4.6 was redissolved in distilled/deionized water at pH 7 to produce a protein concentration of 5%. The solution was subjected to two different treatments to obtain aggregates: 1) the pH was adjusted to 4.6 (SPI-25), and 2) calcium chloride was added to make a final calcium concentration of 20mM (SPCa-25). The curds thus obtained were centrifuged at 1,740 xg for 10 min, and then immersed in a 10% solution of ethanol without any chemical fixation. Specimens were placed on stubs, frozen promptly with liquid nitrogen, and set on a cryounit fitted to a JSM-50A scanning electron microscope. After cutting into small pieces (less than 2mm x 3mm x 3mm) in the cryounit, they were evaporated instantaneously.

No network structure was observed, being dissimilar to heated curds (5). SPI-25 showed denser microstructure, while microstructure of SPCa-25 was more rough. According to Lee and Rha (5), compression force of SPI-25 was considerably higher than that of SPCa-25; SPI-25 had higher water holding capacity determined at centrifugal force of 280 xg, than SPCa-25. The rough microstructure of SPCa-25 seems to be responsible for the lower water holding capacity and compression force.

III. GELS FROM WHEY PROTEIN CONCENTRATES

A. Effects of Salts on Elastic Modulus and Water Holding Capacity

Desalinated sulfuric acid whey protein concentrates (WPC) was dissolved in 50mM imidazole buffer (pH 6.8) containing required amounts of calcium chloride or sodium sulfate at a protein concentration of 11% (w/w). The solution was placed in impermeable heat-stable plastic tubing and heated in boiling water for one hour to form gels. Figure 2 illustrates the dependence of elastic modulus (E_1) on the salt concentration, computed from stress relaxation tests at room temperature under a strain of 10%. Addition of sodium sulfate caused E_1 to increase significantly with a maximum at 0.30%, while E_1 was independent of the concentration of calcium chloride.

TABLE I indicates water holding capacities of the gels in the absence of salt, in the presence of 0.26% of calcium chloride, and of 0.30% of sodium sulfate. Although each gel demonstrated sufficiently high water holding capacity, sodium sulfate-added gel had the highest value with statistical significance. Addition of calcium chloride into the gel, however, resulted in no influence to the water holding capacity of the gel. The different behavior of calcium chloride and sodium sulfate towards the elastic modulus and water holding

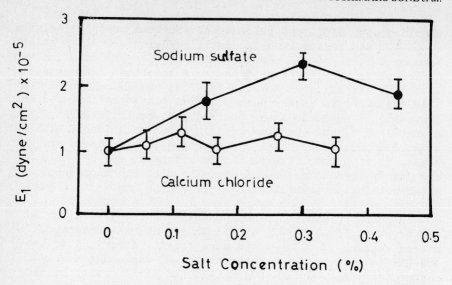

Figure 2. Effects of salt concentrations on elastic modulus of WPC gels. Vertical lines indicate 95% confidence interval.

capacity strongly suggests that these salts play separate role in the gel formation and/or protein-solvent or protein-protein interactions.

B. Microstructure

Figure 3 shows microstructures of the gels in the absence of salt, in the presence of 0.30% calcium chloride, and of 0.30% sodium sulfate. Specimens were immersed in the 10% of

TABLE I. Water Holding Capacities of WPC Gels[a]

Salt	Water Holding Capacity
Salt-free	91.7 ± 0.6
0.26% $CaCl_2$	91.7 ± 0.7
0.30% Na_2SO_4	93.1 ± 0.2
	(mean $\pm$ s.d.)

[a]Unit: %

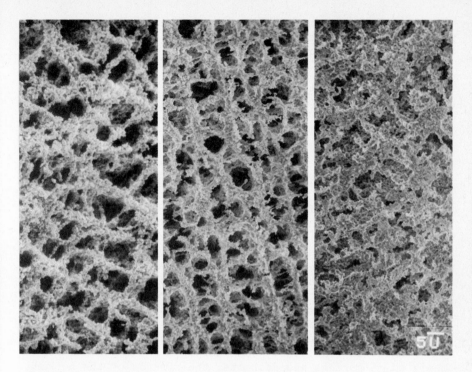

Figure 3. Microstructures of WPC gels in the absence of salt (left), in the presence of 0.30% $CaCl_2$ (middle), and of 0.30% Na_2SO_4 (right), observed through a scanning electron microscope equipped with a cryounit (x 1,000). Specimens were immersed in a 10% solution of ethanol without any chemical fixation.

ethanol solution prior to observation through the cryo-SEM as described earlier. Each gel showed well-defined network structure, although the sodium sulfate-added gel exhibited a distinct feature: the microstructure was denser than the other gels. The calcium chloride-added and salt-free gels essentially had similar microstructure. Again, as observed in the soybean curds (Figure 1), the dense microstructure was compatible with the higher elastic modulus and water holding capacity. This suggests a relatively high affinity of the protein or gel matrix for water; water can be evenly distributed throughout the matrix, limiting the number of different configurations which the strands of the network can assume. Since the elasticity of a three-dimensional network of flexible strands is attributable to a decrease in entropy upon deformation (8), a less "random" network structure like the sodium sulfate-added gel may demonstrate higher elastic

modulus. The rough microstructure, on the other hand, can release water with ease upon compression and is more random in
the network structure, resulting in lower elastic modulus.

IV. DISCUSSION ON THE RHEOLOGICAL PROPERTIES
OF PROTEIN GELS AND THEIR MICROSTRUCTURE,
BASED ON THEIR MOLECULAR BEHAVIOR

Heat-induced gels such as from soybean protein (2,5,6,9)
and from WPC (Figure 3) have meshy network structure. In
these proteins disulfide bonds are primary driving force for
gelation (10,11,12), although the mechanism of gelation is
still under debate. Since cysteine and cystine are hydrophobic in nature, they are, in general, buried within a protein
molecule. Once proteins are heated, cysteine and cystine are
exposed to water, associated with the unfolding, leading to
interchange reaction between SH and S-S groups. If the protein concentration is sufficiently high for intermolecular
cross-linking, proteins can form gels with network structure
through the cross-linking such as disulfide bonds and/or other
non-covalent interactions. Considering a rheological model
consisted of springs and dashpots, the strands of the network
structure appears to play the role of spring since such a network involves a decrease in entropy (8). On the contrary, in
the gels without network structure, in which the non-covalent
interactions of energy being low are the predominant factors
for gelation, the contribution of spring may be much less.
Good evidence has been given by Lee and Rha (5), where spring-
iness of heat-induced gels from soybean protein is appreciably
higher than that of unheated curds.
Rheological properties as well as microstructure of gels
having network structure vary with additives. Addition of
salt is usually expected to increase in gel solvation due to
decreased protein-protein interaction and increased protein-
solvent interaction (12). The high elastic modulus (Figure 2)
and water holding capacity in sodium sulfate added gel (TABLE
1) seem to be interpreted by the increased protein-solvent
interaction. Gels containing a certain amount of added sodium
sulfate have dense microstructure, that is, less random network structure, which holds water tightly in the gel matrix
due to the protein-water interaction endowing elastic nature
to the gel.
Calcium chloride, such as α_{s1}-casein, β-casein, and soybean protein, causes the proteins to precipitate or aggregate
due to salt bridges and/or hydrophobic interaction. Calcium
ions bound to carboxylic groups and phosphate attached to

serine residues neutralize charges on α_{s1}-casein, causing it
to associate with each other through hydrophobic interaction
(13); association of β-casein takes place through calcium-
phosphate bridges (14). Since whey proteins involve no phosphate, calcium ions bound to carboxylic groups seem to enhance
hydrophobicity of the proteins; resultant protein-protein
interaction excludes water from around the protein molecules,
being responsible for rough or porous microstructure.

V. SOME PROBLEMS ENCOUNTERED WITH THE OBSERVATION OF MICROSTRUCTURE THOUGH SEM

An artifact attributed to specimen preparation for SEM is
-occasionaly encountered with the observation of microstructure. Instantaneous freezing with liquid nitrogen can

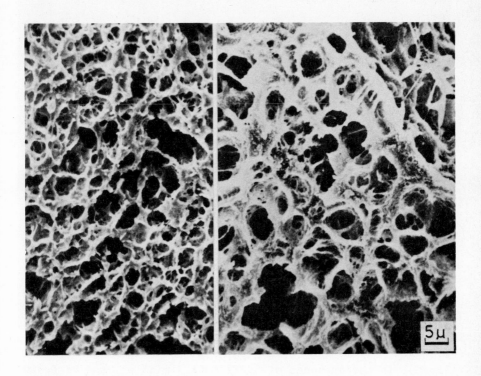

Figure 4. SEM-images of *tofu*, a soybean curd. The left was
observed through a cryo-SEM, and the right was lyophilized
specimen after instantaneous freezing. Used with permission
of JEOL News and by courtesy of Dr. K. Saio.

possibly cause the overall structure to destroy due to growth of icecrystals, unless the size of specimen is sufficiently small. An example for this was demonstrated by Gallant et al. (9) as shown in Figure 4, who observed microstructure of *tofu*, a soybean curd. The microstructure observed through the cryo-SEM demonstrated fine network structure, whereas damage in overall microstructure was shown in the specimen lyophilized after instantaneous freezing, probably due to the growth of icecrystals.

An alteration of native structure can occur, in some cases, to a considerable extent, when specimens are fixed with organic solvent like glutaraldehyde. Heat-induced soybean protein curds were either fixed or unfixed with glutaraldehyde; the unfixed specimen showed a large and better defined network structure (5). Figure 5 shows another series of

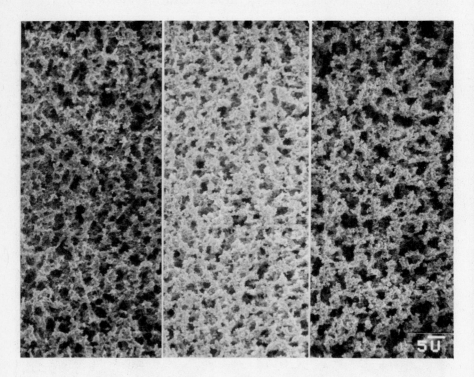

Figure 5. SEM-images of WPC gels in the absence of salt (left), in the presence of 0.30% $CaCl_2$ (middle), and of 0.30% Na_2SO_4 (right) fixed with glutaraldehyde (x 1,000).

TABLE II. Water Holding Capacity of WPC Gels[a]
Fixed with Glutaraldehyde

Salt	Water Holding Capacity[b]
Salt-free	88.3 + 0.3
0.26% $CaCl_2$	86.7 + 0.8
0.30% Na_2SO_4	87.7 + 0.7

[a]Unit: %

[b]mean + s.d.

SEM-image of WPC gels fixed with glutaraldehyde, exemplifying the microstructure not being related to their rheological properties. As already shown in Figure 2 and TABLE I, the elastic modulus and water holding capacity of sodium sulfate-added WPC gel were significantly higher than those of other gels, nevertheless, the microstructure of sodium sulfate-added gel fixed with glutaraldehyde was similar to that of other gels; the dense structure (Figure 3) disappeared. TABLE II indicates the water holding capacities of the fixed WPC gels. Compared with TABLE I, the water holding capacities of the fixed gels decreased to a certain extent; the statistical significance for sodium sulfate-added gel was completely lost. These results strongly suggested the fixation with glutaraldehyde was responsible for the alteration of microstructure.

VI. CONCLUSIVE REMARKS

Relationship between the rheological properties of protein gels and their microstructure was discussed, by using the gels from soybean protein and WPC as examples.

Heat-induced gels of primary driving force for gelation being disulfide bonds have high elasticity and well-defined network structure; the relationship between them has been well explained by the kinetic theory of rubber-like elasticity (8). Microstructure and rheological properties of heat-induced gels vary with salts; sodium sulfate-added WPC gel shows higher elastic modulus and water holding capacity and denser microstructure than calcium chloride-added gel. The denser microstructure is probably due to strong protein-water interaction, leading to less randomness that endows more elastic nature to the gel. Calcium ions, on the other hand, enhance hydrophobic

interaction excluding water from around protein molecules.
In order to observe the native microstructure, care must be
taken to remove artifacts produced during specimen preparation
for SEM. It is strongly recommended to employ cryo-SEM which
is appropriate for the observation of native microstructure.

ACKNOWLEDGMENTS

Careful proofreading throughout this paper by Mr. H.Ohkado
was greatly appreciable. The authors thank Prof. C.K.Rha,
Massachusetts Institute of Technology, for her helpful advice.

REFERENCES

1. Stanley,D.W., and Tung,M.A., *in* "Rheology and Texture in
   Food Quality" (J.M.deMan, P.W.Voisey, V.F.Rasper, and
   D.W.Stanley, ed.), p. 28. The Avi Publishing Co, Inc.,
   Westport, Conn, (1979).
2. Furukawa,T., *in* "SHOKUHIN NO BUSSEI" (Y.Matsumoto, ed.),
   Vol. 4, p.9. Shokuhin Shizai Kenkyu Kai, Tokyo, (1978)
   (in Japanese).
3. Kalab,M., and Harwalkar,V.R., *J. Dairy Res. 41,* 131
   (1974).
4. Harwalkar,V.R., and Kalab,M., *Scanning Electron Microsc.*
   part III, p. 503 (1981).
5. Lee,C.H., and Rha,C-K., *J. Food Sci. 43,* 79 (1978).
6. Saio,K., *Scanning Electron Microsc.* part III, p. 553.
   (1981).
7. Nei,T., Yotsumoto,H., and Hasegawa,Y., *J. Electron
   Microscopy 20,* 202 (1971).
8. Ferry,J.D., *Advances in Protein Chem. 4,* 1 (1948).
9. Gallant,D.J., Saio,K., and Ogura,K., *JEOL News 14(2),* 6
   (1976).
10. Wolf,W.J., and Tamura,T., *Cereal Chem. 46,* 82 (1969).
11. Circle,S.J., Meyer,E.W., and Whitney,R.W., *Cereal Chem.
    41,* 157 (1964).
12. Schmidt,R.H., and Illingworth,B.L., *Food Product
    Development 12(10),* 60 (1978).
13. Ono,T., Kaminogawa,S., Odagiri,S., and Yamauchi,K.,
    Agric. Biol. Chem. 40, 1725 (1976).
14. Yoshikawa,M., Tamaki,M., Sugimoto,E., and Chiba,H.,
    Agric. Biol. Chem. 38, 2051 (1974).

THE INFLUENCE OF THE INTERACTION OF MONO- AND DI-GLYCERIDES WITH MILK PROTEINS ON THE RHEOLOGY AND STABILITY OF FOOD EMULSIONS.

G. Doxastakis
P. Sherman

Queen Elizabeth College
University of London
London, UK.

Abstract

The stability and viscoelasticity of 60% (wt/wt) corn oil-in-water emulsions stabilized by sodium caseinate and mono- and di-glycerides were studied over the pH range 3.0 - 9.0.

At pH values above and below the isoelectric point of the sodium caseinate freshly prepared emulsions exhibited the highest viscoelasticity parameters values. The parameter values were also influenced by the mono-/di-glycerides ratio employed.

When the emulsions were stored the rate of change in the instantaneous elastic modulus (E_o) was much influenced by the mono-/di-glycerides ratio. The highest rate of increase was obtained with a 5/2 glycerides ratio and the slowest rate at a 2/9 ratio.

Viscoelasticity parameter increases during the first 3-5 days storage, especially at low pH, and then they decrease.

The glycerides, as apolar lipids, have only a small hydrophilic part relative to the remainder of the molecule which is hydrophobic.

This hydrophilic part enters into association with proteins loops adsorbed on the surface of adjacent oil drops. Changes in viscoelasticity

219

during storage were attributed to further protein
loop interlinkage and drop coalescence.

Introduction

While nutritional value is ultimately very
important when considering proteins as food components, the physico-chemical characteristics and
interactions of proteins with the other ingredients
determine their usefulness and success. These
characteristics, collectively referred to as functional properties, influence processing, preparation
and quality attributes of foods (Kinsella, 1978).
The intrinsic properties of a protein are governed
by the content and disposition of amino acids,
molecular size, shape, conformation, net charge and
protein-protein interactions. Although the properties of a single component are significant, it is
the manner in which they interact with other components in foods, for example water, proteins and
lipids, that ultimately determines their function-
ality and applications.
Proteins perform a variety of functions and
each of these may depend on different molecular
features or interactions (Kinsella, 1979).
Information correlating the structure of proteins with specific functions in foods is limited,
probably because of the diversity of composition,
structure and conformation of food proteins. This
applies, for example to the fundamental rheological properties of O/W emulsions stabilized by
sodium caseinate.
A comparative study has now been made of the
rheological properties of O/W emulsions stabilized
with sodium caseinate and mono-/di-glycerides in
different ratios, while keeping the total emulsifier
concentration constant. Some of the data are
reported here.

Experimental

Materials; general.   The water used for preparing emulsions was double distilled from an all
Pyrex apparatus. Corn oil was used as the oil
phase. It has a density of 913.0 Kgm^{-3} at $20^{o}C$
(Leon Frenkel Ltd., Kent, U.K.).
Sodium caseinate light white, without further
treatment was supplied by Hopkin and Williams,

Biochemical, Essex, U.K.

Mono-glyceride (KA-2017) and di-glyceride (KA-2018), from Kao-Atlas Chemicals, Japan, were used as emulsifiers and they were incorporated in the oil phase.

All other chemicals used were of "Analytical Reagent" Poole.

Emulsion preparation. 60.0% (wt/wt) corn oil-in-water emulsions were prepared in which the total emulsifier was kept constant at 2.0% (wt/wt), over the pH range 3.0 - 9.0. A standardised procedure was used to prepare all emulsions.

120 ml of 1.0% (wt/wt) sodium caseinate solution at 50°C, and at the required pH, were introduced into a 400ml beaker (approx. 7cm diameter) and corn oil at 75°C, with or without 1.0% (wt/wt) glycerides, was then added dropwise and dispersed for 5 min with the aid of a mechanical stirrer. The coarse emulsion was passed twice through a hand operated homogeniser. Both the aqueous and oil phases were placed in thermostatted glass containers until used, and the desired temperature (55°C) was maintained during the pouring of dispersion.

Some emulsions (Table 3) were prepared with the aqueous phase at pH 7.0. Following preparation each of these emulsions was divided into batches and the pH of each batch was adjusted to the desired level with 5M NaOH or HCl. In this way it was possible to examine the rheological properties of fresh emulsions with identical mean drop size and size distribution at different pH's.

All emulsions were stored in a refrigerator at 5° + 1°C, after preliminary examination, until required for further tests.

The following emulsions were prepared at 55.0° + 0.2°C.
a) Emulsions stabilized only by sodium caseinate (1.0% wt/wt) at various pH's (3.0, 5.5, 7.0 and 9.0), (Table 1).
b) Emulsions stabilized by sodium caseinate (1.0% wt/wt)at various pH's (3.0, 5.5, 7.0 and 9.0) plus various mono-/di-glycerides (1.0% wt/wt) ratios (8/2, 5/2, 2/2, 2/6 and 2/9). Some of the data, at pH 5.5 are reported (Table 2).
c) Emulsions stabilized by sodium caseinate (1.0% wt/wt) at pH 7.0 plus mono-/di-glycerides (1.0% wt/wt) ratio 5/2 and then divided into batches and was adjusted to the desired level of pH. (Table 3).

Rheological examination of the emulsions. The
creep compliance-time behaviour of each emulsion
was investigated with a Deer rheometer (Deer Rheometer Ltd.Essex, England) at a constant shear stress
of 41.7 dyne, cm^{-2}. This stress was within the
linear region of the stress-strain relationship
for each of the emulsions examined.
Drop size enalysis. The size distribution of
the oil drops in each emulsion was determined with
a disc centrifuge photosedimentometer MKIII (Joyce
Loebl, Newcastle, England) as described previously
(Sherman and Benton, 1980), and the mean volume
diameter was calculated.
Electrophoretic mobility of oil drops in the
O/W emulsions: The electrophoretic mobility of the
oil drops in the corn oil-in-water emulsions was
determined, with the Rank Mark II microelectro-
phoresis apparatus (Rank Brothers, Cambridge,
England), as described elsewhere (Vernon Carter and
Sherman, P. 1980).

Results

All the emulsions, irrespective of whether they
were stabilized by sodium caseinate with, or without
glycerides and irrespective of storage time, exhibited viscoelastic behaviour. This behaviour was
characterised (Inokuchi, 1955) by six parameters
(Tables 1-3) viz. an instantaneous elastic modulus
(E_o), two retarded elastic moduli (E_1) and (E_2) and
their associated viscosities (n_1) and (n_2), respectively and a Newtonian viscosity n_N.

Emulsions stabilized by sodium caseinate only.
Table 1 shows quite clearly that the viscoelasticity
parameters values were influenced by pH. In general,
the optimum values, at all storage times were
obtained at pH 5.5. The isoelectric point for
sodium caseinate is 4.6.
During storage all the emulsions showed an
increase in the parameter values over 3-5 days
until optimum values were attained. This was then
followed by a decrease in all parameter values. The
rate at which the parameters values increased initially was influenced by both pH and particle size
distribution. The closer the pH was to 5.5 the
faster the initial increase in instantaneous elastic
modulus (E_o) and the slower the subsequent rate of

decrease. This effect was even more pronounced for E_1 and E_2.

Emulsions stabilized at a pH removed from the isoelectric point showed certain differences in viscoelasticity during storage. For example, at pH 5.5 the E_0 values for 3 days old emulsions were approximately 26%, 9% and 33% higher than at pH 3.0, 7.0 and 9.0 respectively. With the other viscoelastic parameters the trends were even more pronounced than for E_0.

Emulsions stabilized by sodium caseinate and mono-/di-glycerides: The data, table 2, show how pH and glyceride ratio influence the viscoelastic properties of freshly prepared emulsions. The viscoelastic values are higher for the 5/2 mono-/di-glycerides ratio and these differences are more pronounced as we go further to 2/6 and 2/9 ratio. For example, in a 5 days old emulsion with a 5/2 glycerides ratio, the values obtained were 47% and 52% higher for the 8/2 and 1/2 ratio respectively and 97% for both 2/6 and 2/9 ratios.

Emulsions with identical initial mean drop size distribution. Table 3 clearly indicates that the aqueous phase pH influences the viscoelastic parameters. As before, the E_0 values are higher at pH 5.5. The other viscoelastic parameters more or less follow the same trends but more pronounced. It seems that adjustment of the pH away from the isoelectric point, even after the emulsion has been prepared, has a dramatic effect. At pH 3.0 and 9.0, the emulsions were broken down earlier.

The rate of coalescence was found lower with the presence of 5/2 mono-/di-glycerides ratio.

The oil drops in all the emulsions exhibited minimum electrophoretic mobility at a pH around 4.6, (Table 4). The mobility increased as the pH increased or decreased from this value. At any pH, emulsions drops stabilized by a 5/2 mono-/di-glycerides ratio had the highest mobility and drops stabilized by a 2/9 ratio had the lowest mobility.

TABLE 1

Influence of pH on the Viscoelasticity Parameters and Mean Drop Size of O/W Emulsions Stabilized by Sodium Caseinate (1.0% wt./wt.) Temperature $25.0 \pm 0.2^{\circ}\text{C}$.

Emulsion pH	Ageing time (days)	E_o (dyne cm^{-2})	n_N (Poise x 10^5)	E_1 (dyne cm^{-2})	E_2 (dyne cm^{-2})	n_1 (poise x 10^{-4})	n_2 (poise x 10^{-4})	d (μm)
9.0	2/24	6.74	16.17	41.47	70.77	0.59	0.77	0.283
	1	9.38	13.81	70.81	140.47	0.82	0.92	0.311
	3	13.37	9.77	90.89	110.27	1.07	1.37	0.336
	5	20.68	9.04	180.21	105.31	1.19	1.51	0.358
	10	9.08	2.31	130.57	170.41	1.21	1.87	0.374
	20	3.01	1.21	127.30	101.22	0.66	1.21	0.389
	30	2.51	0.47	70.42	197.36	0.71	0.71	0.401
7.0	2/24	9.67	6.51	81.08	60.51	0.61	0.66	0.267
	1	12.38	8.22	107.41	110.74	0.71	0.82	0.288
	3	18.13	10.37	120.37	200.81	0.77	1.44	0.305
	5	14.08	10.71	180.31	156.43	1.33	2.31	0.321
	10	12.03	6.37	107.47	102.44	1.21	1.27	0.332
	20	10.18	2.36	120.37	137.89	0.44	0.45	0.340
	30	6.73	2.07	107.32	67.81	0.28	0.31	0.347
5.5	2/24	14.27	2.37	70.52	110.71	0.21	0.32	0.223
	1	15.31	2.71	100.81	174.81	0.20	0.28	0.236
	3	20.01	3.41	124.32	134.41	0.37	0.37	0.248
	5	22.36	3.99	130.59	153.74	0.41	0.91	0.257
	10	11.71	4.01	117.21	109.18	0.80	1.17	0.265

TABLE 1 Continued

Emulsion pH	Ageing time (days)	E_0 (dyne cm^{-2})	n_N (poise $\times 10^{-5}$)	E_1 (dyne cm^{-2})	E_2 (dyne cm^{-2})	n_1 (poise $\times 10^{-4}$)	n_2 (poise $\times 10^{-4}$)	d (μm)
	20	12.75	3.21	102.34	123.41	0.63	1.04	0.270
	30	8.02	2.19	64.31	112.52	0.51	0.77	0.274
3.0	2/24	10.13	2.17	61.42	80.14	0.47	0.87	0.257
	1	13.72	1.34	70.41	90.21	0.61	0.98	0.295
	3	14.82	1.28	88.27	100.27	0.77	1.71	0.328
	5	17.43	1.41	40.16	147.33	1.03	2.15	0.354
	10	12.18	2.37	18.23	82.23	1.07	2.31	0.366
	20	9.26	2.07	24.73	59.16	0.68	1.69	0.377
	30	4.07	1.71	13.18	46.72	0.51	1.07	0.388

TABLE 2

Influence of Sodium Caseinate (1.0% wt./wt.) mono-/di-glyceride ratio (1.0%wt./wt.) on Viscoelasticity Parameters and Mean Drop Size of O/W Emulsions. Temperature 25.0 ± 0.2°C, pH = 5.5.

mono-/di-glycerides ratio	Ageing time (days)	E_0 (dyne cm^{-2})	n_N (poise x 10+5)	E_1 (dyne cm^{-2})	E_2 (dyne cm^{-2})	n_1 (poise x 10^{-4})	n_2 (poise x 10^{-4})	d (μm)
8/2	2/24	113.37	0.27	184.57	227.32	1.04	2.07	0.337
	1	210.23	0.23	217.40	475.31	0.92	2.73	0.359
	3	885.42	0.20	219.20	485.28	0.87	2.99	0.377
	5	1959.51	0.18	735.09	1528.79	1.20	3.26	0.389
	10	874.28	0.32	531.44	1324.47	2.37	2.17	0.399
	20	254.71	0.46	232.81	681.56	2.39	1.92	0.406
	30	160.65	0.74	177.4	341.19	2.35	1.70	0.409
5/2	2/24	154.19	0.21	115.21	217.32	3.27	5.71	0.315
	2	622.32	0.18	321.37	714.27	5.31	12.19	0.330
	3	1843.63	0.17	713.34	1231.52	11.05	22.15	0.339
	5	3687.21	0.22	1074.17	2341.26	17.34	37.48	0.345
	10	2176.39	0.29	931.42	2024.37	16.21	43.79	0.349
	20	1175.50	0.62	704.39	1837.43	7.28	20.64	0.353
	30	934.42	0.70	532.31	1237.46	6.36	19.71	0.355
2/2	2/24	80.14	0.27	170.47	257.51	5.28	6.54	0.353
	2	178.71	0.24	400.51	1238.27	6.46	13.67	0.381
	3	647.24	0.22	932.26	1314.15	8.45	19.51	0.402
	5	1787.19	0.58	1241.27	1415.21	10.21	23.31	0.417
	10	951.44	0.97	1037.14	938.44	12.17	25.37	0.420
	20	197.30	0.83	842.12	732.13	11.21	20.17	0.425
	30	139.02	0.78	437.73	604.13	8.32	18.03	0.431

226

TABLE 2 continued

mono-/di-glycerides ratio	Ageing time (days)	E_0 (dyne cm^{-2})	n_N (poise x 10^{+5})	E_1 (dyne cm^{-2})	E_2 (dyne cm^{-2})	n_1 (poise x 10^{-4})	n_2 (poise x 10^{-4})	$\bar{d}$ (μm)
2/6	2/24	26.15	0.07	13.18	30.32	1.51	3.27	0.374
	1	61.39	0.15	27.35	54.39	2.47	5.81	0.409
	3	84.43	0.25	51.29	103.35	4.72	9.41	0.433
	5	97.67	0.47	63.41	157.16	8.51	20.52	0.450
	10	52.34	0.41	47.54	127.82	8.07	22.74	0.461
	20	23.41	0.37	20.32	71.33	6.31	16.34	0.466
	30	14.06	0.35	20.03	48.23	4.37	15.21	0.466
2/9	2/24	18.31	0.11	10.51	31.57	0.08	1.19	0.417
	1	30.24	0.26	17.28	42.34	2.28	3.45	0.464
	3	70.41	0.35	33.21	78.21	7.74	12.37	0.505
	5	89.10	0.21	41.34	107.24	7.49	15.28	0.526
	10	40.59	0.18	39.24	91.52	7.00	14.39	0.540
	20	17.31	0.32	28.51	34.37	6.31	10.28	0.547
	30	9.21	0.54	19.34	28.39	4.37	9.04	0.553

TABLE 3

Influence of pH on the Viscoelasticity Parameters and Mean Drop Size of O/w Emulsions Stabilized by mono-/di-glycerides ratio 5/2 (1.0% wt./wt.) and Sodium Caseinate (1.0% wt./wt.) Temperature 25° ± 0.2°C.

Emulsion pH	Ageing time (days)	E_0 (dyne cm^{-2})	n_N (poise x 10^{+5})	E (dyne cm^{-2})	E_2 (dyne cm^{-2})	n_1 (poise x 10^{-4})	n_2 (poise x 10^{-4})	d (μm)
9.0	2/24	12.71	3.14	23.71	40.21	5.44	6.41	0.346
	1	31.74	3.07	33.51	53.47	6.33	8.26	0.377
	3	71.81	5.81	58.76	106.39	8.47	11.31	0.406
	5	64.77	4.67	100.25	168.75	8.56	10.44	0.429
	10	21.29	6.21	133.74	240.82	9.15	6.86	0.446
	20	16.32	7.03	121.23	131.27	10.79	4.57	0.463
	30	—	—	—	—	—	—	—
7.0	2/24	15.63	2.37	5.89	12.84	0.03	0.06	0.346
	1	17.09	3.71	6.38	15.71	0.03	0.06	0.362
	3	18.08	4.27	10.74	17.32	0.07	0.14	0.275
	5	16.17	5.11	11.05	19.79	0.09	0.16	0.384
	10	17.32	5.07	9.41	20.21	1.11	2.02	0.392
	20	8.64	5.34	7.42	17.54	0.08	2.18	0.399
	30	6.36	5.76	5.74	14.43	0.06	2.30	0.405
5.5	2/24	25.47	5.71	9.14	50.19	1.14	3.15	0.346
	1	32.21	10.84	18.71	71.32	4.57	9.79	0.358
	3	34.46	20.31	37.14	80.44	20.38	18.16	0.369
	5	51.27	22.14	41.52	110.27	29.76	25.67	0.376
	10	21.33	18.12	36.24	91.21	9.21	23.71	0.381
	20	13.77	16.31	19.31	74.67	7.37	13.48	0.385
	30	9.44	13.14	18.37	57.34	4.00	12.07	0.386

TABLE 3 continued

Emulsion pH	Ageing time (days)	E (dyne cm^{-2})	n_N (poise x 10^{+5})	E_1 (dyne cm^{-2})	E_2 (dyne cm^{-2})	n_1 (poise x 10^{-4})	n_2 (poise cm^{-4})	d (μm)
3.0	2/24	4.37	0.02	0.05	0.23	0.67	1.58	0.246
	1	2.21	0.07	0.21	0.54	0.91	2.04	0.388
	3	1.24	0.09	0.51	1.18	1.37	3.17	0.416
	5	0.91	1.03	0.87	2.27	1.74	3.34	0.438
	10	–	–	–	–	–	–	–

TABLE 4.

Electrophoretic mobility data for O/W emulsions, stabilized by sodium caseinate (1.0% wt/wt) of various pH's and mono-/di-glyceride ratios.

Emulsion pH	Electrophoretic mobility ($m\mu$ sec^{-1} at Vol/cm^{-1}* (mono-/di-glycerides ratio)				
	8/2	5/2	2/2	2/6	2/9
9.0	6.12	6.09	6.24	6.51	6.80
7.0	5.39	5.28	5.73	6.18	6.69
5.5	2.48	2.21	2.72	3.28	3.35
4.6	-	No movement			-
3.0	3.73	3.17	5.06	5.25	5.86

* Initial emulsions' aqueous phase pH 7.0.
Emulsions were 3 hrs old.
Measurements took place at $25^{\circ} \pm 0.1^{\circ}C$.

Discussion

Since the stability of sodium caseinate stabilized O/W emulsions to coalescence is greater when also using mono-glycerides and di-glycerides than when using either emulsifier on its own, it is necessary to review the nature and structure of the film formed around the oil drops. This film is mainly responsible for the resistance to drops coalescence and for emulsion stability.

The mono-glyceride and di-glyceride molecules are likely to arrive at the oil-water interface more quickly than the protein molecules because they have lower molecular weights. Therefore they diffuse to the interface more readily. There, they arrange themselves in the form of a monolayer. The excess of glyceride molecules that do not absorb remain in the oil phase.

According to Dervichian, D.G. (1958) Van der Waals' forces between the hydrocarbon chains are primarily responsible for the regular arrangement of emulsifier molecules in mixed monolayers. In addition, mono-glyceride and di-glyceride molecules associate easily because their hydrocarbon chains contain more than 8 C atoms.

The monoglycerides and diglycerides, as apolar lipids, project their relative small hydrophilic

groups into the water phase and the remainder of
each molecule which is hydrophobic, peojects into
the oil phase.

According to Kako, M and Kondo, S. (1978)
because di-glyceride has a strong attraction for
oil, its molecules must be, at least to some extent,
in a dissolved state in the oil phase. Most of each
di-glyceride hydrocarbon chain lies deeply within
the oil phase, and a small portion of the hydrophilic
part is located at the oil-water interface. On the
other hand, mono-glyceride is somewhat more attracted to the water phase than di-glycerides. Thus the
molecules of mono-glycerides are oriented at the
oil-water interface in such a way that each hydrocarbon chain does not lie so deeply within the oil
phase as the hydrocarbon chains of di-glycerides.
When mono- and di-glycerides are used together, the
hydrophilic parts of mono-glycerides may be located
deeper within the aqueous phase then the hydrophilic parts of di-glyceride molecules.

The main conclusion is, therefore, that when
molecules of both emulsifiers are absorbed at the
oil-water interface they may be packed more closely
together than when either emulsifier is present
alone and they enter into some form of association.

Sodium caseinate (proteins) molecules are
attached to the oil phase only by the trains (Fig.
1) and numerous molecular segments (loops) project
outwards into the aqueous phase and pretend close
approach of adjacent oil drops. Within the train
areas there is likely to be some form of association
with the previously absorbed glyceride molecules.
The resultant, heterogeneous, film on the surface of
drops, consists in some regions of associated mono-
and di-glycerides on their own and in other parts
of a mono- and di-glycerides/sodium caseinate
complex. This heterogeneous film provides better
surface elasticity and viscosity and improves
emulsion stability.

Furthermore, the longest trains of sodium
caseinate projecting into the aqueous phase oppose
close approach of oil drops; interlink via hydrophobic bonding, by ion-dipole and dipole-dipole
bonds into a network with a weak gel-like structure
(Whitney, R. McL. 1977; Van Vliet et al. 1978;
Sonntag et al. 1982) and greatly increase the
viscoelasticity of the emulsions as compared with
emulsions in which stabilisation derives wholly
from electrical repulsion and attraction forces.

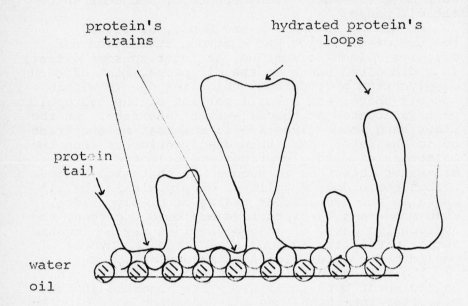

Fig. 1. Schematic representation of
orientation of mono-glyceride and
di-glyceride molecules in interaction with a protein (sodium casein-
ate/molecule in mixed films adsorbed
at the oil-water interface.

By altering the environmental conditions, such
as pH, temperature and concentration, the molecules
unfold and so increase the number of residues
available for association and interlinking.

At their respective isoelectric points the
sodium caseinate's proteins are in their most
compact configuration in aqueous solution and
consequently, they diffuse more quickly to the drop
surfaces and form an absorbed layer which has a
higher protein concentration than at any other pH.
Under these conditions there is increased interlinking through hydrated loops and tails leading to
more pronounced viscoelasticity. The degree of

interlinking in freshly prepared emulsions is obviously higher around the isoelectric point and less at all the other pH's investigated. Adjustment of the pH away from the isoelectric point, even after the emulsion has been prepared, has a dramatic effect on interlinking and reduces the viscoelasticity and emulsion stability (Table 3).

When the emulsions are stored there is further interlinking of adsorbed protein loops as the drops move closer together and the loops compress and overlap (Napper, 1977). During the first 3-5 days storage this process, which increases the viscoelasticity parameters values, exerted a greater influence than drop coalescence, which reduces the parameter values. At longer storage times drop coalescence was the dominant process. Away from the isoelectric point exhibited a relatively smaller increase in E_o, and in the first 3-5 days storage. this can be attributed to one of two possible mechanisms.

(a) Interlinking of adsorbed protein loops process more slowly immediately after emulsion preparation away from the isoelectric point than in emulsion near the isoelectric point and is almost complete after a relatively short time.

(b) Steric stabilisation by the longer protein loops does not prevent drops near the isoelectric point from moving as close together as in emulsion at other pH's.

The changes in parameters values of sodium caseinate/glycerides emulsions when stored can be attributed to mechanism (α).

Finally, the sodium caseinate and glycerides association provides an interfacial layer with a structure and rheological properties which are more effective in resisting drop coalescence than when mono-glycerides or di-glycerides are used on their own with sodium caseinate proteins. Only the longest loops and tails influence the resistance to flocculation of the O/W emulsions (Sonntag, H. et al, 1982).

Therefore, regions where mono-glycerides/di-glycerides and protein associate are unlikely to influence this process, since the hydrophilic regions of the mono-glyceride and di-glyceride molecules project into the aqueous medium to a far lesser extent than the loops and tails of protein. However, they should exert a significant influence on the rate of drop coalescence.

ACKNOWLEDGMENTS

One of the authors (G.D.) gratefully acknowledges
the financial assistance provided by the "State
Scholarships Foundation", Athens, Greece, which
made this study possible.

REFERENCES

1. Aschaffenberg, R., J. Dairy Sci., 49, 792 (1966).
2. Becher, P., "Emulsions: Theory and Practice"
   p. 196 Reinhold, New York, (1957).
3. Boyd, J., Parkinson, C. and Sherman, P.J.
   Colloid Interface Sci., 41, 359 (1972).
4. Boyd, J.V., Krog, N. and Sherman, P., Paper No.7
   at Symposium on Theory and Practice of Emulsion
   Technology, Brunel Univ. London, (1974).
5. Bigelow, C.C., J. Theor. Biol. 16, 187 (1967).
6. Colaccio, G., J. Colloid Interface Sci., 29,
   345 (1969).
7. Doxastakis, G., Ph.D. thesis, University of
   London "The interaction of the glycerides with
   proteins and their influence on the rheological
   properties of O/W emulsions", (1983).
8. Dervichian, D.G. in "Surface Phenomena in
   Chemistry and Biology"
   (J.F. Danielli, K.G.A. Pankhurst and A.C.
   Riddiford, eds), p.70 Pergamon, London (1958).
9. Friberg, S., and Janssen, P.O., J. Colloid
   Interface Sci., 55, 614 (1976).
10.Graham, D.E. and Phillips, M.C., J. Colloid
   Interface Sci., 70, 403 (1979).
11.Graham, D.E. and Phillips, M.C., J. Colloid
   Interface Sci., 70, 415 (1979).
12.Graham, D.E. and Phillips, M.C., J. Colloid
   Interface Sci., 70, 427 (1979).
13.Graham, D.E. and Phillips, M.C. J. Colloid
   Interface Sci., 76, 227 (1980).
14.Graham, D.E. and Phillips, M.C. J. Colloid
   Interface Sci., 76, 240 (1980).
15.Graham, D.E. and Phillips, M.C. in "Theory and
   Practice of Emulsions Technology" (A.L. Smith,
   ed) p. 75-98, Academic Press, London (1976).
16.Inokuchi, K.P., Bull. Chem. Soc. Japan, 28,
   453-465 (1955).
17.Ottewill, R.H., J. Colloid Interface Sci., 58,
   357 (1977).

18. Krog, N. Fett Seif. Anstrichm, 77, 267 (1975)
19. Karel, M., J. Food Sci., 38, 756 (1973).
20. Kinsella, J.E. Crit. Rev. Food Sci., and Nutr. 10, 147 (1978).
21. Kako, M. and Kondo, S. J. Colloid Interface Sci., 69, 163 (1979).
22. MacRitchie, F., in "Advance in Protein Chemistry", (Afinsen, C.B., Edsall, J.T. and Richards, F.M. eds) Vol. XXXII, p. 283, Academic Press, New York, (1978).
23. Pearson, A.M., Spooner, A.M., Hogarty, G.R. and Bratzler, L.J., Food Technol., 19, 1841 (1965).
24. Rahman, A. and Sherman, P., Colloid and Polymer Sci., 260, 1035 (1982).
25. Rivas, H.J., Ph.D. thesis, University of London "A comparison of the effectiveness of soybean and meat proteins in stabilising food emulsions and their influence on the rheological properties" (1982).
26. Sherman, P., J. Texture Studies, 1, 43 (1969)
27. Shama, F. and Sherman, P. J. Food Sci., 31, 699 (1967).
28. Sherman, P. in "Proc. Int. Cong. Surface Activ., 4th Vol 2, p. 1199 (1967).
29. Sherman, P. J. Phys. Chem. 67, 2531 (1963).
30. Sherman, P. and Benton, M., J. Texture Studies 11, 1 (1980)
31. Sonntag, H., Ehmke, B., Miller, R. and Knapschinsky, L. in "The Effect of Polymers on Dispersion Properties" (Tadros,Th. F. ed) p. 207, Academic Press, London (1982).
32. Napper, H.P. in "Colloid and Interface Science" Vol. 1 (M. Kerker, R.L. Rowell and A.C. Zettlemoyer eds) p. 470, Academic Press, New York, (1977).
33. Van Kliet, Lyklema, J. and Van den Tempel, M. J. Colloid Interface Sci., 65, 505 (1978).
34. Vernon Carter, F.J. and Sherman, P. J. Texture Studies 11, 351 (1980).
35. Whitney, R. McL., in "Food Colloids" (H.D. Graham ed.) p. 88, AVI Publishing Co., Westport, Conn. U.S.A. (1977).

ANALYSIS OF DRIED MILK AND CHEESE POWDERS BY NEAR INFRARED REFLECTANCE SPECTROSCOPY

K.I. Ereifej[1]
Pericles Markakis

Department of Food Science and Human Nutrition
Michigan State University
East Lansing, Michigan

I. ABSTRACT

A Neotec near infrared reflectance analyzer, Model GQA-41, was calibrated to estimate the content in moisture, protein, fat, ash and lactose-by-difference in the following samples: a) 41 milk powders prepared by spray drying mixtures of skim and whole milk; b) 31 milk powders prepared by mixing predried skim and whole milk; c) 11 commercial milk powders; d) 40 commercial cheese powders. The correlation coefficients (r) between NIRS and conventional analysis data were in the range .68-.85 for moisture, .94-.99 for protein, .98-.99 for fat, .98-.99 for ash, and .95-.99 for lactose-by-difference. The lowest r's were associated with cheese powders, probably because some contained non-dairy ingredients. The NIR prediction was best for commercial milk powders when also commercial milk powders were used for calibration of the instrument. NIRS analysis of dehydrated milk products is promising.

II. INTRODUCTION

Norris and Hart (1965) are credited with the idea of using near infrared reflectance spectroscopy (NIRS) for determining moisture in agricultural products. A few years later, NIRS was introduced to the grain industry as a means of rapid analysis not only for moisture, but also for protein and oil content (Rosenthal, 1971). Since then, three US manufacturers (Neotec[2], Technicon, and Dickey-john) have made available instruments of increasing sophistication for the compositional analysis of grains and oilseeds. Efforts for predicting by NIRS additional food or feed constituents (e.g. starch, fiber, lignin, lysine) have also been made (Norris et al., 1976; deGroen, 1980; Rubenthaler and Bruinsma, 1978).

This communication presents the results of an on-going effort to predict the moisture, protein, fat, ash and lactose

[1]*Present address: Food Science Department, King Saud University, Riyadh, Saudi Arabia*
[2]*Neotec is now a division of Pacific Scientific Co.*

237

content of dehydrated dairy products using modern NIRS
instrumentation. Previously, Casado et al. (1979) reported
good agreement between conventional and NIRS analyses for the
moisture, protein and fat content of milk powders. Also,
Giangiacomo et al. (1979) found a good correlation between
conventional and NIRS methods for the protein, ash, and free
lysine content of freeze-dried blue cheese, but a rather poor
correlation for the pH and fat content.

III. MATERIALS AND METHODS

A. *Dried Milk Samples*

Three groups of dried milk samples were prepared. In
group A there were 41 samples prepared by mixing whole
pasteurized milk (3.06% fat) and skim milk (0.11% fat) at
various proportions and spray-drying the mixtures. In group B
there were 31 samples prepared by mixing commercially dehydrated whole milk (26.49% fat) with commercially dehydrated
skim milk (1.03% fat) at various proportions. The purpose of
wet-mixing (group A) and dry-mixing (group B) was a) to obtain
samples with a wide range of fat concentrations for calibrating
the NIRA instrument, and b) to see if the mixing procedure had
any influence on the calibration values which would be used to
predict the composition of unknown samples. Group C was
composed of 11 commercially dried skim milk and whole milk
samples. All samples were ground for two minutes in a Mitey-
Mill and passed through a 100-mesh sieve.

B. *Cheese Powders*

Forty samples of commercial cheese powders which are used
for snack seasoning were also available. They were all
provided by one manufacturer.

C. *Conventional Analysis*

All samples were analyzed for moisture, ash, fat and
total nitrogen contents. Moisture, ash and nitrogen contents
were determined by the oven method, dry ashing, and the micro-
Kjeldahl method, respectively, according to AOAC (1980). Fat
content was determined by the Mojonnier procedure (1925).

D. *NIRR Analysis*

The Neotec Grain Quality Analyzer Model 41, or GQA-41,
interfaced with the main computer of Michigan State University
by teletype, was used. In this instrument, the sample is

scanned in the 1800-2320 nm region and the computer selects
four wavelengths which best represent the reflectance characteristics of a food constituent. These wavelengths are used
to calibrate the instrument by means of samples of known composition. Calibration results in a polynomial regression
equation on the basis of which the composition of an unknown
sample may be predicted.

IV. RESULTS AND DISCUSSION

The correlation coefficients (r), the slopes and the y-
intercepts of the regression relationships established between
the conventional analysis data and the NIRS values of the
dehydrated dairy products used for calibration of the instrument are shown in Table 1.

Table 1. Correlation coefficients (r), slopes (a) and y-
          intercepts (b) of linear regressions between NIRS
          values (y) and conventional analysis data (x) of de-
          hydrated dairy products (n=number of samples
          analyzed.

		Water	Protein	Fat	Ash	Lactose by dif.
Milk powders, wet-mixed n=41	r	.793	.965	.989	.823	.953
	a	.804	.932	.935	.621	.998
	b	.589	1.747	.953	2.459	.333
Milk powders, dry-mixed n=31	r	.814	.985	.995	.985	.996
	a	.574	.976	.991	.445	.998
	b	1.190	.426	-.083	4.036	.140
Commercial milk powders n=11	r	.851	.988	.995	.995	.989
	a	1.002	1.025	1.009	1.044	1.012
	b	.001	-.908	-.197	-.288	-.979
Commercial cheese powders n=40	r	.680	.936	.980	.836	-
	a	.496	.868	1.014	.791	-
	b	1.995	2.016	.087	1.592	-

In a comparison among the five constituents, the higher r's are associated with the content in protein, fat and lactose-by-difference. The r's for the ash content are slightly lower and those for moisture considerably lower.

Among commodities, the commercial milk powders displayed the highest r's, while the commercial cheese powders showed the lowest r's. The low r's of the cheese powders are probably due to the presence of non-dairy ingredients in some (but not all) of these powders.

When the GQA-41 was used to predict the composition of six commercial milk powders, applying the calibration obtained first by the 41 wet-mixed samples and then by the 31 dry-mixed samples, the following results were obtained. It was first observed that the deviations of the predictions from the real (conventional analysis) values were not very different for the two calibrations. For this reason the following comments refer to all 12 samples, independently of calibration method.

The predictions for moisture content were off by an average of 30% and biased downward; the mean moisture content of the 12 samples was 4.1% as measured by the oven method, and 2.9% by NIRS. The predictions for protein content were considerably better, deviating by about 3%, and they were unbiased; the mean protein content was 31.1% by Kjeldahl and 31.5% by NIRS. The predictions for fat content were off by about 10% and were unbiased; the mean fat content was 9.7% by Mojonnier and 9.3% by NIRS. The predictions for samples of low fat content (1-5% fat) were less accurate than those for high fat content (15-20% fat) samples. The predictions for ash content were off by about 5% and they were unbiased; the mean ash content was 7.9% by combustion and 7.3% by NIRS. The predictions for lactose-by-difference content were off by about 4% and unbiased; the average value based on conventional methods was 46.9%, and the corresponding NIRS average 48.8%.

In an experiment in which the GQA-41 was first calibrated by eleven commercial milk powders and then two commercial samples, other than those used for calibration, were analyzed, the results of Table 2 were obtained.

Table 2.   Analysis of two commercial milk powders by conven-
           tional and NIRS methods following calibration of the
           GQA-41 by other commercial milk powders.

Sample	H_2O		Protein		Fat		Ash		Lactose-by-dif.	
	Conv.	NIRS	Conv.	NIRS	Conv.	NIRS	Conv.	NIRS	Conv.	NIRS
1	3.7	2.9	31.6	31.3	1.2	1.1	7.8	7.6	55.7	57.1
2	4.1	2.1	31.5	31.2	1.1	0.7	7.7	7.9	55.6	58.1

The predictions, in this case, for protein content are excellent, differing by only 1% from the analytical values. Very good are also the predictions for ash content (3% deviation) and lactose-by-difference content (3.5% deviation). The constituents that were present in very small quantities, water and fat, were predicted poorly. The reasons for improved predictability of the composition of commercial dry milk samples when commercial rather than laboratory samples were used for calibrating the instrument are probably differences in the physical characteristics of the two sample types. It is known that the physical condition of powders affect their reflectance properties.

It may be concluded from this study that NIRS holds excellent promise for the prediction of protein content in milk powders and relatively good promise for the prediction of ash and lactose content. Much additional work would be necessary, however, before NIRS methodology can be recommended for official purposes. Variabilities between subsamples for the same analyst, between analysts, between units of the same instrument type, and between instruments of different types and/or manufacturers should be explored and evaluated. And the rapidly advancing instrument technology should be also fully exploited.

ACKNOWLEDGEMENTS

Thanks are due to Prof. D. Reicosky for placing the GQA-41 instrument at the disposal of the authors and helping with its use, and to Commercial Creamery, Spokane, WA for providing the cheese powders.

REFERENCES

AOAC. Official Methods of Analysis. Assn. Off. Anal. Chem., Washington, D.C. (1980).

Casado, P., Blanco, C. and Rozas, A. Rev. Isp. Tech. 108, 97 (1978).

deGroen, A. Monatschr. f. Braurei 33, 131 (1980).

Giangiacomo, R., Torregianni, D., Frank, J.F., Loewenstein, M. and Birth, G.S. J. Dairy Sci. 62 (Suppl. 1) 39 (1979).

Mojonnier Bros. Co. Mojonnier Milk Tester Instruction Manual. Chicago, IL (1925).

Norris, K.H., Barnes, R.F., Moore, J.E. and Shenk, J.S. J. Animal Sci. 43, 889 (1976).

Norris, K.H. and Hart, J.R.   Proc. 1963 Intern. Symp. on
    Humidity & Moisture.   Vol. 4, p. 19.   Reinhold Publ.,
    N.Y. (1965).

Rosenthal, R.D.   Proc. Ann. Meeting, Kansas Assn. Wheat
    Growers, Hutchinson, KS (1971).

Rubenthaler, G.L. and Bruinsma, B.L.   Crop Sci. 18, 1039 (1978).

BOVINE, CAPRINE, AND HUMAN MILK XANTHINE OXIDASES: ISOLATION, PURIFICATION, AND CHARACTERIZATION[1]

John P. Zikakis
Michael A. Dressel
Mark R. Silver

Department of Animal Science and Agricultural Biochemistry
University of Delaware
Newark, Delaware

For the purification of bovine milk xanthine oxidase, virtually all methods employ proteolytic and lipolytic enzymes and denaturing organic reagents. Although it has been shown that such treatments do not appear to adversely affect enzyme activity and cofactor composition, these treatments modify the enzyme's molecular structure. A non-proteolytic purification method was developed which produces high purity native xanthine oxidase with a PFR of 4.1, a specific activity of 7.8 IU/mg, a symmetric peak by ion-exchange chromatography, a single active protein band by polyacrylamide disc gel electrophoresis, and a yield of 23%. Comparative kinetics showed significant differences between proteolytically and non-proteolytically prepared enzyme. These and other findings were used to explain how limited proteolysis causes structural alterations leading to enzyme with reduced molecular weight,

---

[1]Published with the approval of the Director of the Delaware Agricultural Experiment Station as Miscellaneous Paper No. 73, Contribution 816 of the Department of Animal Science and Agricultural Biochemistry, University of Delaware, Newark.

243

diminished stability, reduced catalytic efficiency,
and lower affinity for both substrate and competitive
inhibitors. Caprine milk xanthine oxidase was isolated
and purified for the first time. Amino acid analysis
of the purified preparation indicated that goat's milk
xanthine oxidase contains higher amounts of aspartic
acid, glutamic acid, proline, and glycine and lower
amounts of serine than cow's milk xanthine oxidase.
The goat milk enzyme had a pH optimum of 8.35 and FAD
was identified as one of its cofactors. Furthermore,
free and membrane-bound xanthine oxidase were isolated
from human colostrum for the first time and partially
purified and characterized. The molecular weight, pH
optimum, and activation energy for free and membrane-
bound xanthine oxidases were: 318,810, 8.2, and 17.2
kcal/mole and 4×10^7, 8.5-10.0, and 15.1 kcal/mole,
respectively. The free enzyme had a Km of 2.79×10^{-5}M
and a Vmax of 21.23×10^{-3} IU/ml for xanthine and a Km
of 1.34×10^{-5}M and a Vmax of 8.47×10^{-3} IU/ml for
hypoxanthine. The membrane-bound enzyme had a Km of
2.64×10^{-5}M and a Vmax of 14.83×10^{-3} IU/ml for
xanthine and a Km 9.24×10^{-6}M and a Vmax of $3.50 \times
10^{-3}$ IU/ml for hypoxanthine. Both free and membrane-
bound enzymes contained FAD.

INTRODUCTION

In 1902, Schardinger noted the ability of methylene
blue to oxidize formaldehyde in the presence of bovine milk
(88). Some 20 years earlier, Horbaczewski (52) and Spitzer
(93) described the oxidation of xanthine and hypoxanthine
to uric acid in mammalian tissue homogenates. For a long
period of time, these two observations remained unrelated.
After years of extensive research by many, it is now
generally accepted that xanthine oxidase (xanthine: O_2
oxidoreductase, E.C.1.2.3.2) catalyzes both reactions
(14,15,36,37). Xanthine oxidase is widely distributed in
animals, plants, and microorganisms. Bovine milk is a rich
source of xanthine oxidase (2,5,43,117) whereas the milk of
sheep and goats contains lower concentrations of the enzyme
(20,29,64,65,79,91). Xanthine oxidase activity in the milk
of monogastric mammals also varies. The milk of mouse, rat,
guinea pig, and donkey contains high activity of xanthine
oxidase (112) while milk from mare, cat, dog, patas monkey,

and human contains moderate to low activity (110,112).
Another rich source of the enzyme is body organs, especially
the liver of mammals and birds (23,31). However, due to its
abundance and availability, xanthine oxidase from cow's milk
has been extensively studied in the last 60 years or so.
Bovine milk xanthine oxidase is a conjugated iron-sulfur-
molybdenum flavoprotein and was one of the first flavo-
proteins to be purified (4,5,27) and crystallized (2,3).
Bovine milk xanthine oxidase and xanthine oxidase from
animal tissue were the first mammalian enzymes found to
contain molybdenum (32,44) and soon thereafter it was
discovered that they also contained iron (82). In most
subsequent studies, bovine milk xanthine oxidase has been
used as a model for detailed study into the enzyme
mechanisms and biological electron transport systems.

Since its discovery, the biological role of xanthine
oxidase remains unclear although recent studies, discussed
below, indicate some important roles for the enzyme. Its
main function is the oxidation of purines but due to its
broad specificity, it oxidizes a wide variety of aldehydes,
pteridines, and other heterocyclic compounds by a variety
of electron acceptors. Xanthine oxidase has been implicated in the control of various redox reactions in the cell.
For example, there is evidence that oxidation of reduced
glutathionine and other thiols (53,73) and the oxidation
of glyoxalate to oxalate (42) are coupled to xanthine
oxidation. The mechanism of these couplings is thought to
occur via the production of hydrogen peroxide and super-
oxide radicals which serve as oxidants for coupled biological oxidations (40,81). Recently Tophan et al. (95)
demonstrated that intestinal xanthine oxidase plays a major
role in the absorption of dietary iron by promoting the
oxidation and incorporation of iron into transferrin.

The interconversion of bovine milk xanthine oxidase from
an oxidase to a dehydrogenase has also been studied. When
the enzyme is bound to the milk-fat globule membrane (MFGM),
it acts as an NAD^+-dependent dehydrogenase while free
xanthine oxidase appears as an oxidase (7,19). The oxidase
form can be converted into an NAD^+-dependent dehydrogenase
by treatment with dithioerythritol. This form of the
enzyme can be converted, irreversibly, back again to the
oxidase by treatment with proteolytic enzymes such as
chymotrypsin, papain, and subtilisin (7,19). A similar
oxidase-dehydrogenase interconversion occurs with liver
xanthine oxidase (94).

The biological function of xanthine oxidase in milk is
equally unclear. It is assumed that nature has provided
enzymes (and other nutrients) in milk for the benefit of

the young who lack a complete digestive system and are un-
able to produce their own enzymes for the digestion of
milk.   However, it seems unlikely that the main function of
xanthine oxidase in milk would be in the catabolism of
purines for the survival of the young.   As mentioned
earlier (95), intestinal xanthine oxidase is important in
the absorption of dietary iron.   Since intestinal xanthine
oxidase in the young is not well developed, one of the major
functions of milk xanthine oxidase may be to assure absorption of iron from the gut.   Another function of xanthine
oxidase in milk may be its coupling antibacterial effect via
the lactoperoxidase system.   The lactoperoxidase system in
milk requires a source of hydrogen peroxide to oxidize
thiocyanate to hypothiocyanate, an antibacterial agent (12).
Since milk contains no hydrogen peroxide, the lactoperoxidase
system will be inoperative without a source of hydrogen
peroxide.   The presence of xanthine oxidase in milk provides
the needed hydrogen peroxide for the system as it degrades
purines (12) and other substrates in milk.   From a practical
point of view for the dairy industry, xanthine oxidase has
been implicated with the spontaneous development of
undesirable oxidized flavor in market milk and other dairy
products (1,89).

Renewed interest has centered on bovine milk xanthine
oxidase because of the theory which involves the enzyme in
the initial development of atherosclerosis in humans (75,76,
77,84).   Oster and his associates postulated that ectopic
xanthine oxidase from bovine milk destroys the palmitalde-
hydes liberated from the cell-membrane-based plasmalogens
(aldehydogenic phospholipids) as they are metabolically
turned over.   This causes the initial damage to the cell
membrane of the arterial intima and the myocardium.   The
resulting histochemical injury is followed by cell pro-
liferation and scar formation in the affected myocardium,
local deposition of cholesteryl esters, and ultimate
development of typical atherosclerotic lesions in the
arteries.   This concept holds that the enzyme survives
passage through the gastrointestinal tract and is absorbed
enzymatically active into the lymphatic system finally
entering the general circulation.   Since homogenization of
milk increases the bioavailability of xanthine oxidase (by
converting the membrane-bound enzyme from dehydrogenase to
oxidase, by micronization of fat globules, and by forming
liposomes, this point is discussed later on), consumption
of homogenized dairy products may be a predisposing factor
in the development of atherosclerosis (77).

In the 1970's this subject received considerable
publicity (30,63,99,102).   In 1975 the Food and Drug

Administration commissioned the Life Sciences Research Office of the Federation of American Societies for Experimental Biology (FASEB) to review the evidence and evaluate the theory. The same year this group issued a technical report (22) concluding that the evidence for or against the theory is inconclusive. Because of the lack of knowledge in certain areas and the urgency of elucidating this disease, the FASEB group recommended more research to be done in specific areas.

Following the release of the FASEB report, a considerable number of studies were carried out, most of which support the concept. Contrary to criticisms (11,58) of the xanthine oxidase theory, it was demonstrated in vivo with rats that xanthine oxidase in ingested market milk persisted from 2-7 hours in the stomach (113). The transit gastric time of intubated milk for the rat was between 30-40 minutes. Similar observations were made in vitro using simulated gastric juice (113). Furthermore, in in vitro experiments simulating the small intestine, it was shown that xanthine oxidase in milk persisted for up to 24 hours after successive incubation with artificial gastric juice and pancreatin (113). These findings were confirmed by Ho et al. (51) using human gastric juice and pancreatin digestion.

In a survey of market dairy products, Zikakis and Wooters (117) found xanthine oxidase activity in the majority of about 200 products (including some 110 domestic and imported cheeses) assayed. An earlier study with a limited number of dairy products tested reported similar observations (24). Intradermal and subcutaneous inoculations of guinea pigs and miniature pigs with purified xanthine oxidase proved that the milk enzyme is antigenic (108,109). Oster et al. (78) showed that individuals with clinically manifested atherosclerosis had higher titers to bovine milk xanthine oxidase than did individuals with no clinical evidence of atherosclerosis. In another study, antibodies to purified bovine milk xanthine oxidase were found in the sera of 73 of the 94 human volunteers tested (86). Furthermore, as the mean consumption of volume of whole milk and milkfat increased, so did the enzyme antibody levels. This dose response supports the claim of dietary bovine milk xanthine oxidase entering the bloodstream via absorption or persorption (98) from the gut.

Other studies supporting the uptake of dietary milk xanthine oxidase include the following. Clark et al. (25) found that serum xanthine oxidase activity increased in rats 2 hours after intubation with processed cow's cream (half and half). Ho et al. (50) estimated that of 100 mg xanthine oxidase in fresh raw milk, 41 mg survived processing, 27 mg

entered the intestine active, and 20 ng were absorbed
enzymatically active. Gandhi and Ahuja (41) dosed rabbit
cubs orally with purified milk xanthine oxidase and
sacrificed them after 2, 3, and 5 hours. Then, they
determined xanthine oxidase activity in blood, stomach, and
small intestine and found a direct relationship. As the
stomach enzyme activity decreased, the blood activity
increased while activity in the small intestine remained
relatively constant. In vitro studies confirmed their
results (41). Furthermore, Gandhi and Ahuja found that
xanthine oxidase decreased the plasmalogen content of the
heart muscle after daily oral and intravenous administration to rabbits with purified xanthine oxidase for 14
days. They also found an increase in the total cholesterol.
Finally, Ross et al. (85) reported that the process of
homogenization creates liposomes. Using column chromatography, they purified liposomes from homogenized market
milk and found that these intact liposomes contained trace
or no xanthine oxidase activity. However, after treatment
with a detergent, the liposomes burst open releasing
entrapped xanthine oxidase and exhibiting high enzyme
activity. They concluded that liposomes entrapped xanthine
oxidase and acted as carriers of the enzyme into the body.
This observation was confirmed in our laboratory (87).
These "trojan horses" are well known to pass undetected the
immuno-surveillance system and enter the body (46,47,103).
These findings suggest another mechanism by which active
xanthine oxidase may enter the body.

     In this study, we will present isolation and purifica-
tion methodologies for obtaining undenatured high purity
bovine milk xanthine oxidase. Such a high purity native
preparation (unavailable commercially) is needed for on-
going studies described above, especially for the elucidation of dietary xanthine oxidase and its implication in the
etiology of heart diseases. Similarly, we will discuss
methods developed for the isolation, purification, and
partial characterization of xanthine oxidase from goat's
cream for the first time. Finally, we will show that
xanthine oxidase is present in human milk and colostrum and
describe methods for its isolation, purification, and
characterization from colostrum, a feat never accomplished
before.

I. BOVINE MILK XANTHINE OXIDASE

Nearly half of xanthine oxidase in cow's milk is closely associated with the MFGM (66). This membrane has a protein-aceous surface that interfaces with the milk plasma phase on the exterior and the globule lipids on the interior (21). In order to increase the yield in nearly all methods published since 1939, bovine milk xanthine oxidase has been purified using proteolytic and lipolytic enzymes and organic reagents (2,5,43,48,60,61). Lipolytic enzymes have been used in xanthine oxidase purifications to break up the MFGM and release membrane-bound xanthine oxidase. On the other hand, pancreatin has been used to degrade casein micelles to lower molecular weight components so that they may be eluted behind xanthine oxidase in chromatographic fractionations. However, pancreatin is not specific for casein degradation and has a proteolytic effect on all proteins including xanthine oxidase. Hart et al. (48) and Nelson and Handler (72) have shown that purified xanthine oxidase differs according to the purification method and that proteolysis adversely affects the enzyme. Waud et al. (100) and Nagler and Vartanyan (68) demonstrated that purification procedures using pancreatin yield xanthine oxidase and subunits with lower molecular weight and that xanthine oxidase migrates faster on polyacrylamide gel electrophoresis than enzyme prepared by a non-proteolytic treatment. Furthermore, Nathans and Hade (70) showed that xanthine oxidase isolated in the presence of pancreatin co-purifies with proteases from pancreatin.

Our objective in this section is to describe isolation and purification procedures for obtaining high purity native xanthine oxidase from cow's milk. The main features are the use of a mild non-ionic detergent, maintaining a low temperature to remove a maximum amount of casein, and use of ultrafiltration and multiple chromatographic columns to concentrate the enzyme and remove non-xanthine oxidase proteins with lower or higher molecular weights. The final purified enzyme has been kinetically characterized and compared with a proteolytically derived enzyme preparation (92,114,121). A preliminary account of parts of this research has appeared (107,119). A more elaborate account was given in a series of 3 United States patents (115,118, 120).

A. MATERIALS AND METHODS

Sodium acid pyrophosphate was purchased from Alfa-Inorganics-Ventron, Beverly, Mass. Xanthine was from Eastman Organic Co., Rochester, N. Y. Hypoxanthine was purchased from ICN Pharmaceuticals, Inc., Cleveland, Ohio. Folic acid, pterin-6-carboxylic acid, xanthine oxidase, bovine serum albumin, and neotetrazolium chloride were obtained from Sigma Fine Chemicals Co., St. Louis, MO. Triton X-100, polyoxyethylene sorbitan (Tween 80), and flavin adenine dinucleotide (FAD) were from U.S. Biochemical Corp., Cleveland, Ohio. Sephadex G-75, G-200, Sepharose 6B, DEAE Sepharose, and DEAE Sephadex A-50 were obtained from Pharmacia Fine Chemicals, Piscataway, NJ. Acrylamide (electrophoresis grade) was from Bio-Rad Laboratories, Richmond, CA. Disodium ethylene diamine tetraacetate (EDTA), potassium phosphate (mono- and dibasic), sodium phosphate (primary and secondary), sodium salicylate, trichloroacetic acid, and tris (hydroxyamino) methane were purchased from Fisher Scientific Co., King of Prussia, PA. Ultrafiltration membrane filters XM-50 and XM-100A were purchased from Amicon Corp., Lexington, Mass. Other reagents and solvents were of reagent grade. Glass distilled-deionized water was used throughout. Fresh raw milk was obtained from the University of Delaware dairy Guernsey and Holstein herds.

1. Gel Filtration Chromatography

Sephadex gels G-75, and G-200 were swollen in 0.1M pyrophosphate buffer pH 7.1 at 4°C for the appropriate time. Sepharose 6B was purchased pre-swollen. Prior to the addition of a protein sample, gel columns were equilibrated at 4°C with 4-5 volumes of elution buffer or until eluate had a constant absorbance at 280 nm. A constant head pressure was maintained with a peristaltic pump and 3 ml fractions were collected.

2. Ion Exchange Chromatography

DEAE Sephadex-A50 or DEAE Sepharose CL-6B was equilibrated overnight at 25°C in 0.1M sodium pyrophosphate buffer pH 8.6. The gel was washed copiously with 0.005M pyrophosphate buffer until it reached equilibrium. Elution of the enzyme from the column was accomplished on a linear continuous salt gradient from 0.005M to 0.1M sodium pyrophosphate buffer, pH 8.6.

3. Ultrafiltration and Total Protein Concentration

Three stirred Amicon ultrafiltration systems and Amicon membrane filters XM-50 and XM-100A were used to concentrate protein fractions from chromatography. Ultrafiltrations were performed at 4°C at pressures between 3.5-4.2 kg/cm^2 with XM-50 membrane and 1.1-1.4 kg/cm^2 with XM-100A membrane using nitrogen or helium. Total protein in samples was determined according to Lowry et al. (56) with bovine serum albumin as standard.

4. Polyacrylamide Disc Gel Electrophoresis (PAGE)

Analytical PAGE was performed using a Buchler 18 tube Polyanalyst according to the methods of Ornstein (74) and Davis (33). Pore sizes in the gels were based on the use of 3.5% acrylamide in the stacking gel and 10% acrylamide in the separating gel. Gels were run at pH 8.3 using Tris/glycine buffer at 4°C. Gels which were to be stained for protein, first were fixed in 12.5% trichloroacetic acid. Following fixing, gels were stained for protein in a mixture containing 1% Coomassie Brilliant Blue and 12.5% trichloroacetic acid in a ratio of 1:20 Coomassie Blue to trichloroacetic acid and destained in 7.5% acetic acid. Detection of enzymatically active xanthine oxidase in gels was performed according to the method of Zikakis (120) using neotetrazolium chloride.

5. Enzyme Assays

Enzyme activity of purified preparations was measured spectrophotometrically using either a Beckman DB or Gilford Model 250 spectrophotometer. One International Unit (IU) of enzyme activity was defined as that amount of enzyme which catalyzes the conversion of one micromole of xanthine to uric acid per minute at 295 nm, 23°C, and pH 8.3. The assay was carried out in 0.1M pyrophosphate (pH 8.3) containing 0.005% EDTA. Changes in absorbance were converted to IU of activity by using the difference of extinction coefficient (9.5×10^3 cm^{-1} M^{-1}) determined experimentally at 295 nm between substrate and uric acid. The extinction coefficients for uric acid and xanthine were 1.2×10^4 and 2.3×10^3 cm^{-1} M^{-1}, respectively, and for hypoxanthine negligible. Xanthine oxidase activity in samples which were too turbid (such as whole milk, skim milk, samples containing sodium salicylate, and ammonium sulfate) for spectrophotometric analysis were assayed polarographically by the method of Zikakis and Treece (106) and polarographic units were converted to IU according to Zikakis (116).

6. Spectrophotometric Analysis

Avis et al. (2,3) developed three parameters for determining the purity of xanthine oxidase. These parameters
are: Protein-flavin ratio (PFR), activity-flavin ratio, and
activity-protein ratio. The PFR is by far the most
sensitive of the three indicators and it was the parameter
used in this study to monitor the progressive purification
of the enzyme, in addition to electrophoresis.

The PFR (E_{280}/E_{450}) is an absorbance reading of protein
measured at 280 nm divided by the absorbance of flavin
measured at 450 nm. The PFR depends largely on the relative
amounts of colored material and the total protein in the
sample. Once xanthine oxidase becomes the major constituent
of a preparation, a decrease in the PFR will represent an
increase in purity. In this study the PFR was determined by
either of two methods. In the first method, each eluant
fraction from chromatography was read at 280 and 450 nm.
In the second method, 25 μl of a concentrated-pooled sample
was read in 2.5 ml of 0.1M pyrophosphate buffer, pH 8.3, at
both wavelengths.

7. Isolation and Purification of Xanthine Oxidase from
   Fresh Raw Cow's Milk

This method treats the starting sample (whole milk or
cream) under the mildest conditions to prevent denaturation
of xanthine oxidase. Denaturing substances (such as
proteolytic enzymes and organic reagents) are not used in
this procedure. Also, in all published methods of xanthine
oxidase extraction, no attempt has been made to measure the
enzyme activity of the starting milk or cream. Thus, a lot
of time and effort are wasted on starting material containing
very low activity of xanthine oxidase which results in
disappointingly low enzyme yield. This fact makes any method
of xanthine oxidase purification undependable. To overcome
this difficulty, in this method the starting milk sample is
assayed for activity polarographically (106) before subject-
ing it to isolation and purification procedures. As a rule,
good yield is expected when enzyme activity in starting fresh
milk (assayed 10-30 minutes from milking) is between 60-100
μl O_2/ml/hr or raw milk which has been refrigerated (for
1-12 hours at 4°C) 140-210 μl O_2/ml/hr. In IU, the
corresponding activity values for raw fresh warm milk and
refrigerated raw milk are 39-65 ImU/ml and 92-137 ImU/ml,
respectively. Also, polarography was used to measure

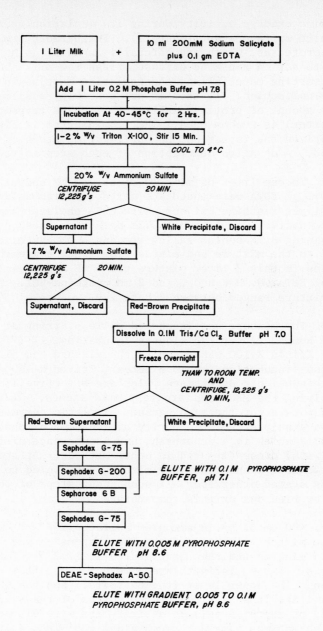

Figure 1. Flow diagram for isolation and purification of bovine milk xanthine oxidase.

xanthine oxidase activity during its isolation up to the 7%
ammonium sulfate step (Figure 1). Furthermore, to ensure
that proteases, which are naturally in milk, do not have
time to act on xanthine oxidase, fresh raw milk should be
the starting material and the isolation must begin soon
after collection. If that is not possible, addition of a
small amount of protease inhibitor (such as trypsin
inhibitor) is recommended. The flow diagram in Figure 1
summarizes the isolation and purification steps of this
method. A detailed description of each step follows.

a. To one liter of milk 10 ml of 200 mM sodium salicylate
and 0.1 gm EDTA were added and mixed. Sodium salicylate
stabilizes xanthine oxidase while EDTA chelates heavy metal
contaminants. One liter of 0.2M potassium phosphate buffer,
containing 8 mM sodium salicylate and 4 mM cysteine-HCl was
added to the mixture and mixed. The final concentration of
solutes in this 2 liter mixture was 5 mM sodium salicylate,
0.005% EDTA, 0.1M K_2HPO_4, and 2 mM cysteine-HCl. The pH of
the mixture ranged between 7.8 to 7.9.

b. The mixture was incubated while stirring at 40 to 45°C
for 2 hours. After 105 minutes incubation, 1% (V/V) Tween
80 or Triton X-100 was added to the mixture and allowed to
continue incubation for 15 minutes. Triton X-100 and Tween
80 are non-ionic detergents which are effective in
dissolving the MFGM and are very mild agents. This is
a substitute for the much harsher lipolytic enzymes (which
may adversely effect the purity of xanthine oxidase) and
butanol (which is a denaturant and a substance difficult to
work with) presently used in other methods. At the end of
the two hour incubation, the mixture was cooled to 4°C and,
unless stated otherwise, all subsequent steps of the method
were carried out at this temperature.

c. 400 gm of solid ammonium sulfate (20% W/V) was added
to the mixture with stirring. The suspension was stirred
for 15 minutes and then centrifuged at 12,225 g for 20
minutes. Three distinct layers were formed after centrifugation. The upper layer (the butterfat) and the white
precipitate (the caseins) at the bottom of the tubes were
devoid of xanthine oxidase activity and were discarded. The
supernatant was passed through glass wool into a graduated
cylinder. The filtrate appeared as an opalescent yellow
fluid which contained all the xanthine oxidase activity.

d. The concentration of ammonium sulfate in the filtered supernatant was adjusted from 20% to 27% with solid ammonium sulfate, the mixture stirred for 15 minutes, and centrifuged at 12,225 g for 20 minutes. The resultant brownish-red precipitate was dissolved in 10 to 15 ml of 0.1M Tris/CaCl$_2$ buffer (pH 7.0) containing 2 mM sodium salicylate, 0.07M CaCl$_2$, and 0.005% EDTA and stored at -20°C. The objective of this step was the precipitation of caseins (5). About 80% of the total protein in cow's milk is casein which precipitates at 20 to 26.4% (W/V) ammonium sulfate (62). This fractionation range is close to the 27% W/V (20% in step c and 7% in this step) ammonium sulfate used to precipitate xanthine oxidase in this procedure. Therefore, the inclusion of caseins in the above precipitations was unavoidable.

e. Upon thawing the mixture to 22°C, it yielded a coarse white precipitate of caseins. Upon centrifugation at 12,225 g for 20 minutes, the mixture yielded a reddish-brown super-natant and a slightly brown precipitate. The precipitate was redissolved in 0.1M Tris/CaCl$_2$ buffer and recentrifuged. The supernatant from both centrifugations was combined and showed high activity of xanthine oxidase while the white precipitate of caseins had negligible activity. It was found that the longer the preparation was frozen, the more caseins can be removed. Maximum casein precipitation occurred after about 3 to 4 weeks of storage at -20°C. Since precipitation of caseins is very low after 7 days and some decomposition of xanthine oxidase will occur even at -20°C and time is of the essence, most batches were stored from 15 hours to 1 week.

f. The active reddish-brown supernatant obtained in step e was concentrated to 5 ml on an Amicon ultrafiltration system using a XM-50 membrane designed to retain molecules of 50,000 daltons and greater. This concentrate was then applied to a Sephadex G-75 superfine column (1.5 x 125 cm) equilibrated and eluted with 0.1M pyrophosphate buffer, pH 7.1. The purpose of this chromatographic step was to desalt (remove the ammonium sulfate) and remove low molecular weight (< 75,000 daltons) impurities from the sample. All fractions were analyzed individually at 280 nm for protein and at 450 nm for FAD on either a Beckman DB or a Gilford Model 250 spectrophotometer. From this point on the enzyme activity in each fraction was measured spectrophoto-metrically at 295 nm at 23°C. Fractions with activity were pooled and concentrated by ultrafiltration to 5 ml. The pooled sample was analyzed for activity, absorption spectra, total protein, and purity by electrophoresis.

g. The pooled concentrated sample was applied to a
Sephadex G-200 column (2.5 x 100 cm) equilibrated with 0.1M
pyrophosphate buffer pH 7.1 and eluted with the same buffer.
All fractions were analyzed as in step f.

h. The fractions from the Sephadex G-200 step above
showing activity at 295 nm and flavin at 450 nm, were pooled,
concentrated, and applied to a Sepharose 6B column (2.5 x
100 cm) equilibrated and eluted with 0.1M pyrophosphate
buffer, pH 7.1. Following analyses of eluted fractions,
those with xanthine oxidase activity were pooled and concentrated to 3-5 ml.

i. The concentrated sample of step h was desalted by
passing it through a Sephadex G-75 column (0.9 x 60 cm)
equilibrated and eluted with a 0.005M sodium pyrophosphate
buffer, pH 8.6. Fractions containing the enzyme were pooled
and concentrated.

j. The concentrated sample was applied to a DEAE
Sephadex A-50 or to a DEAE Sepharose CL-6B anionic exchange
column (1.6 x 20 cm) which was equilibrated with 0.005M
sodium pyrophosphate buffer, pH 8.6. Initial elution of the
column was with the 0.005M phosphate buffer. At this pH and
salt concentration, xanthine oxidase effectively bound to
the exchanger as was apparent from the appearance of a dark
brown band in the upper 2 to 4 cm of the column and its
failure to elute in 0.005M salt. Elution of xanthine
oxidase from the column was accomplished on a linear
continuous salt gradient from 0.005M to 0.1M sodium pyrophosphate buffer, pH 8.6.
This enzyme preparation is designated as a non-
proteolytically derived xanthine oxidase (NPDXO). The
kinetic characteristics of this and a proteolytically
derived xanthine oxidase (PDXO) were determined and compared
(92,114,121).

B.   RESULTS

1. Isolation and Purification

The isolation and purification scheme shown in Figure 1
was followed to prepare NPDXO and the results obtained at
the various stages of purification are listed in Table 1.

Table 1. Average Values Obtained From Six Runs at Various Stages of Xanthine Oxidase (XO) Purification From Cow's Milk.

Procedure	Total volume (ml)	XO activity (IU/ml)	Protein (mg/ml)	Specific activity (IU/mg)	PFR[1]	Fold purification	Recovery (%)
Whole milk	1000	0.034*	32.66	0.0010	**	0	–
After buffer addition	2000	0.017*	11.32	0.0010	**	0	–
After digestion	2000	0.104*	12.60	0.0083	**	6	100
20% cut	1862	0.110*	2.26	0.0480	**	34	99
7% cut	6.5	31.30	18.90	1.6560	24.5	1125	98
G-75	6.0	18.43	18.60	0.9910	16.0	661	54
Sephacryl S-200 or Sephadex G-200	6.0	17.71	12.40	1.4280	10.0	952	51
Sepharose 6B	3.5	20.71	12.40	1.6700	8.0	1113	35
DEAE+	3.5	13.34	1.70	7.8230	4.1	4784	23

[1]Protein flavin ratio.

*Activity of samples prior to the 7% ammonium sulfate cut was determined polarographically.

**The PFR could not be calculated for samples prior to the 7% ammonium sulfate cut due to the turbidity of the sample.

Figures 2, 3, 4, and 5 are the elution profiles from the
G-75, G-200, Sepharose 6B, and DEAE Sephadex columns,
respectively. Figure 2 shows that xanthine oxidase,
emerging in the void volume, was the first eluting protein
and was followed by trailing lower molecular weight
impurities. Fractions showing activity following the
Sephadex G-75 column were pooled, concentrated, and applied
to a Sephadex G-200 column. The elution profile from this
column (Figure 3) shows that xanthine oxidase, again
emerging in the void volume, was the first protein to
elute followed closely by a shoulder of lower molecular
weight impurities. Active fractions were pooled, concentrated, and applied to a Sepharose 6B column. Figure 4
depicts the elution profile from the Sepharose 6B column.
Xanthine oxidase emerged beyond the void volume while a
protein peak of higher molecular weight proteins began
eluting prior to the enzyme and a larger peak of lower
molecular weight proteins followed the xanthine oxidase
peak. Active fractions were again pooled, concentrated,
and desalted on a Sephadex G-75 column and applied to a
DEAE Sephadex A-50 ion exchange column equilibrated in

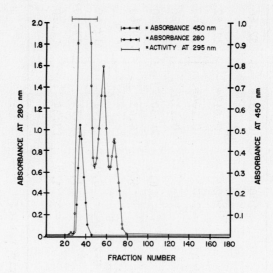

    Figure 2. Elution profile of xanthine oxidase from the
Sephadex G-75 column. Elution was with 0.1M sodium pyrophosphate buffer, pH 7.1. The volume of the fractions was
3 ml and the flow rate was 0.2 ml/min.

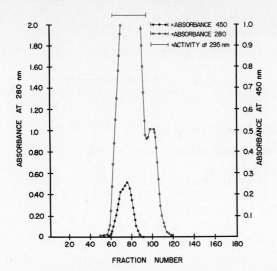

Figure 3. Elution profile of xanthine oxidase from Sephadex G-200 column. The enzyme was eluted with 0.1M sodium pyrophosphate buffer, pH 7.1. The volume of the fractions was 3 ml and the flow rate was 0.06 ml/min.

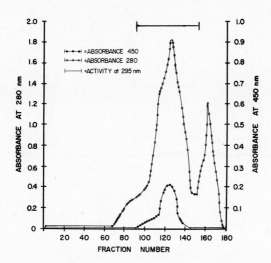

Figure 4. Elution profile of xanthine oxidase from Sepharose 6B column. Elution was with 0.1M sodium pyrophosphate buffer, pH 7.1. The volume of the fractions was 3 ml and the flow rate was 0.45 ml/min.

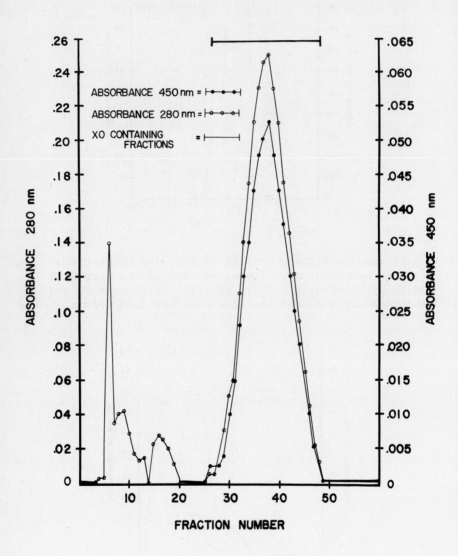

Figure 5. Elution profile of xanthine oxidase from the DEAE Sephadex A-50 column. Elution was made by a linear continuous salt gradient from 0.005M to 0.1M sodium pyrophosphate buffer, pH 8.6. The volume of the fractions was 3 ml and the flow rate was 0.75 ml/min.

0.005M pyrophosphate buffer, pH 8.6. Proteins were eluted
with a continuous linear gradient of 0.005M to 0.1M pyrophosphate buffer, pH 8.6. As shown in Figure 5, xanthine
oxidase eluted as one symmetric peak separated distinctly
from all protein impurities which eluted prior to the enzyme.

Table 1 lists the average results from six runs. The
specific activity increased as the purification advanced,
except after the Sephadex G-75 column step. The final
specific activity of the enzyme ranged from 6.2 to 11.8 and
averaged 7.8. The PFR decreased from the 7% ammonium
sulfate cut to the final chromatography on the ion exchange
column. This decrease in PFR implies a loss of 280 nm
absorbing non-xanthine oxidase proteins. The PFR of the
final pooled xanthine oxidase preparation of six runs
ranged from 2.7 to 4.8 and averaged 4.1, the lowest PFR
value ever reported. Similarly, the yield of the method for
the final pooled preparation ranged from 18-26% and
averaged 23%. Therefore, this method produces enzyme which
is on the average 20% purer (about 4800-fold purity) and its
yield is about 130% higher than the best method in the
literature (100).

2. Polyacrylamide Disc Gel Electrophoresis

PAGE was carried out on pooled fractions after
chromatography on Sephadex G-75 and G-200, Sepharose 6B, and
ion exchange columns. Purified bovine milk xanthine oxidase
from Sigma was also analyzed for comparison. Figure 6 shows
typical results of PAGE of samples from various stages in
the purification. As the purification advanced, there was a
gradual loss of non-xanthine oxidase proteins. The sample
after the ion exchange shows a single band. Additional PAGE
analyses with sample concentration increased up to 100-fold
still yielded only a single protein band. Figure 6 also
contains a sample of purified xanthine oxidase from Sigma,
which was used as a "standard". As you can see, the Sigma
preparation contained at least 14 bands making it difficult
to determine which of these is xanthine oxidase. This led
to the development of a visualization procedure to stain and
identify specifically for enzymatically active xanthine
oxidase in polyacrylamide gels using neotetrazolium chloride
(120). Using this procedure it was shown that only one of
the 14 bands in the Sigma preparation was active xanthine
oxidase; the rest were impurities. Aside from being highly
impure, Sigma xanthine oxidase migrated faster in PAGE than
did enzyme as prepared in this study (Figure 7).

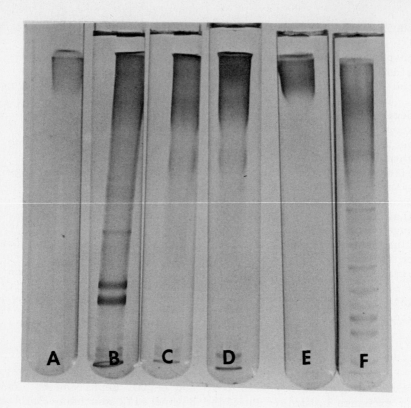

Figure 6. Polyacrylamide disc gel electrophoresis after: A, ion-exchange column; B, G-75 column; C, G-200 column; D, Sepharose 6B column; E, ion-exchange column; and F, Sigma xanthine oxidase. Gel A was stained for enzyme activity with neotetrazolium chloride. Gels B through F were stained for protein with Coomassie Blue.

C. DISCUSSION

1. Isolation

In this investigation the purification of bovine milk xanthine oxidase was accomplished using the mildest conditions possible to obtain a preparation with properties as close as possible to the native enzyme. All proteolytic purification schemes use pancreatin to degrade large casein micelles which co-elute with xanthine oxidase in subsequent purification steps. Pancreatin is a mixture of serine proteases (elastase, trypsin, chymotrypsin), leucine aminopeptidase, carboxypeptidases, and other uncharacterized

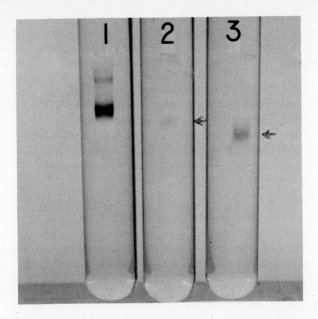

Figure 7. Polyacrylamide disc gel electrophoresis. Gel 1 is non-proteolytically derived bovine milk xanthine oxidase; Gel 2 is free human colostral xanthine oxidase; and Gel 3 is Sigma xanthine oxidase. The location of light xanthine oxidase bands in gels 2 and 3 is indicated with arrows. Gels were run simultaneously and stained for activity using neotetrazolium chloride (original color of bands is pink). The heavy molecular weight band behind the xanthine oxidase band in Gel 1 demonstrates enzymatically active xanthine oxidase which aggregates during storage; this may be a dimer, tetramer, or a larger polymer of xanthine oxidase.

proteases (49,96). These proteases are not specific for the degradation of casein micelles but exert their effect on all proteins. Furthermore, it was shown (70) that xanthine oxidase isolated in the presence of pancreatin co-purified with proteases from pancreatin thereby making it very difficult to purify xanthine oxidase.

The presence of large quantities of caseins in the final ammonium sulfate precipitation of xanthine oxidase is expected for two reasons. First, caseins display properties similar to detergents and are capable of micelle formation (101). In effect caseins act as large (polypeptide) detergent-like molecules and undergo non-ionic interactions. Since xanthine oxidase in milk is partly membrane-bound, it

is probable that there exist sites on the enzyme molecule
which are capable of undergoing non-ionic interactions with
the caseins. Second, the ammonium sulfate concentrations
used in xanthine oxidase isolation overlap with the concentration used to isolate caseins (62). However, we have
demonstrated in this study that the removal of casein
micelles can be accomplished without proteolysis. The
caseins were removed in the first step by the phosphate
buffer containing cysteine-HCl. The remaining caseins were
removed in step d by adding to the brownish-red precipitate
$0.1M$ Tris/CaCl$_2$ and storing the mixture at $-20°C$ for a
period from 15 hours to 1 week. In step e the thawed mixture
yielded a coarse white precipitate of caseins. Lipolytic
enzymes and butanol, which have been used to remove xanthine
oxidase from MFGM, were replaced by cysteine-HCl and Triton
X-100 (or Tween 80). The effectiveness of these mild agents
in increasing the yield of the enzyme is shown in Table 1.
After digestion, the yield of xanthine oxidase increased by
a factor of approximately 6-fold (2000 ml x 0.017 IU/ml =
34 total units vs. 2000 ml x 0.104 IU/ml = 207 total units).
Without Triton X-100 and cysteine-HCl, there was only a
slight increase in the yield of enzyme after the digestion
step. Therefore, Triton X-100 and cysteine-HCl were
effective agents for the release of xanthine oxidase from
the MFGM.

2. Purification

In virtually all purification methods in the literature,
desalting of the final ammonium sulfate precipitate is
performed through Sephadex G-25. In this investigation
desalting was done on a Sephadex G-75 column. The reason
for this change was that the use of G-75 extends the range
of separation from 5000 to 80,000 daltons. Thus, non-
micellar caseins (17,000 to 27,000 daltons) and other low
molecular weight impurities could be separated from xanthine
oxidase which elutes in the void volume.

Samples assayed after each chromatographic step showed
a lower PFR than the preceding step. This indicates that
non-xanthine oxidase protein contaminants absorbing at 280
nm were gradually and effectively removed through the large
fractionation range employed in the purification steps.
The elution profiles from Sephadex G-75, G-200, and Sepharose
6B chromatography indicate that the majority of the non-
xanthine oxidase contaminants eluted beyond xanthine oxidase
and were of lower molecular weight than xanthine oxidase.
As increased amounts of contaminants were removed, the PFR
decreased and reached its lowest point after the ion

exchange step. The final non-xanthine oxidase proteins from
the ion exchange column eluted ahead of the enzyme. Since
elution was done in a linear continuous salt gradient,
proteins which eluted ahead of xanthine oxidase had a lower
net negative charge and, therefore, a lower attraction for
the DEAE$^+$ exchanger at pH 8.6. These non-xanthine oxidase
proteins were very near their isoelectric point, and even a
slight increase in the salt concentration would interfere
with binding to the ion exchanger, allowing these proteins
to elute early in the elution gradient.

3. Electrophoresis

The results of PAGE on samples tested at each
chromatographic step in the purification of xanthine
oxidase are shown in Figure 6. The majority of the non-
xanthine oxidase proteins which eluted with the enzyme in
the void volume after the Sephadex G-75 step were removed
after chromatography on the Sephadex G-200 column. This
correlates well with the large decrease in the PFR (from
16.0 to 10.0) after passage of the pooled G-75 fractions
through the Sephadex G-200 column. A further reduction in
the PFR (from 10.0 to 8.0) after the Sepharose 6B column
also correlates well with the removal of non-xanthine
oxidase proteins at this step as shown by PAGE (Figure 6).
Further analysis by PAGE of the pooled sample after the ion
exchange column showed no residual impurities (Figure 6E).
In order to highlight possible non-xanthine oxidase protein
impurities, the protein concentration of the pooled sample
after ion exchange chromatography was increased nearly 100-
fold and still yielded a single protein band in PAGE. This
single protein band was identified by the neotetrazolium
chloride reaction to be active xanthine oxidase (Figures
6A and 7).

PAGE of a commercially available purified bovine milk
xanthine oxidase (obtained from Sigma Fine Chemicals),
demonstrated that this preparation was highly impure
(Figure 6F). In addition, the Sigma enzyme migrated faster
in PAGE than did the enzyme from this study (Figure 7).
Sigma would not divulge information pertaining to their
purification of the milk enzyme (59). However, since
previous research (68,100) has shown that PDXO yields
xanthine oxidase that migrates faster in PAGE than milk
enzyme prepared by non-proteolytic means, it can be assumed
that the Sigma enzyme was obtained by proteolytic digestion
and was impure.

In summary, the present purification method provides
high purity xanthine oxidase with high yield. The PFR and

the specific activity of the final pooled preparation were
4.1 and 7.8 IU/mg, respectively, had one symmetric peak by
ion exchange chromatography, and a single protein band by
PAGE which was shown by a differential staining technique
to be enzymatically active xanthine oxidase. The yield of
this method ranged from 18 to 26% and averaged 23%. Therefore, this method produces xanthine oxidase which is on the
average about 20% purer (about 4800-fold purity) and yields
about 130% more enzyme than the best available method in
literature (100).

4. The Effect of Proteolysis on Xanthine Oxidase

From previous studies (48,67,68,72,100,115,118) it has
been established that purified bovine milk xanthine oxidase
differs according to the purification method, and that
proteolysis adversely affects the enzyme. Furthermore,
PDXO migrated faster in PAGE, and in SDS electrophoresis it
resolved in subunits with lower molecular weight than
NPDXO. In addition to the above enzyme modifications
induced by proteolysis, in recent studies Zikakis and
Silver (121) have found significant kinetic differences
between PDXO and NPDXO. In this section, we will use these
kinetic data and data from the literature to explain how
proteolysis may alter the molecular structure of the milk
enzyme.

In 1976, Nagler and Vartanyan (67,68) proposed a
structural model of a native molecule of xanthine oxidase
(Figure 8) based on a comparison study of PDXO and NPDXO
preparations. These investigators demonstrated (68) that
pancreatin (used in PDXO) cleaved four segments of the
primary structure of xanthine oxidase with a total molecular
weight of about 12,000 daltons. Although this modification
does not appear to change significantly the catalytic
activity of the enzyme, it reduces total molecular weight
and alters its subunit structure. Using SDS electrophoresis,
Nagler and Vartanyan (68) found that NPDXO resolved in a
single subunit of 150,000 molecular weight whereas PDXO
showed a molybdenum containing subunit approximately 92,000
daltons, an FAD containing subunit of about 42,000 daltons,
and a third subunit of about 20,000 daltons.

Aside from the migrational differences observed between
PDXO and NPDXO, there are several noteworthy kinetic
differences which may be attributed to molecular changes
resulting from proteolysis. In kinetic comparisons between
PDXO and NPDXO preparations (121) it was found that in sub-
saturating concentration of substrate in 0.1M Tris/HCl
buffer, the PDXO showed significantly suppressed activity

relative to activity found in 0.1M pyrophosphate buffer. In contrast, NPDXO activity was unaffected in Tris/HCl buffer. This suggests that proteolysis, by modifying the enzyme's primary structure, has reduced the enzyme's conformational stability. Furthermore, substantial differences were observed between the two preparations in Km and Vmax. With xanthine at pH 8.3, the Km and Vmax values for NPDXO and PDXO were 4.1×10^{-5}M and 4.65 IU/ml and 6.2×10^{-5}M and 1.88 IU/ml, respectively. This indicates that NPDXO had a higher affinity (51%) and Vmax (147%) than PDXO. A similar trend was observed at pH 8.7 with the two preparations. Moreover, when hypoxanthine was the substrate, at pH 8.3 again NPDXO demonstrated higher affinity (lower Km) and a higher Vmax value. This increase in affinity (lower K_I) with NPDXO was also observed with three competitive inhibitors (folic acid, allopurinol, and pterin-6-carboxylic acid).

These kinetic observations may be directly related to the arrangement of the globules described in Nagler and

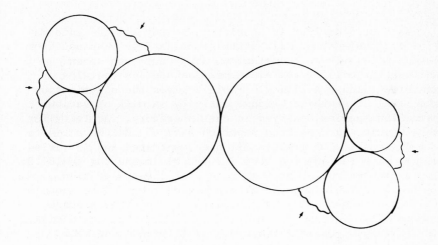

Figure 8. Molecular model for native xanthine oxidase proposed by Nagler and Vartanyan (68). Arrows indicate 4 protein segments removed by proteolysis.

Vartanyan's molecular model (68), shown in Figure 8. For optimum catalytic efficiency, the electron transport chain at the active site of the enzyme requires that the molybdenum and FAD containing globules must be within a favorable proximity of each other. Such an arrangement is consistent with the model generated from the magnetic interaction measurements proposed by Barber et al. (6). If the protein segments described in Nagler and Vartanyan's model are represented as segments of the enzyme's primary structure which closely interact with the amino acid side chains of the globules, then it is possible that these segments might have a more effective structure-to-function relationship to the intact enzyme molecule than realized heretofore. This hypothesis suggests that NPDXO is as close as possible to the native milk enzyme. Such a conclusion is supported by NPDXO's greater catalytic efficiency and higher affinity for both substrates and competitive inhibitors.

II.   CAPRINE MILK XANTHINE OXIDASE

As stated earlier in Section I, nearly all the information available about milk xanthine oxidase is obtained from bovine milk. In most instances, it is not even known whether the milk of certain mammalian species contains xanthine oxidase activity. Table 2 gives the distribution and level of xanthine oxidase activity in milk of various mammalian species, assayed in our laboratory. Activity for most species has not been reported before. Enzyme activity was present in the milk of all species tested and it varied widely with the species from 3.3 ImU/ml in the dog to 187.2 ImU/ml in the rat. The goat is one of the few species included in Table 2 whose milk has been reported to contain xanthine oxidase activity (29,64,65,79). On the average, goat's milk contains 10-fold less activity than cow's milk.

In view of Oster's theory (75-77) implicating milk xanthine oxidase as an initiating cause of atherosclerosis and the fact that a greater portion of the world's population consumes goat's milk than cow's milk (26), it is important to know more about goat's milk xanthine oxidase. The objective of this section is to develop isolation and purification procedures for caprine milk xanthine oxidase and partially characterize it. A preliminary account of parts of this study has appeared (20).

Table 2. Distribution of Xanthine Oxidase (XO) Activity in Milk of Various Mammalian Species.

Species	Xanthine oxidase activity[1] (μl O_2/ml/hr)		(ImU/ml)
	Range	$\bar{X}$	$\bar{X}$
Rat	161–385	281	187.2
Guinea Pig	120–326	225	149.9
Cow	94–261	156	103.9
Rabbit	93–256	149	99.2
Donkey	89–169	121	80.6
Mouse	80–189	119	79.3
Horse	11–32	23	15.3
Goat	10–28	16	10.7
Sheep	14–31	15	9.9
Man (milk)	0.2–21	11	7.3
Man (colostrum)	12–45	31	20.6
Cat	0.1–29	12	7.9
Patas monkey	2–16	11	7.3
Dog	2–9	5	3.3

[1]Raw whole milk or colostrum was collected fresh and stored at -20°C for 24 hrs before assaying polarographically. The mean activity (from at least 2 animals) was converted to ImU/ml (116) and shown in the second $\bar{X}$ column.

A. MATERIALS AND METHODS

In addition to the reagents listed under Materials and
Methods, Section I, the following materials were used.
Sephadex G-25 was purchased from Pharmacia Fine Chemicals,
Piscataway, NJ. Ultrafiltration membrane filters type
DP045 were purchased from Amicon Corp., Lexington, Mass.
Fresh raw goat's milk was obtained from AMYR Farms,
Oxford, PA.

1. Chromatography, total protein determination, PAGE,
enzyme assays, and spectrophotometric analyses were performed as described under Materials and Methods, Section I.

2. Ultrafiltration

This procedure was performed as described under
Materials and Methods, Section I. However, Amicon membrane
filters DP045 (with pore size of 0.45 μm) were used in
addition to XM-100A.

3. Determination of the pH Optimum

Xanthine oxidase activity in pH range of 4.0-6.5 was
surveyed with 0.1M acetate buffer. The pH range of 6.5-7.5
was surveyed with 0.1 sodium phosphate buffer. A 0.1M
sodium pyrophosphate buffer was employed to assay activity
in the pH range of 7.5-10. These buffers were prepared by
making the acid and basic components of each buffer at the
proper molarity and titrating them until the desired pH was
obtained. Enzyme samples were incubated at the desired pH
for 10 minutes at 23°C before adding saturating amounts of
xanthine. The final concentration of xanthine was 2.5 x
10^{-4}M. Every 0.1 pH interval was assayed from 4.0-10.0.

4. Amino Acid Analyses

The purified goat milk xanthine oxidase preparation was
analyzed for amino acid composition according to the methods
of Blackburn (13) and Benson (8). Analysis was performed
in a Durrum D-500 Amino Acid Analyzer.

5. Isolation and Purification of Xanthine Oxidase from
   Fresh Raw Goat's Cream

Enzyme activity of whole milk and blended buffer-fat
fraction was determined polarographically. After this
point, enzyme activity was monitored polarographically

and spectrophotometrically.

a. 1100 ml of fresh pooled whole milk was cooled to 4°C
and centrifuged at 12,225 g for 20 minutes. The fat layer
formed was removed and added to 500 ml of 0.2M potassium
phosphate buffer, pH 7.8. The total volume of the mixture
was brought up to 662 ml by adding 32.5 ml of 0.2M sodium
salicylate. Then, 33.0 mg of EDTA and 1.1636 gm of cysteine-
HCl were added to the mixture. The final concentration of
solutes in this mixture was 9.8 mM salicylate, 0.005% EDTA,
10 mM cysteine-HCl, and 0.15M K_2HPO_4. The pH of the mixture
was 7.5. This mixture was designated as fat-buffer fraction
and contained very high xanthine oxidase activity.

b. This fraction was incubated while stirring at 40–45°C
for 2 hours. Following incubation, the mixture was cooled
to 4°C and blended in a Shetland blender at high speed for
30 seconds. This step was used to rupture the MFGM and
release the enzyme into the aqueous phase. Unless stated
otherwise, all subsequent steps of the method were carried
out at this temperature. Following blending, the mixture
was centrifuged at 12,225 g for 20 minutes. The mixture
separated into an upper fat layer, a clear brown supernatant, and a small white precipitate composed chiefly of
caseins. The fat layer was lifted with a spatula and the
supernatant containing the enzyme was filtered through a
plug of glass wool and sterile cheesecloth. The filtrate
was concentrated to 10 ml by two consecutive ultrafiltrations, first through a DPO45 membrane filter and then
through an XM-100A. This fraction was designated as xanthine
oxidase isolate.

c. The concentrated xanthine oxidase isolate was applied
to a Sepharose 6B column (2.5 x 100 cm) equilibrated and
eluted with 0.1M sodium pyrophosphate buffer, pH 7.1. All
fractions were analyzed individually at 280 nm for protein
and at 450 nm for FAD. Activity in each fraction was
measured spectrophotometrically at 295 nm at 23°C.
Fractions with activity were pooled and concentrated by
ultrafiltration. The enzyme activity of the concentrated
fraction was also measured.

d. The concentrated sample from the Sepharose 6B column
was desalted by passage through a Sephadex G-25 column
(2 x 50 cm) equilibrated and eluted with 0.005M pyrophosphate buffer, pH 8.6. Active fractions were assayed,
pooled, and concentrated as before.

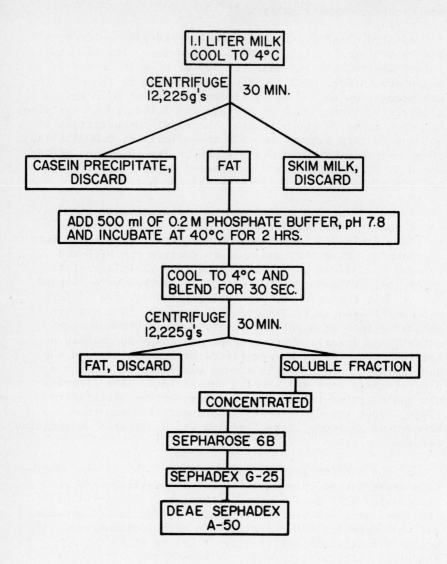

Figure 9.   Flow diagram for the isolation and purification of xanthine oxidase from fresh goat's raw cream.

e. The concentrated desalted xanthine oxidase sample
was applied to a DEAE Sephadex A-50 column (1.6 x 20 cm)
equilibrated with 0.005M pyrophosphate buffer, pH 8.6. The
column was washed with the same buffer to elute non-xanthine
oxidase proteins. Elution of bound xanthine oxidase from
the column was accomplished by a linear continuous salt
gradient from 0.005M to 0.1M sodium pyrophosphate buffer,
pH 8.6.

B. RESULTS AND DISCUSSION

1. Isolation and Purification

Figure 9 shows the flow diagram scheme followed for the
isolation and purification of xanthine oxidase from goat's
cream. The results from a typical run are summarized in
Table 3. Figure 10 represents the elution profile from
Sepharose 6B column showing several protein peaks, two of
which had strong absorbance at 450 nm. The first of these
two peaks eluted in the void volume at or near fraction
tube 40 and although absorbed at 450 nm, it had no xanthine
oxidase activity. Since this peak had low protein concentration, it was assumed that it contained the residual
MFGM material which was released during the isolation
procedure. Xanthine oxidase emerged beyond the void volume
followed by three lower molecular weight protein impurities.
The enzyme was contained in tube fractions 82-94 as a clear
brown fluid, characteristic of xanthine oxidase solutions.
The concentrated active fraction contained 60% of the
recovered enzyme activity and its purity was increased from
302-fold (of the isolate) to 1211-fold. The specific
activity was also increased from 18.6 to 74.8 ImU/mg. This
is indicative that the Sepharose 6B column was effective in
separating xanthine oxidase from several lower and higher
molecular weight protein impurities.

The concentrated fraction from the Sepharose column was
desalted by passage through Sephadex G-25 and eluted with
0.005M pyrophosphate buffer, pH 8.6. Active fractions were
pooled and concentrated. This step reduced slightly the
purity and recovery was 53%. Figure 11 shows the elution
profile from DEAE Sephadex column. In the initial washing
with 0.005M pyrophosphate, a few minor protein impurities
were eluted. Following washing with the gradient buffer,
xanthine oxidase emerged as a symmetric peak. Near the end
of the xanthine oxidase elution peak, protein impurities
emerged as a shoulder. Active fractions 45-54 were pooled,
concentrated, and analyzed. All other active fractions were

Table 3. Characterization of the Isolation and Purification Procedures for Xanthine Oxidase (XO) From Whole Goat's Milk.

Procedure	Total volume (ml)	XO activity (ImU/ml)	Protein (mg/ml)	Specific activity (ImU/mg)	PFR[1]	Fold Purifi- cation	Recovery (%)
Whole milk	1100.0	1.08*	17.55	0.062	**	0	–
Fat-buffer	662	25.50*	7.50	3.400	**	123	–
Isolate	10.0	453.00	24.30	18.600	19.64	302	100
Sepharose 6B	6.3	437.30	5.85	74.800	10.40	1211	60
G-25	3.0	795.20	11.70	68.000	7.70	1101	53
DEAE[+]	4.5	237.70	1.93	123.000	3.71	1996	24

[1]Protein flavin ratio.

*Activity of samples prior to the isolate was determined polarographically.

**The PFR could not be calculated for samples prior to the isolate due to the turbidity of the sample.

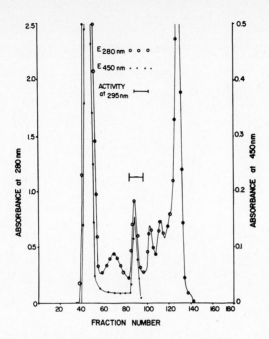

Figure 10. Elution profile of goat xanthine oxidase from Sepharose 6B column. The volume of the fractions was 3 ml and the flow rate was 0.1 ml/min.

excluded from this preparation to avoid fractions with undesirable low specific activity and PFR. Analyses showed that the DEAE^{+} fraction had a specific activity of 123 ImU/mg, a PFR of about 3.7, an increase in purity of nearly 2000-fold, and a recovery of 24%.

PAGE with samples from various purification steps, showed a progressive removal of protein impurities. The final fraction from the DEAE^{+} column contained a single broad band which was identified by neotetrazolium chloride differential staining to be active xanthine oxidase. Even though the concentration of the protein sample applied on polyacrylamide gels was increased 100-fold, only a single protein band resolved. The data indicate that the method is effective in removing the non-xanthine oxidase impurities. Samples assayed after each step (Table 3) showed higher specific activity, higher fold purification, and lower PFR than the preceding step (except the desalting step). The method has a final yield of 24% and the product is of high purity as judged by its PFR, analysis on PAGE,

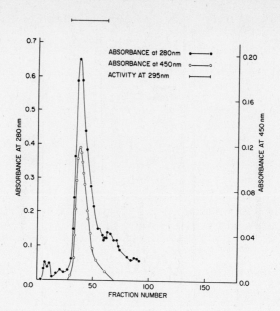

Figure 11. Elution profile from the DEAE Sephadex A-50 column. Desorption of xanthine oxidase was by a linear continuous salt gradient from 0.005M to 0.1M pyrophosphate buffer, pH 8.6. The volume of the fractions was 2.5 ml and the flow rate was 0.1 ml/min.

purification, and specific activity. Furthermore, this was accomplished without use of proteolytic and lipolytic enzymes and denaturing organic reagents.

2. Amino Acid Composition

Table 4 contains the amino acid composition of the final purified goat milk enzyme along with the composition of bovine milk xanthine oxidase taken from Nelson and Handler (72). There are many similarities between the two enzyme sources. Caprine milk xanthine oxidase contained higher amounts of aspartic acid, glutamic acid, proline, and glycine, and a lower amount of serine than bovine milk xanthine oxidase.

3. Determination of the pH Optimum

The pH profile of goat xanthine oxidase is shown in Figure 12. The pH optimum for this enzyme was 8.35 which is

Table 4. Amino Acid Composition of Caprine and Bovine Milk Xanthine Oxidase.

Amino acid[1]	Amino acid content (mole %)	
	Goat[2]	Nelson & Handler 1968 Bovine[3]
Lysine	7.0	6.9
Histidine	2.5	2.3
Arginine	5.0	4.7
Aspartic acid	9.1	8.6
Threonine	6.7˙	7.0
Serine	5.4	6.5
Glutamic acid	10.7	10.0
Proline	6.6	5.5
Glycine	8.7	8.2
Alanine	7.7	7.6
Valine	6.8	6.8
Methionine	2.0	2.2
Isoleucine	5.1	4.8
Leucine	9.3	8.9
Tyrosine	2.3	2.4
Phenylalanine	5.0	4.9

[1] Cysteine and tryptophan were not determined.

[2] Values represent averages from duplicate analyses.

[3] Original values were given in number of residues per mole which were recalculated to moles % for comparison with data from this study.

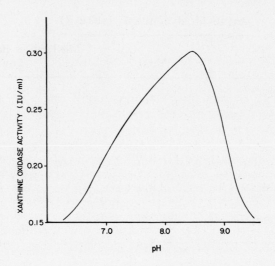

Figure 12.   pH profile for goat's cream xanthine oxidase.

close to the bovine milk xanthine oxidase, 8.3, and
slightly higher than the free human colostral enzyme, 8.2.

4. Enzyme Cofactor and Stability

FAD was identified as a cofactor of the goat milk
xanthine oxidase. This cofactor was identified spectrophotometrically at 450 nm and by differential staining in
PAGE using neotetrazolium chloride.

To check enzyme stability, purified samples were stored
at 4°C and -20°C and periodically checked for loss of
activity. The enzyme was stored in 0.1M pyrophosphate
buffer, pH 7.1, containing 5 mM sodium salicylate, 3.0 mM
cysteine-HCl, and 0.005% EDTA. At -20°C, 27% of the activity
was lost after 2 weeks, 51% after 4 weeks, and 89% after 12
weeks of storage. At 4°C, activity was lost at a faster
rate. 31% of activity was lost in 6 days, 54% in 12 days,
and 72% in 16 days of storage at 4°C.

III. HUMAN MILK XANTHINE OXIDASE

Prior to 1976, there was considerable debate over whether or not xanthine oxidase existed in human milk. Rodkey and Ball (83) found that the enzyme was present in low levels in human milk while Modi (65) reported that it was absent. Bradley and Gunther (16), using a more sensitive assay, found the enzyme present in milk. However, their work was criticized because bacteria normally in milk contain xanthine oxidase (34,39,97) and, therefore, the xanthine oxidase activity measured by Bradley and Gunther may have been of bacterial origin. Because of these discrepancies, Zikakis et al. (110) assayed polarographically and radiochemically 59 individual human milk and colostrum samples for xanthine oxidase activity. They found activity in all samples; activity varied widely among individuals and colostrum showed the highest activity. Nearly all of the enzyme was contained in the cream fraction. From diagnostic bacteriological tests, the pH optimum of the milk enzyme, and its stability at -20°C, they demonstrated conclusively that the enzyme is not of bacterial origin. Furthermore, it was shown that xanthine oxidase activity is very low immediately after birth, rises to a sharp peak at day 3 to 4 postpartum, rapidly declines to a very low base level at day 6 to 7, and thereafter remains at base levels throughout the lactation (110). Although the significance of this cycle is not known, the purpose of the high xanthine oxidase activity in human colostrum may be to effect maximum absorption of dietary iron from the undeveloped infant gut (refer to INTRODUCTION pp. 3 and 4).

Notwithstanding the above evidence for the presence of xanthine oxidase in human milk and colostrum, some investigators remained skeptical and unconvinced. To settle this question it would be necessary to isolate the enzyme from human milk or colostrum. Our objective in this section is to develop methodologies for the isolation and partial purification of xanthine oxidase from human colostrum (since it is a richer source than milk). Because most of the enzyme is contained in the cream fraction (110), we will isolate the free and membrane-bound enzymes and partially characterize them. A preliminary account of parts of this study has appeared (38).

A. MATERIALS AND METHODS

In addition to the reagents listed under Materials and

Methods, Section I, the following items were used. The following reagents and materials were purchased from Fisher Scientific Co., King of Prussia, PA: Standard buffer solutions and seamless cellulose dialyzer tubing. P-nitrophenol phosphate was from Sigma Fine Chemicals Co., St. Louis, MO. Sepharose 2B (pre-swollen), DEAE Sephacel, and high molecular weight standard proteins were purchased from Pharmacia Fine Chemicals, Piscataway, NJ. Ultrafiltration membrane filters type DP02 were obtained from Amicon Corp., Lexington, Mass. Other reagents and solvents were of reagent grade. Glass distilled-deionized water was used throughout. Human colostrum samples were obtained from the Mother's Milk Bank of Wilmington Medical Center, Wilmington, DE. The Milk Bank collected colostrum from individual donors and pooled the samples in approximately 300 ml aliquots. Samples which were designated for this study were not pasteurized and were stored at -20°C until used, usually within 10-50 days.

1. Gel Filtration Chromatography

Sepharose 6B was equilibrated in 0.1M pyrophosphate buffer, pH 7.1. A flow adaptor was inserted and samples were eluted with an upward flow. A 3-way valve was used to facilitate sample application. As with all columns, a peristaltic pump was used to control the flow rate. Sepharose 2B was also equilibrated and eluted with 0.1M pyrophosphate buffer, pH 7.1. In this case, no flow adaptor was used and samples were eluted with a downward flow. The volume of fractions for both columns was 5 ml.

2. Ion Exchange Chromatography

DEAE Sephacel was initially washed with 0.1M pyrophosphate buffer, pH 8.6, in order to exchange the original chloride counter ions for new phosphate counter ions. The gel was then equilibrated with the starting buffer, 0.005M pyrophosphate pH 8.6. Samples were dialyzed against the starting buffer and applied to the column. The column was then washed with 4-5 column volumes of the starting buffer and non-adsorbed proteins were eluted. Selective desorption of bound proteins was achieved by a linear continuous salt gradient from 0.005M to 0.1M sodium pyrophosphate, pH 8.6. The volume of fractions collected was 3 ml.

3. Ultrafiltration and Total Protein Concentration

These steps were carried out as described under
Materials and Methods, Section I. In addition to the XM-50,
Amicon membrane filter DPO2 was used under a pressure of
0.7 kg/cm^2,

4. Polyacrylamide Disc Gel Electrophoresis

The purity, electrophoretic mobility, and detection of
enzymatic activity were monitored by PAGE as described under
Materials and Methods, Section I.

5. Enzyme Assays

Polarographic and spectrophotometric assays for enzyme
activity were as described under Materials and Methods,
Section I. The only exceptions were: In the spectrophotometric method, the assay temperature was 30°C and the pH
at the optimum of the enzyme.

6. Isolation of Human Colostral Xanthine Oxidase

a. Preparation of Membrane Material. Frozen colostrum
was thawed, mixed with a solution containing 0.01% EDTA and
1% 0.2M sodium salicylate, and centrifuged at 5000 g for
10 minutes at 20°C. The resulting fat layer was dispersed
in 30-40 ml of 0.1M pyrophosphate buffer, pH 7.1. The
mixture was churned for 10 minutes in a tissue grinder and
recentrifuged as before. This process of churning and
centrifuging was repeated until little or no enzyme activity
remained in the fat. The supernatants from each centrifuging
step were pooled and centrifuged at 12,000 g for 20 minutes
at 4°C. The resulting supernatant was filtered through a
glass wool plug and ultrafiltered using a DPO2 membrane
until the retentate reached a volume of 30 ml or less. The
ultrafiltration cell was filled with 0.1M pyrophosphate
buffer, pH 7.1, and again ultrafiltered. The retentate was
washed twice in this manner and dispersed in 50-100 ml
pyrophosphate buffer. This fraction was referred to as
membrane material.

b. Separation of Free and Membrane-Bound Xanthine.
Oxidases from Membrane Material. A 10 ml fraction of
the membrane material was applied to a Sepharose 2B column
(2.5 x 100 cm), eluted with 0.1M pyrophosphate buffer, and
collected in 5 ml fractions. All fractions were analyzed
individually at 280 nm for protein and for xanthine oxidase

activity spectrophotometrically at 295 nm at 30°C.
Fractions with activity were pooled and concentrated by
ultrafiltration using an XM-50 membrane.

After separation of free from membrane-bound xanthine
oxidase, the concentrated active fraction of free enzyme
was further purified through a Sepharose 6B (2.5 x 100 cm).
The most active fractions were dialyzed and applied to a
DEAE Sephacel column (2.5 x 20 cm). Xanthine oxidase
adsorbed to the column was eluted by a continuous linear
salt gradient from 0.005M to 0.1M pyrophosphate buffer, pH
8.6. Active fractions were pooled and concentrated. This
preparation was designated as free xanthine oxidase.

7. Characterization of Colostral Xanthine Oxidase

a. Determination of Molecular Weight by Gel Filtration.
A Sepharose 6B column (2.5 x 100 cm) was standarized with
the following molecular weight standards: Aldolase 158,000,
catalase 232,000, and thyroglobulin 669,000. Each standard
protein was dissolved to a concentration of 20 mg/ml in
0.1M pyrophosphate buffer, pH 7.1. The standards were
applied to the column and eluted with the same buffer in
an upward flow. The peak of activity of xanthine oxidase
eluting from this column was used to calculate its Kav from
the formula: Kav = Ve-Vo/Vt-Vo where, Ve = elution volume,
Vt = total volume, and Vo = void volume. The molecular
weight of the enzyme was determined from a plot of the Kav
versus the logarithm of the molecular weight for the
above standard proteins.

b. Determination of the pH Optimum. Solutions of 0.1M
pyrophosphate buffer were made up for every 0.5 pH interval
between pH 4.0 and 10.0. Xanthine oxidase samples were
incubated at the desired pH for 10 minutes at 30°C before
adding saturating amounts of xanthine. The final concentration of xanthine was 2.5×10^{-4}M. Critical areas were
assayed at every 0.1 pH interval. The optimum pH and the
pH at 1/2 Vmax were determined from a plot of pH versus
Vmax. The pH at 1/2 Vmax was used to estimate the pK's of
the amino acids which may be involved in catalysis.

c. The Effect of Temperature. Xanthine oxidase samples
were incubated at every 4°C intervals between 15°C and 35°C
for 10 minutes at the previously determined pH optimum of
the enzyme before adding saturating amount of xanthine. The
final concentration of xanthine was 2.5×10^{-4}M. Arrhenius
plots, log Vmax versus 1/°K, were constructed and the
activation energy (Ea) and the Q10 were determined.

Activation energies were calculated from the equation: Ea =
-b (2.3)R where, R = 1.98. The Q10 was calculated from the
equation: Q10 = Vmax t°/Vmax t° + 10.

   d. Determination of Km for Hypoxanthine and Xanthine.
Initial velocities were determined at final substrate concentrations ranging from 2.5 x 10^{-6}M to 10.0 x 10^{-6}M for
hypoxanthine and 1.0 x 10^{-5}M to 10.0 x 10^{-5}M for xanthine.
Assays were performed at 30°C at previously determined pH
optima, 8.2 for free xanthine oxidase and 8.5 for membrane-
bound enzyme. Lineweaver-Burk plots were constructed by
plotting 1/v_i versus 1/[S]. A linear regression program
which uses the method of least squares was employed to
determine the Km and Vmax. Vmax was determined from the
Y-intercept, and Km from the slope according to the
relationship of b = Km/Vmax. Standard error for Km and
Vmax was calculated as described by Wilkinson (104).

                  B. RESULTS AND DISCUSSION

   1. Isolation of Humam Colostral Xanthine Oxidase

   a. Preparation of Membrane Material. Table 5 contains
the results from a representative isolation run starting
with 275 ml colostrum. The isolated membrane material
contained nearly 3.5-fold more xanthine oxidase activity
than did the original colostrum. This observation is in
agreement with bovine milk xanthine oxidase studies that
have shown when the enzyme is associated with MFGM, its
catalytic action is restricted (5,45,64) and that certain
treatments can disrupt the MFGM and thereby increase enzyme
activity (43,45,105,117). It appears that some xanthine
oxidase molecules may be associated with the inner surface
of the MFGM and become accessible for catalysis only after
this portion of the membrane is exposed by the removal of
the central core of triglycerides. Another similarity with
bovine milk is that freeze and thaw increased xanthine
oxidase activity by releasing the enzyme from the MFGM.
Approximately 98% of the initial accountable activity was
found in the isolated membrane material, 1.8% in the
discarded fat,and none in the filtrate. Following thawing
and centrifuging of the frozen colostrum, about 55% of
xanthine oxidase activity appeared in the aqueous phase
and 45% in the lipid phase.
   More than 95% of the lipids in human milk, and other
milks, are triglycerides which are part of the MFGM (80).
These triglycerides must be removed before chromatography

Table 5. Characterization of the Isolation Procedure for Membrane Material From Human Colostrum.

Fraction	Total volume (ml)	Total protein (mg/ml)	XO activity[1] (ImU/ml)	Total activity (ImU)	Specific activity (ImU/mg)
Colostrum	275	15.00	4.69	1291.40	.31
Supernatant	234	15.00	6.39	1497.36	.42
Fat	122	9.50	11.65	1421.42	1.22
Combined supernatants	425	10.50	10.23	4348.60	.97
Discard fat	24	1.93	3.33	80.01	1.73
Centrifuge 12,000 g	418	8.00	10.41	4354.30	1.30
Discard filtrate	965	1.13	0.00	0.00	0.00
Membrane material	85	25.50	52.23	4439.89	2.04

[1]Activities were determined polarographically.

in order to avoid column clogging and to facilitate
separation of proteins. In this study, we avoided chemical
methods (which use reagents that denature proteins) for
the removal of triglycerides. Instead, we used such
physical methods as freeze-thawing and churning. After
centrifugation at 20°C, the destabilized fat emulsion rose
to the surface while the membrane material remained in the
aqueous phase. The high temperature for centrifugation was
chosen so that the membranous material would not be
entrained in the lipid phase and to induce the lipid to
rise to the surface as an oil.

Assuming that the diameter of membrane material ranges
from 1-10 µm (80), the Amicon DPO2 membrane filter with
pore size of 0.2 µm should retain the membrane material in
the retentate while allowing free xanthine oxidase to pass
in the filtrate. Since all xanthine oxidase activity
remained in the retentate, apparently the enzyme did not
dissociate from the membranous material under these conditions. An alternative explanation is that the enzyme
may have been prevented from passing the DPO2 filter
because of non-covalent interactions between caseins and
xanthine oxidase (71).

b. Separation of Free and Membrane-Bound Xanthine
    Oxidases. The results from the chromatographic
purification of free and membrane-bound xanthine oxidases
from membrane material are listed in Table 6. The elution
profile (Figure 13) of membrane material through Sepharose
2B column shows two protein peaks both of which contained
xanthine oxidase activity. The first peak (Peak 1)
emerged in the void volume, was cloudy, consisted of intact
membrane material, and contained about 20% of the recovered
xanthine oxidase activity. The second peak (Peak 2) was
clear, consisted of free xanthine oxidase and other lower
molecular weight proteins which passed through the DPO2
filter, and contained about 78% of the recovered enzyme
activity. This indicates that the Sepharose 2B column was
effective in dissociating about 78% of xanthine oxidase
from the membrane material or other complex. Other
investigators (9,19) have described similar results with
bovine buttermilk xanthine oxidase. The fact that 78% of
xanthine oxidase eluted as free enzyme suggests that the
majority of the enzyme is bound to the membrane as an
extrinsic protein.

The peak 2 fraction of free xanthine oxidase was
further purified by passing it through Sepharose 6B. The
elution profile (Figure 14) from this column contained two
major protein peaks with the peak for xanthine oxidase

Table 6. Characterization of the Purification Procedure for Free and Membrane-Bound Xanthine
Oxidase (XO) from Membrane Material.

Fraction	Total volume (ml)	Total protein (mg/ml)	XO activity[1] (ImU/ml)	Total activity (ImU)	Specific activity (ImU/mg)	Recovery (%)
Membrane material	10.0	25.50	62.75	627.58	2.46	100
Sepharose 2B:						
Peak 1	7.6	1.33	16.18	123.01	12.21	20
Peak 2	10.0	21.50	49.20	492.02	2.28	78
Further Purification of Fraction Under Peak 2						
Peak 2:						
Sepharose 6B	10.0	5.59	32.30	323.01	5.77	52
Dialysis	12.0	4.66	26.91	322.92	5.77	52
DEAE-Sephacel	8.5	2.12	23.98	203.89	11.28	42

[1] Activities were determined spectrophotometrically.

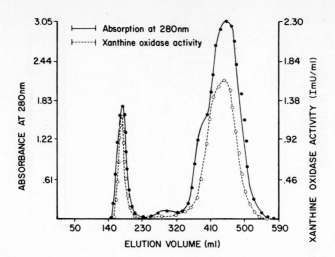

Figure 13. Elution profile from Sepharose 2B. The solid line represents absorbance at 280 nm while the dashed line is xanthine oxidase activity. The flow rate was maintained at 15 ml/hr.

activity eluting in the valley between the two protein peaks. Approximately the middle one-third of the enzyme activity peak fractions (fractions with an activity to protein ratio of at least 0.4) were pooled and concentrated. The recovery of this step was 52% of the amount applied to the Sepharose 6B column. All other active fractions were excluded from this preparation to avoid including fractions with undesirable low specific activities.

Figure 15 shows the elution profile of free xanthine oxidase from the DEAE Sephacel column. Two minor non-xanthine oxidase impurities were eluted far ahead of the enzyme followed by one protein peak slightly offset from the peak of enzyme activity. This is indicative that the fraction under the enzyme peak is not homogeneous. However, the primary goal of this study was not to achieve homogeneity. Rather, it was to separate free from the membrane-bound xanthine oxidase and to partially purify (while conserving sufficient activity) and characterize the free enzyme. Although we are presently developing methods for obtaining homogeneous colostral enzyme, homogeneous preparation is not necessary for determination of kinetic and other parameters. The recovery of this step was 42% thereby providing

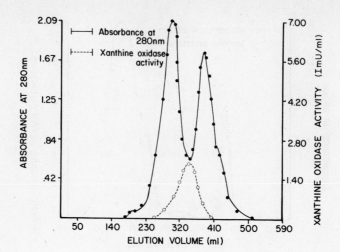

Figure 14.   Elution profile from Sepharose 6B.   The
solid line represents absorbance at 280 nm while the dashed
line is xanthine oxidase activity.   Flow rate was maintained
at 15 ml/hr.

good quantities for further testing and characterization.

   2. Characterization of Free and Membrane-Bound
      Colostral Xantine Oxidases

   Results from the characterization of free and membrane-
bound xanthine oxidases are summarized in Table 7.   Furthermore, a comparison of the characteristics of free human
colostral and bovine milk (values obtained from the
literature) xanthine oxidases is given in Table 8.

   a. Molecular Weights.   The molecular weight of free
colostral xanthine oxidase was determined by gel filtration
through Sepharose 6B, as described under Materials and
Methods, Section I.   The K_{av} for the free enzyme corresponded to a molecular weight of 310,000 daltons.   This value
is close to the molecular weight for bovine milk xanthine
oxidase (100) although in the literature the weight of
active bovine milk enzyme varies widely from 75,000 to
400,000 (10,18,55,69) depending on the method used for its
purification.   Membrane-bound xanthine oxidase eluted in the
void volume of the Sepharose 2B column and, therefore, had

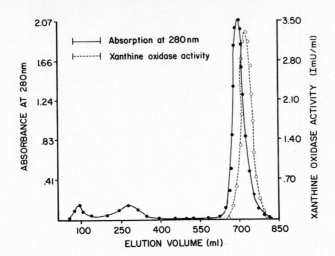

Figure 15.   Elution profile from the DEAE Sephacel column.   The solid line represents absorbance at 280 nm and the dashed line is xanthine oxidase activity.   Flow rate was maintained at 10 ml/hr.

a molecular weight of 4×10^7 daltons or greater, which is the exclusion limit of this column for proteins.   This weight is more of a function of the size of the membrane material the enzyme is associated with rather than the molecular weight for xanthine oxidase.

b. The Effect of pH on Free and Membrane-Bound Xanthine
    Oxidase Activities.   Figures 16 and 17 are the pH profiles of free and membrane-bound xanthine oxidases, respectively.   Free xanthine oxidase had a pH optimum of 8.2 and a 1/2 Vmax at pH 6.23 and 9.94.   This optimum compares close to an optimum of 8.3 for free bovine milk xanthine oxidase (18).   Using human raw whole milk samples and a radiochemical assay, we reported earlier (110) a pH optimum of 7.5.   This lower optimum for human milk may be attributed to the fact that the enzyme is not purified, to the radio-chemical assay, or both.   In this study with the free colostral enzyme, the pH profile was bell shaped (Figure 16). For the bovine, oxidation of xanthine appears to be regulated by at least three pK values.   The pK's for xanthine are 7.5 and 11.8 and one group on the enzyme with a pK of 10.7.   According to Bray (18), the substrate bears a single negative charge whereas the enzyme group has a positive

Table 7. Summary of the Characteristics of Human Colostrum Xanthine Oxidase (XO)

Characteristic	Free XO	Membrane-Bound XO
Molecular Weight	310,800	4×10^7
pH Optimum	8.2	8.2 - 10.0
Ea	17.2 kcal/mole	15.1 kcal/mole
Q10	2.7	2.4
Km:		
Xanthine	$2.79 \pm 0.37 \times 10^{-5}$ M	$2.64 \pm 0.08 \times 10^{-5}$ M
Hypoxanthine	$1.34 \pm 0.20 \times 10^{-5}$ M	$9.24 \pm 0.29 \times 10^{-6}$ M
Vmax:		
Xanthine	$21.13 \pm 0.72 \times 10^{-3}$ IU/ml	$14.83 \pm 0.16 \times 10^{-3}$ IU/ml
Hypoxanthine	$8.47 \pm 0.93 \times 10^{-3}$ IU/ml	$3.5 \pm 0.65 \times 10^{-3}$ IU/ml

Table 8. Comparison of Free Human Colostral and Bovine Milk Xanthine Oxidases

	Human colostrum	Bovine milk[1]
Molecular weight	310,800	303,000 (100)
pH optimum	8.2	8.3 (17)
Ea	17.2 kcal/mole	14.5 kcal/mole (60)
Km:		
Xanthine	$2.79 \pm 0.37 \times 10^{-5}$M	4.10×10^{-5}M (121)
Hypoxanthine	$1.34 \pm 0.20 \times 10^{-5}$M	7.00×10^{-6}M (121)
Vmax:		
Xanthine	$21.23 \pm 0.72 \times 10^{-3}$ IU/ml	4.65 IU/ml (121)
Hypoxanthine	$8.47 \pm 0.93 \times 10^{-3}$ IU/ml	1.95 IU/ml (121)

[1]Values from the literature.

charge.   From the pH values at 1/2 Vmax of 6.23 and 9.94 for
the human enzyme, these pK's would represent the groups
histidine and lysine or tyrosine, respectively (99).

Membrane-bound xanthine oxidase had a constant maximum
activity between pH 8.5 and 10.0 and a 1/2 Vmax at pH 5.85.
These differences between free and membrane-bound xanthine
oxidases may be due to dissociation-association phenomena
in the membrane-bound enzyme which occur in bovine membrane-
bound xanthine oxidase (10,71), known to affect pH profiles
(28).   The membrane may afford some protection from
denaturation at high pH values which may explain the
leveling off phenomenum shown in Figure 17.   These results
indicate that the catalytic properties of human colostral
xanthine oxidase depend on whether the enzyme is membrane-
bound or in the free form.   The same has been shown to be
true for bovine milk xanthine oxidase.   Briley and Eisenthal
(19) have demonstrated that the oxidase activity of membrane
-bound enzyme toward NADH is enhanced relative to that
toward xanthine.   The same effect can be mimmicked by free
xanthine oxidase as the pH is lowered from 9.0 to 6.0.   The
catalytic difference between bound and free enzymes (from
bovine and human colostrum) may result from the enzyme
binding to the membrane in a microenvironment of low pH
(19).   The results from this investigation supports this
explanation.   The optimum for the bound enzyme shifted
towards higher pH values relative to free xanthine oxidase.
According to Laidler and Bunting (54), this effect is
characteristic for enzymes attached to supports in a
microenvironment of low pH.

c. The Effect of Temperature on Free and Membrane-
Bound Xanthine Oxidases.   From Arrhenius plots, free
xanthine oxidase was found to have an activation energy of
17,241 cal/mole and a Q10 of 2.7.   These values for the
membrane-bound enzyme were 15,125 cal/mole and 2.4,
respectively.   Thus, the activation energy for the free
enzyme was 2,116 cal/mole higher than that for the membrane-
bound enzyme.   Apparently, the association of xanthine
oxidase with the membrane allows the transition state to be
achieved more readily and therefore requires less energy
than the free enzyme.   The activation energy for free
bovine milk xanthine oxidase was found to be 14,100 cal/
mole (57) and 14,500 cal/mole (60).   On the other hand,
Bray (18) reported that the activation energy decreased
slightly above 20°C.   His values for the range of 5-20°C
and 20-25°C were 17,800 and 16,000 cal/mole, respectively.

The Arrhenius plots of free and membrane-bound xanthine
oxidases demonstrated that the relationship between log

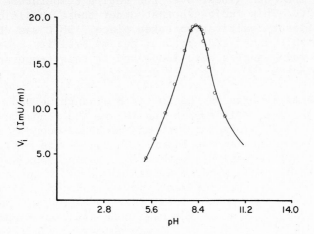

Figure 16. pH profile for free xanthine oxidase.

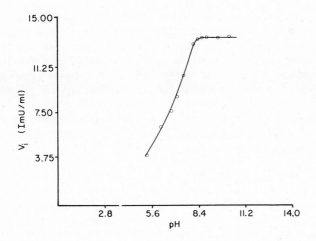

Figure 17. pH profile for membrane-bound xanthine oxidase.

Vmax and $1/K°$ was linear over the entire temperature range of 15-35°C. This indicates that under these conditions no significant inactivation takes place. 30°C was chosen as a suitable assay temperature; enzyme activity remained constant when samples were incubated at this temperature for 30 minutes.

    d. Determination of the Km for Hypoxanthine and
       Xanthine. Figures 18 and 19 are the respective Lineweaver-Burk plots for free and membrane-bound xanthine oxidases for xanthine and Figures 20 and 21 are similar plots for the two enzyme preparations using hypoxanthine. Free xanthine oxidase had a Km of 2.79×10^{-5}M and a Vmax of 21.23×10^{-3} IU/ml for xanthine and a Km of 1.34×10^{-5}M and a Vmax of 8.47×10^{-3} IU/ml for hypoxanthine. When hypoxanthine was used at the same concentration range as xanthine, substrate inhibition was observed. This is in agreement with previous reports with bovine milk xanthine oxidase (35,121). This inhibition was circumvented by reducing the hypoxanthine concentration. Membrane-bound xanthine oxidase had a Km of 2.64×10^{-5}M and a Vmax of 14.83×10^{-3} IU/ml for xanthine and a Km of 9.24×10^{-6}M and a Vmax of 3.50×10^{-3} IU/ml for hypoxanthine.

    There are many differences between the Km values for human colostral and bovine milk xanthine oxidases (see Tables 7 and 8). For example, the Km for free human xanthine oxidase was lower than that for the bovine enzyme for xanthine whereas the Km for hypoxanthine was higher. The human membrane-bound xanthine oxidase had a lower Km for xanthine than the bovine membrane-bound enzyme with a reported value of 5.41×10^{-5}M (19). For both free and membrane-bound xanthine oxidases, the Km for xanthine is higher than for hypoxanthine. Also for each enzyme, the Vmax is higher for xanthine than for hypoxanthine. This same pattern is followed for bovine milk xanthine oxidase (Table 8).

    e. Enzyme Cofactor, Stability, and Electrophoretic
       Mobility. FAD was identified as a cofactor of the human colostral xanthine oxidase. This cofactor was detected spectrophotometrically at 450 nm and by neo-tetrazolium chloride differential staining in PAGE. FAD is also present in bovine (18) and caprine (as shown in Section II) milk xanthine oxidases. The free human colostral enzyme contained one minor and two major protein bands, one of which was identified by neotetrazolium chloride as active xanthine oxidase. In parallel gels containing PDXO and NPDXO from bovine milk and free human colostral xanthine

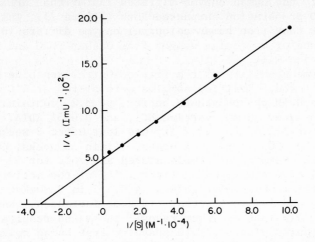

Figure 18. Lineweaver–Burk plot for free xanthine oxidase using xanthine as substrate.

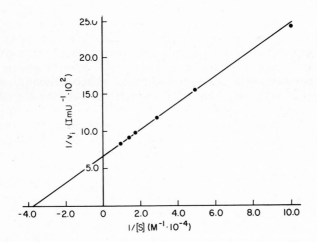

Figure 19. Lineweaver–Burk plot for membrane-bound xanthine oxidase using xanthine as substrate.

oxidase, the human enzyme migrated faster than NPDXO, but
the PDXO preparation surpassed both (Figure 7). This
suggests that free human colostral enzyme differs in net
charge and/or molecular weight from either PDXO and NPDXO
from bovine milk.

Enzyme stability of the free colostral xanthine oxidase
was monitored. Purified samples were stored at 4°C and
-20°C in 0.1M pyrophosphate buffer, pH 8.2, containing 5
mM salicylate, 30 mM cysteine-HCl, and 0.005% EDTA. We
found that 11% of the activity was lost after 2 weeks in
storage at -20°C, 18% in 4 weeks, 34% in 10 weeks, and
57% in 22 weeks. One sample stored at -20°C for 59 weeks
lost 81% of its original activity. 4% of the activity was
lost after 1 week in storage at 4°C, 7% in 2 weeks, 11% in
3 weeks, and 17% in 6 weeks. In some instances, enzyme
activity was lost at lower rates. These and results from
the Arrhenius plot indicate that the free human colostral
enzyme is stabler than either the bovine or caprine milk
xanthine oxidase, but not as stable as undiluted whole
human milk and colostrum samples (110).

IV. CONCLUSION

Native bovine milk xanthine oxidase was obtained by a
streamline method without the use of proteolytic enzymes
and other denaturing organic reagents. Electrophoretic and
kinetic comparisons between PDXO and NPDXO demonstrated
that proteolysis modifies milk xanthine oxidase by cleaving
its four interconnecting polypeptide loops (68). Besides
reducing its molecular weight and increasing its electrophoretic mobility, this treatment yields enzyme with
reduced stability, catalytic efficiency, and affinity for
both substrate and competitive inhibitors. Thus, the use
of pancreatin or other proteolytic agents in the purification of milk xanthine oxidase should be avoided, especially
when the enzyme is intended for use in critical studies.
Likewise, using non-proteolytic methods, xanthine oxidase
was purified for the first time from caprine cream and
human colostrum. Although we found some differences between
xanthine oxidases from bovine, caprine, and human, overall
xanthine oxidase from these sources is not very different.
Finally, with the results presented in Section III we have
confirmed our previous findings (110) and proven conclusively that xanthine oxidase is present in the human

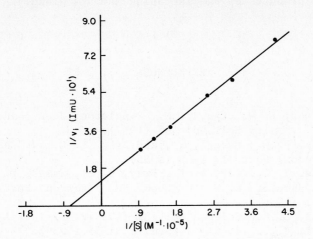

Figure 20.   Lineweaver–Burk plot for free xanthine oxidase using hypoxanthine as substrate.

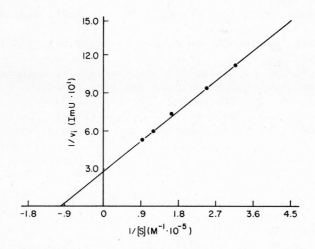

Figure 21.   Lineweaver–Burk plot for membrane-bound xanthine oxidase using hypoxanthine as substrate.

secretory products of the lactating mammary gland.

REFERENCES

1. Aurand, L. W., Chu, T. M., Singleton, J. A., and Shen, R. J. Dairy Sci. 50: 465 (1967).
2. Avis, P. G., Bergel, F., and Bray, R. C. J. Chem. Soc. (London) Part II: 1100 (1955).
3. Avis, P. G., Bergel, F., Bray, R. C., James, D. W. F., and Shooter, K. V. J. Chem. Soc. (London) Part I: 1212 (1956).
4. Ball, E. G. Science 88: 131 (1938).
5. Ball, E. G. J. Biol. Chem. 128: 51 (1939).
6. Barber, M. J., Salerno, J. C., and Siegel, L. M. Biochemistry 21: 1648 (1982).
7. Battelli, M. G., Lorenzoni, E., and Stirpe, F. Biochem. J. 131: 191 (1973).
8. Benson, J. R. Am. Lab. Volume 4, October (1972).
9. Bhavadasan, M. K. and Ganguli, N. C. J. Dairy Sci. 63: 362 (1980).
10. Biasotto, N. O. and Zikakis, J. P. J. Dairy Sci. 58: 1238 (1975).
11. Bierman, E. L. and Shank, R. E. J. Amer. Med. Ass. 234: 630 (1975).
12. Bjorck, L. and Claesson, O. J. Dairy Sci. 62: 1211 (1979).
13. Blackburn, S. Amino acid determination, methods and techniques, Marcel Dekker, New York (1968).
14. Booth, V. H. Biochem. J. 29: 1732 (1935).
15. Booth, V. H. Biochem. J. 32: 494 (1938).
16. Bradley, P. L. and Gunther, M. Biochem. J. 74: 15P (1960).
17. Bray R. C., in "The Enzymes" (P. D. Boyer, H. Lardy, and R. Myrback, eds.), Vol. 7, 2nd Ed., Academic Press, New York (1963).
18. Bray, R. C., in "The Enzymes" (P. D. Boyer, H. Lardy, and R. Myrback, eds.), Vol. 12, 3rd Ed., Academic Press, New York (1975).
19. Briley, M. S., and Eisenthal, R. Biochem. J. 143: 149 (1974).
20. Brown, M. L. and Zikakis, J. P. Presented at the 174th Annual Meeting Am. Chem. Soc., Biol. Div., Aug. 31, Chicago, IL (1977).

21. Brunner, J. R., in "Fundamentals of Dairy Chemistry" (B. W. Webb, A. H. Johnson, and J. A. Alford, eds.), p. 474, 2nd Ed., AVI Publishing Co., Westport, Conn. (1974).

22. Carr, C. J., Talbot, J. M., and Fisher, K. D. Life Sciences Research Office, Federation of American Societies for Experimental Biology, Contract No. FDA 223-75-2090, Bethesda, MD (1975).

23. Catignani, G. L., Chytil, F., and Darby, W. J. Biochim. Biophys. Acta 377: 34 (1975).

24. Cerbulis, J. and Farrell, H. M. J. Dairy Sci. 60: 170 (1977).

25. Clark, A. J., Pratt, D. E., and Chambers, J. V. Life Sci. 19: 887 (1976).

26. Colby, B. E., Evans, D. A., Lyford, S. J., Nutting, W. B., and Stern, D. N. Dairy goats breeding, feeding, and management, Am. Dairy Goat Assn. (1972).

27. Corran, H. S., Dewan, J. G., Gordon, A. H., and Green, D. E. Biochem. J. 33: 1694 (1939).

28. Cousin, C. L. Clin. Biochem. 9: 160 (1976).

29. Crossland, A., Owen, E. C., and Proudfoot, R. Brit. J. Nutr. 12: 312 (1958).

30. Davidson, H. F. Evening J., Wilmington, Del., March 1 (1972).

31. Della Corte, E., Gozzetti, G., Novello, F., and Stirpe, F. Biochim. Biophys. Acta 191: 164 (1969).

32. DeRenzo, E. C., Kaleita, E., Heytler, P., Oleson, J. J., Hutchings, B. L., and Williams, J. H. J. Am. Chem. Soc. 75: 753 (1953).

33. Davis, B. J. Ann. N. Y. Acad. Sci. 121: 404 (1964).

34. Dikstein, S., Bergmann, F., and Henis, Y. J. Biol. Chem. 224: 67 (1957).

35. Dixon, M. and Thurlow, S. Biochem. J. 18: 971 (1924).

36. Dixon, M. Biochem. J. 20: 703 (1926).

37. Dixon, M. Enzymologia 5: 198 (1938-39).

38. Dressel, M. A. and Zikakis, J. P. Presented at the 181st National Mtg. Am. Chem. Soc., AFC Div., April 2, Atlanta, GA (1981).

39. Franke, W. and Hahn, G. E. Z. Physiol. Chem. 301: 90 (1955).

40. Fried, R., Fried, L. W., and Babin, D. R. Eur. J. Biochem. 33: 439 (1973).

41. Gandhi, M. P. S. and Ahuja, S. P. Zbl. Vet. Med. A. 26: 635 (1979).

42. Gibbs, D. A. and Watts, R. W. E. Clin. Science 31: 285 (1966).

43. Gilbert, D. A. and Bergel, F. Biochem. J. 90: 350 (1964).

44. Green, D. E. and Beinert, H. Biochim et Biophys. Acta 11: 599 (1953).
45. Gudnason, G. V. and Shipe, W. F. J. Dairy Sci. 45: 1440 (1962).
46. Gregoriadis, G. and Neerunjun, E. D. Biochem. Biophys. Res. Commun. 65: 537 (1975).
47. Gregoriadis, G. New Eng. J. Med. 295: 704 (1976).
48. Hart, L. I., McGartoli, M. A., Chapman, H. R., and Bray, R. C. Biochem. J. 116: 851 (1970).
49. Hartnett, J. C. and Gjessing, E. C. J. Biol. Chem. 237: 2201 (1962).
50. Ho, C. Y. and Clifford, A. J. J. Nutr. 106: 1600 (1976).
51. Ho. C. Y., Clifford, A. J., and Hill, F. W. Fed. Proc. 35: 538 (1976).
52. Horbaczewski, J. Monatsh. Chem. 12: 221 (1882).
53. Jocelyn, P. C. Nature 202: 1115 (1964).
54. Laidler, K. J. and Bunting, P. S. The chemical kinetics of enzyme action. Second Edition, Oxford University Press, Ely House, London, England (1973).
55. Lowry, O. H., Bessey, O. A., and Crawford, E. J. J. Biol. Chem. 180: 399 (1949).
56. Lowry, O. H., Rosebrough, N. J., Farr, A. L., and Randall, R. J. J. Biol. Chem. 193: 265 (1951).
57. Mangino, M. E. and Brunner, J. R. J. Dairy Sci. 60: 841 (1977).
58. Mangino, M. E. and Brunner, J. R. J. Dairy Sci. 59: 1511 (1976).
59. Marx, J. (Technical advisor for Sigma). Personal communication.
60. Massey, V., Brandy, P. E., Komai, H., and Palmer, G. J. Biol. Chem. 244: 1682 (1969).
61. Massey, V., Edmondson, D., Palmer, G., Beacham, L. M., and Elion, G. B. Biochem. J. 127: 10P (1972).
62. McKenzie, H. A., in "Milk Protein Chemistry and Molecular Biology" (H. A. McKenzie, ed.), Vol. 2, p. 87, Academic Press, New York (1971).
63. Michelini, C. B. Nat. Enquirer p. 1, March 11 (1975).
64. Mittal, V. K. and Mathur, M. P. Indian J. Dairy Sci. 28: 296 (1975).
65. Modi, V. V., Owen, E. C., and Proudfoot, R. Proc. Nutr. Soc. 18: i (1959).
66. Morton, R. K. Biochem. J. 57: 231 (1954).
67. Nagler, L. G. and Vartanyan, L. S. Biokhimika 38: 561 (1973).
68. Nagler, L. G. and Vartanyan, L. S. Biochim. Biophys. Acta 427: 78 (1976).

69. Nathans, G. R. and Hade, E. P. R. Fed. Proc. 34: 606 (1975).
70. Nathans, G. R. and Hade, E. P. R. Biochem. Biophys. Res. Commun. 66: 108 (1975).
71. Nathans, G. R. and Hade, E. P. R. Biochim. Biophys. Acta. 526: 328 (1978).
72. Nelson, C. A. and Handler, P. J. Biol. Chem. 243: 5368 (1968).
73. Olinescu, R. Rev. Roum. Biochim. 8: 303 (1971).
74. Ornstein, L. Ann. N. Y. Acad. Sci. 121: 321 (1964).
75. Oster, K. A. Cardiol. Digest 3: 29 (1968).
76. Oster, K. A. Am. J. Clin. Res. 2: 30 (1971).
77. Oster, K. A., in "Myocardiology" (E. Bajusz and G. Rona, eds.), Vol. 1, p. 803. University Park Press, Baltimore, MD (1972).
78. Oster, K. A., Oster, J. B., and Ross, D. J. Amer. Lab. 6: 41 (1974).
79. Owen, E. C. and Dundas, I. Proc. Nutr. Soc. 28: 59 (1969).
80. Patton, S. and Keenan, T. W. Biochim. Biophys. Acta 415: 273 (1975).
81. Pederson, T. C. and Aust, S. D. Biochem. Biophys. Res. Commun. 52: 1071 (1973).
82. Richert, D. A. and Westerfield, W. W. J. Biol. Chem. 209: 179 (1954).
83. Rodkey, F. L. and Ball, E. G. J. Lab. Clin. Med. 31: 354 (1946).
84. Ross, D. J., Ptaszynski, and Oster, K. A. Proc. Soc. Exp. Biol. Med. 144: 523 (1973).
85. Ross, D. J., Sharnick, S., and Oster, K. A. Proc. Soc. Exp. Biol. Med. 163: 141 (1980).
86. Rzucidlo, S. J. and Zikakis, J. P. Proc. Soc. Exp. Biol. Med. 160: 477 (1979).
87. Rzucidlo, S. J. and Zikakis, J. P. Unpublished data (1982).
88. Schardinger, F. Z. Untersuch. Nahr. Genussm. 5: 1113 (1902).
89. Schultz, H. W., Day, E. A., and Sinnhuber, R. O. Symposium on Foods: Lipids and their Oxidation. Avis Publishing Co., Westport, CT (1962).
90. Segel, I. H. Enzyme kinetics. John Wiley and Sons, Inc., New York (1975).
91. Shahani, K. M. J. Dairy Sci. 58: 1123 (1966).
92. Silver, M. R. and Zikakis, J. P. Presented at joint Ann. Mtg. Northeastern Region. Am. Soc. Animal Sci. and Am. Dairy Sci. Assn., June 18, Morgantown, WV (1979).

93. Spitzer, W. Arch ges. Physiol. 76: 192 (1899).
94. Stirpe, F. and Della Corte, E. J. Biol. Chem. 244: 3855 (1969).
95. Tophan, R. W., Woodruff, J. H., and Walker, M. C. Biochemistry 20: 319 (1981).
96. Uriel, J. and Avrameas, S. Biochem. J. 4: 1740 (1965).
97. Villela, G. C., Affonso, O. R., and Mitidieri, E. Arch. Biochem. Biophys. 59: 532 (1955).
98. Volkheimer, G., in "Gastroenterologie und Stoffwechsel" (H. Bartelheimer, H. A. Kuhn, V. Becker, and F. Stelzner, eds.), Georg Thieme Verlag Stuttgart (1972).
99. Von Hoffman, N. The Washington Post, April 8 (1974).
100. Waud, W. R., Brady, F. O., Wiley, R. D., and Rajagopalan, K. V. Arch. Biochem. Biophys. 169: 695 (1975).
101. Waugh, D. F., in "Milk Protein Chemistry and Molecular Biology" (H. A. McKenzie, ed.), Vol. 2, p. 3, Academic Press, New York (1971).
102. Webb, J. M. Nat. Enquirer p. 10, July 1 (1975).
103. Weissmann, G., Collins, T., Evers, A., and Dunham, D. Proc. Nat. Acad. Sci. 73: 510 (1976).
104. Wilkinson, G. N. Biochem. J. 80: 324 (1961).
105. Zikakis, J. P. and Treece, J. M. J. Dairy Sci. 52: 644 (1970).
106. Zikakis, J. P. and Treece, J. M. J. Dairy Sci. 54: 648 (1971).
107. Zikakis, J. P. and Biasotto, N. O. Presented at the 172nd National Meeting of the Amer. Chem. Soc., September 29, San Francisco, CA (1976).
108. Zikakis, J. P. and Rzucidlo, S. J. Presented at the 71st Ann. Mtg. of the Am. Dairy Sci. Ass., June 20, Raleigh, NC (1976).
109. Zikakis, J. P. and Rzucidlo, S. J. J. Dairy Sci. 58: 796 (1975).
110. Zikakis, J. P., Dougherty, T. M., and Biasotto, N. O. J. Food Sci. 41: 1408 (1976).
111. Zikakis, J. P. and Rzucidlo, S. J. J. Dairy Sci. 59: 1051 (1976).
112. Zikakis, J. P., Rzucidlo, S. J., and Biasotto, N. O. J. Dairy Sci. 58: 1238 (1975).
113. Zikakis, J. P., Rzucidlo, S. J., and Biasotto, N. O. J. Dairy Sci. 60: 533 (1977).
114. Zikakis, J. P. and Silver, M. R. Presented at the 176th National Mtg. Am. Chem. Soc., September 16, Washington, D. C. (1979).
115. Zikakis, J. P. United States Patent No. 4,172,763, October 30 (1979).

116. Zikakis, J. P. Am. Laboratory 11: 57 (1979).
117. Zikakis, J. P. and Wooters, S. G. J. Dairy Sci. 63: 893 (1980).
118. Zikakis, J. P. United States Patent No. 4,238,566, December 9 (1980).
119. Zikakis, J. P. Dairy Sci. Abstr. 42(12): 953, No. 8234 (1980).
120. Zikakis, J. P. United States Patent No. 4,246,341, January 20 (1981).
121. Zikakis, J. P. and Silver, M. R. J. Agric. Food Chem. (Submitted) (1983).

INTRODUCTION TO WINES AND SPIRITS SECTION

Because of the complexity of its chemical composition, wine presents some analytical problems that are truly difficult. The last decades have been marked by a profound evolution which corresponds to the fine-tuning of increasingly improved methods permitting the analysis of ever smaller concentrations of wine constituents more simply and with greater accuracy. These were conductometric and photometric methods, also the several chromatographic techniques (paper-, thin layer-, column- both gas and liquid, gravity as well as low- and high-pressure) which have completely metamorphosed our knowledge of the chemical composition of wine. Finally, the wine analyst can no longer ignore the enzymatic methods.

Along with the organoleptic evaluation, which it could not possibly replace, analysis remains an essential means of quality control, of production, of marketing and of consumption.

The diversity of increasingly complex techniques necessitates a continuing checking carried out by highly skilled specialists.

This section comprises several articles dealing with recent developments in chromatographic techniques, gas liquid- and high performance liquid chromatography, which now form an indispensable part of the armamentarium of all oenological laboratories. Similarly, neutron activation opens up new possibilities for the determination of certain minerals in wine.

These days, multivariate analysis of great masses of data is used to point out the more relevant analytical differences between grapes and between wines due to the variety of charac-

ter, climate conditions and degree of grape ripeness; similarly
the application of pattern recognition techniques permits the
correct classification of wines. Two important chapters in
this book are devoted to this type of approach which will certainly come into its own in oenological research in the years
to come.

Yet another chapter is devoted to the components of wine
aroma. This most complex matter is in perpetual evolution and
subject to new developments.

Finally, two contributions deal with spirits, sake, rhum
and brandy. In their relation to wine, it is seen that these
beverages have similar analytical problems and that the changes
which are brought about by aging, especially in wooden casks,
call for techniques of the same nature.

There is no doubt that this compilation of research papers
makes a significant contribution to the analysis of wine and
spirits, presenting the application of modern and up-to-date
techniques to actual problems in practice.

                              Professor Pascal Ribereau-Gayon
                              Director of the Institute of Oenology
                              of the University of Bordeaux II

GLASS CAPILLARY GAS CHROMATOGRAPHY
IN THE WINE AND SPIRIT INDUSTRY

Peter Liddle, André Bossard

MARTINI & ROSSI
Saint-Ouen, France

Alcoholic beverages are often complex mixtures of different classes of volatile and non-volatile compounds. Routine analyses of these products, as well as of the associated raw materials, have been greatly facilitated by the technique of capillary gas chromatography using glass or fused silica columns. Whilst the advantage of increased separating power with capillary columns is now well-known, this sometimes overshadows the associated advantages of faster analysis times and increased sensitivity, allowing simpler sample preparation techniques and smaller sample sizes. Applications discussed cover both the volatile compounds of alcoholic beverages, and also the determination of some of the non-volatile constituents, after formation of suitable derivatives. The gains in sensitivity and analysis time are especially useful for a small laboratory which is required to carry out routine quantitative analyses on large numbers of samples.

I. INTRODUCTION

The techniques currently involved in the analysis of alcoholic beverages range from the simple determination of alcoholic strength to the detection and determination of a few parts par billion of certain compounds. The finished products involved range from virtually pure aqueous alcohol, as in the case of vodka, to complex mixtures such as vermouths or spirit-based liqueurs, which contain not only compounds derived from an alcoholic fermentation, but also those extracted from a wide range of plant materials.

Instrumental Analysis of Foods
Volume 2

Between these two extremes are alcoholic beverages obtained
from the fermentation of different raw materials (grapes,
cereals, molasses, etc.) and those in which aqueous ethanol
is used as an extracting solvent for the aroma and flavour
principles of plant materials.

The routine analyses of such products have been greatly
facilitated by capillary gas chromatography. However,
although the increased separating power of this technique is
well-known, what is often important in a practical situation
are the associated advantages of faster analysis times, if
necessary operating under non-optimum conditions, and
increased sensitivity. The latter allows simpler sample-
preparation techniques to be used, along with smaller sample
sizes.

Some of the routine applications of gas chromatography
with glass or fused silica capillary columns in a small
laboratory are described below. These applications concern
the analyses of the volatile compounds of vermouths, liqueurs
and distilled spirits, and also the determinations of some of
their non-volatile constituents, which require a suitable
derivatisation procedure before being analysed by gas chromatography.

II. ANALYSIS OF VOLATILE COMPOUNDS

The volatile fractions of alcoholic beverages generally
account for, in total, less than 1 % of the aqueous-ethanol
"background". The most important compounds, from a quantitative point of view, are other aliphatic alcohols (up to five
carbons) and a few low molecular-weight compounds such as
ethyl acetate and acetaldehyde. These are often still
analysed by direct injection on packed columns, as the
analysis is relatively simple and the use of a packed column
may be stipulated by an official method (1,2). However, the
gains in separation and analysis time are appreciable with
capillary columns (3,4), and furthermore the introduction of
more polar bonded phases in capillary GC allows the use of
direct aqueous injections. This will doubtless result in the
gradual adoption of this type of column in the future.

Most of the volatiles of interest are present at levels
of less than 50 mg/l, and often around or below the level of
a few mg/l. Using packed columns, a preliminary solvent
extraction is generally required, in order to separate the
volatiles from the excess aqueous ethanol, together with a
concentration step to increase sensitivity. In the case of
capillary columns, where the injection of a few nanograms or
less of a compound is often sufficient for a reasonable

quantitative analysis (provided that the entire GC system is sufficiently inert), a simple solvent extraction of an alcoholic beverage, without subsequent concentration, is usually sensitive enough for routine work. Sample volumes of 10-20 ml are thus normally sufficient.

A. General Method

After addition of a suitable internal standard, and if necessary dilution of the sample to an alcoholic strength of less than about 20 % v/v, 10 to 20 ml of the sample containing 1 g sodium chloride are extracted for 1 minute with 1 ml iso-octane (2,2,4-trimethylpentane) in a large test-tube on a "Vortex" type mixer. After separation of the organic phase, the latter is analysed directly by splitless injection of about 1 microlitre on to a fused silica column coated with Carbowax 20M, held at 60°C for 2 minutes, then programmed at 5°C/min up to 220°C.

Despite poor recoveries of certain classes of compounds, we have found iso-octane to be a useful solvent for several practical reasons :
- with a b.p. of 99°C, a good solvent effect is obtained with an initial column temperature of 60-70°C; lower temperatures require longer cooling times for the instrument, and are in any case below the minimum working temperature for Carbowax 20M,
- transfer of a suitable amount of the organic phase to automatic-sampler vials is easier than with more volatile solvents, and there is less risk of premature evaporation in a long automatic series of analyses,
- it is generally available in a high state of purity, and has a low toxicity.

The use of sodium chloride increases the recoveries of some of the poorly-extracted compounds by 10-15 %, but is especially useful in the prevention of emulsions. If these occur, the addition of one or two drops of ethanol will usually give a clear phase separation, without affecting quantitative results. The extraction time of 1 minute is sufficient for equilibrium to be attained between the phases.

The two examples shown below were obtained with a 50 m fused silica column, with hydrogen as carrier gas. Despite initial fears, we have been using hydrogen for many years, with real gains in speed and sensitivity; the advantages of hydrogen have been summarized by Grob (5).

A permanent check can be kept on the inertness of the whole GC system by using two internal standards. One of these should be unaffected by active sites, whilst the other should be sensitive to such sites, which give rise to adsorption or decomposition. Once the ratio of the peak areas of the two compounds have been determined under ideal conditions, it can be monitored for each analysis. Any departure from the initial value indicates undesirable system activity, which is not always shown up by, for example, tailing of certain classes of compounds. Such activity is generally due to septum particles in the injector, or a build-up of non-volatile compounds in the latter or in the first few coils of the column.

For practical purposes (i.e., availability and lack of interference with other compounds occurring in the samples routinely analysed, at least on Carbowax 20M), we have found methyl undecanoate to be a good primary internal standard,

TABLE I. Recoveries[a] and Precision with a Model Solution
(20 % v/v alcohol, n = 10)

Compound	Concentration (mg/l)	Recovery (%)	Relative standard deviation (%)
Limonene	0.5	100	1.6
Ethyl hexanoate	0.5	97	1.8
Tridecane[b]	0.5	100	1.0
Decanal	0.5	101	1.3
Linalool	0.5	77	1.8
Octanol	0.5	67	2.3
Diethyl succinate	5	46	3.0
Methyl undecanoate[b]	0.5	95	1.3
Anethole	0.5	97	1.4
2-Phenylethanol	40	8	3.1
Cinnamaldehyde	0.5	25	2.4
Octanoic acid	5	5	3.8
Carvacrol	5	83	1.9
Ethyl hexadecanoate	0.5	97	1.8

[a]Relative to hexadecane
[b]Possible internal standards

with a free fatty acid as a secondary standard, acting as a check on the system. In many cases, undecanoic acid is useful for the latter, if absent from the product or present only in trace amounts.

In order to check the precision of this type of procedure, a model solution in 20 % v/v alcohol, containing compounds typical of those found in alcoholic beverages, was analysed ten times. The results are shown in Table I, and it can be seen that the extraction is sufficiently reproducible for routine analyses, despite poor recoveries (relative to hexadecane) of certain classes of compounds. The concentrations of some of the latter in the model solution were deliberately higher, reflecting the actual levels found in finished products. However, it may obviously be necessary to resort to a more polar solvent if analyses are required for this type of compound at lower levels.

For a given type of product, standardisation with model solutions and the internal standard(s) chosen are necessary, with due regard to the composition of the blank medium, e.g. sugar content, pH, alcoholic strength. The method of standard additions should be used as a check on the accuracy with a particular type of product, especially if GC-MS is not available to verify that a particular peak represents only the compound to be analysed.

B. Applications

Perhaps one of the main applications of this type of analytical procedure is in the area of quality control. Samples taken at different stages in the production of an alcoholic beverage, from raw materials to the finished product, can be rapidly analysed with a minimum of sample preparation. Thus, for example, many alcoholic beverages are subjected in the final stages of production to a process known as chill-proofing, whereby the product is refrigerated and subsequently filtered. This improves the stability of the finished product, by decreasing the levels of fatty acids and their ethyl esters (generally from 14 carbon atoms upwards), and other compounds which may subsequently cause a problem. A rapid check on the profile of the higher ethyl esters and free fatty acids can avoid the problem known as "chill haze" in distilled spirits, and of rancid off-flavours which may develop in products containing unsaturated (mainly 18-carbon) fatty acids and ethyl esters, derived from the triglycerides of the botanicals used in flavoured alcoholic beverages.

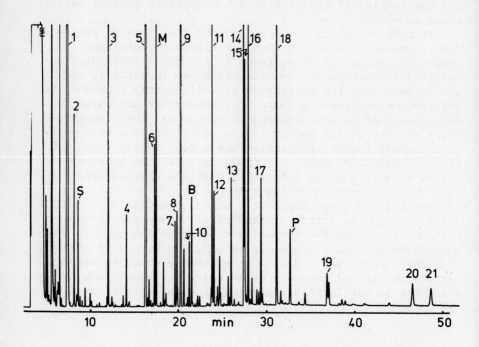

FIGURE 1. Typical chromatogram of a blended whisky.
1: 2- and 3-methylbutanol, 2-6: ethyl hexanoate, octanoate,
nonanoate, decanoate and 9-decenoate, 7: 2-phenylethyl
acetate, 8: hexanoic acid, 9: ethyl dodecanoate, 10: 2-
phenylethanol, 11: octanoic acid, 12: ethyl tetradecanoate,
13: tetradecanol, 14: decanoic acid, 15-16: ethyl hexa-
decanoate and 9-hexadecenoate, 17: hexadecanol, 18-21: do-
decanoic, tetradecanoic, hexadecanoic, and 9-hexadecenoic
acids. Added compounds are styrene (S, 1 mg/ℓ), BHT (B, 1
mg/ℓ), di-isobutyl phthalate (P, 1 mg/ℓ), and methyl undecanoate (M, internal standard, 10 mg/ℓ).

Such analyses will also rapidly show up certain types
of contaminants which may be introduced from bulk transport
or storage containers, hoses, bottle closures etc. Figure 1
shows a typical chromatogram of a blended whisky, to which
have been added trace amounts of examples of such contaminants : a residual monomer (styrene), a plasticiser (di-
isobutyl phthalate) and an anti-oxidant (BHT). Some of the
main volatile compounds are also indicated.

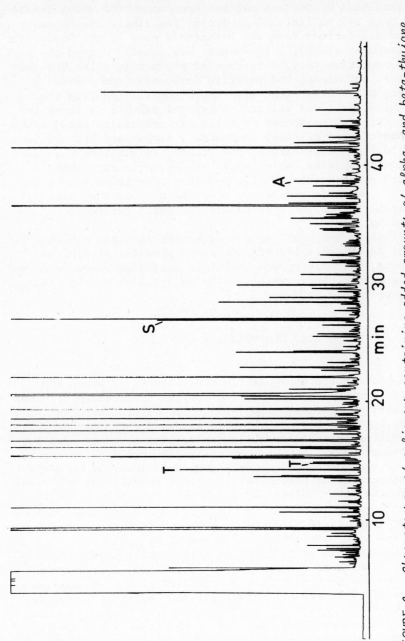

FIGURE 2. Chromatogram of a liqueur containing added amounts of alpha- and beta-thujone (T, 0.07 and 0.02 mg/ℓ, respectively), safrole (S, 0.2 mg/ℓ), and beta-asarone (A, 0.1 mg/ℓ).

Another problem in this industry is the detection and
determination of certain natural compounds, for which limits
have been set in finished products. The limits for these
compounds, derived from the botanicals used in the production
of flavoured alcoholic beverages, are usually around the mg/l
level. Unfortunately, this type of product is often the most
complex, containing the volatile compounds from a wide
variety of plants, along with those contributed from the
base wines or added spirits. A typical example is shown in
Figure 2, a chromatogram of a liqueur containing added trace
amounts of some of these compounds : alpha- and beta-thujone,
safrole, and beta-asarone. The liqueur in question contained
no wine or spirits, which would have given a much more
complex chromatogram. Whilst positive identification of such
limited compounds requires confirmation by another method,
preferably GC-MS, this procedure is useful for rapid
screening.

Finally, the small amount of sample required can be
useful in establishing "fingerprint" profiles of finished
products. This has proved useful in analysing products served
to the public, where substitution or dilution of the beverage
is suspected.

III. NON-VOLATILE COMPOUNDS

As far as the non-volatile constituents of wines and
spirits are concerned, a useful routine application of glass
capillary gas chromatography is in the determination of
sugars, polyols and non-volatile acids. Gas-chromatographic
analysis of suitable volatile derivatives has been used for
some time (6-12), but has generally involved a prior separation of the acidic and neutral compounds (e.g. by column
chromatography or precipitation of the acids). The increased
separating power offered by glass capillary columns enables
a considerable amount of information on the principal carbo-
hydrates and fixed acids to be obtained in a single run,
starting with sample sizes of a few ml or less.

The analysis of amino acids using capillary gas chromatography has been described by many authors, although the
specific application to the analysis of alcoholic beverages
has received little attention (13), and will not be dealt
with here. It remains to be seen whether this technique will
become more popular, as opposed to the traditional use of
specific amino-acid analyzers or the development of HPLC
methods (14).

A. General Method

A complete description of the procedure which involves the formation of trimethylsilyl (TMS) derivatives, is given in reference 15. After suitable dilution in the case of samples with a high sugar content (e.g., wine-based liqueurs), and addition of a solution of internal standards, a small volume (e.g., 25 microlitres) is evaporated under a stream of dry nitrogen at room temperature. The dry residue is silylated with N,O-bis (trimethylsilyl)trifluoro-acetamide (BSTFA), containing 1 % trimethylchlorosilane, for 1 hour at about 80°C, and the reaction mixture analysed on a borosilicate column coated with a bonded methylsilicone phase, prepared by the method of Madani (16). Columns prepared with this type of non-extractable bonded or "immobilized" phase are extremely inert and stable, exhibiting practically no bleeding. Other column preparation techniques have been extensively described, and such columns are now available commercially.

The TMS ester derivatives (i.e., obtained from the acids) are generally more sensitive to the existence of active sites than the TMS ethers (obtained from the sugars and polyols). By choosing a suitable acid and carbohydrate as internal standards, in this case vanillic acid and methylmannoside, a continuous check can again be kept on the inertness of the gas-chromatographic system. The ratio of the peak areas of the two internal-standard derivatives is determined under optimum conditions, and any variation in this ratio in subsequent analyses indicates that the system is no longer inert. The most common site of activity is the injector insert, which requires periodic cleaning and silanization, but it may be necessary, if the activity persists, to remove one or more coils from the injector-end of the column. In extreme cases, the sample preparation may be at fault, but this is usually indicated by the presence of multiple peaks with the standards, or turbidity in the reaction vial.

A chromatogram from a dry white wine is shown as an example in Figure 3. The precision for most of the major compounds, present at levels of 10-20 mg/l or more, is reasonable; some examples are given in Table II. The main disadvantage is the presence of several peaks from the anomeric forms of certain compounds, such as fructose and glucose, making quantitation more difficult, though not impossible for a person used to this particular analysis. In fact, these two compounds are readily determined by HPLC if present in amounts of more than about 1 g/l.

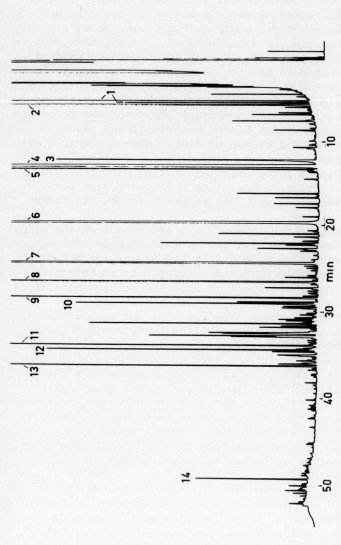

FIGURE 3. Chromatogram from a dry white wine: 1 : 2,3-butanediols, 2: lactic acid, 3: orthophosphoric acid, 4: glycerol, 5: succinic acid, 6: malic acid, 7: tartaric acid, 8: vanillic acid (internal standard, 1 g/ℓ), 9: alpha-methylmannoside (internal standard, 0.5 g/ℓ), 10: citric acid, 11: glucose (one anomeric form), 12: galacturonic acid (one anomeric form), 13: myo-inositol, 14: trehalose. For further peak identifications and conditions, see reference 15.

TABLE II. Replicate Analyses (n = 8) of a Wine

Compound	Peak number[a]	Mean (g/l)	Relative standard deviation (%)
2,3-Butanediol	3	0.33	7.8
Lactic acid	5	1.08	1.9
Phosphoric acid	27	0.24	7.1
Glycerol	33	5.83	2.9
Succinic acid	34	0.50	4.0
Citramalic acid	62	0.020	5.0
Malic acid	66	3.38	2.7
Erythritol	69	0.10	5.0
Tartaric acid	87	2.12	2.8
Arabitol	109	0.037	2.7
Citric acid	122	0.27	3.7
Shikimic acid	121	0.071	2.8
Mannitol	151	0.11	1.9
Myo-inositol	171	0.51	4.4
Trehalose	188	0.14	3.6

[a] See ref. 19

B. Applications

This type of derivatisation method, though not suited to a detailed study of a particular beverage, is very useful for routine control of wines or wine-based products. In addition to providing quantitative information on the principal compounds, such a procedure allows a rapid screening for certain compounds which are normally absent in this type of product, or present only in small amounts, e.g. sorbic, benzoic, salicylic or ascorbic acids, saccharose etc.

The same procedure can be useful in the examination of other alcoholic beverages. Thus, distilled spirits can be analysed rapidly for certain sugars and polyols which are extracted from the wood during the aging process (17-18).

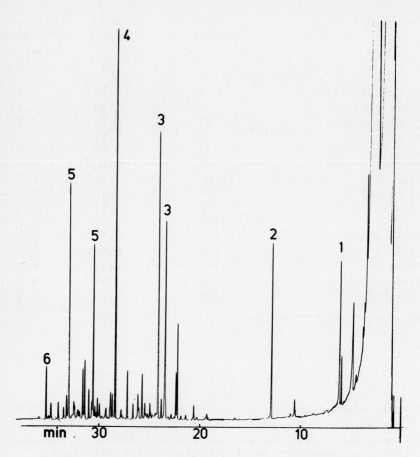

FIGURE 4. Chromatogram from an 8-year old rum. 1: lactic acid, 2: glycerol, 3: arabinose, 4: proto-quercitol, 5: glucose, 6: myo-inositol. Conditions as in reference 15.

Some of these compounds are shown in Figure 4, although it should be pointed out that the lower levels of certain phenolic compounds (extracted from the wood), which may be more important from an organoleptic point of view, usually require a more complex method (19-22). During the ageing process, certain sterols are also extracted from the wood (23-24), and these can subsequently form a flocculent precipitate from a super-saturated or colloidal state, even at ambient temperatures.

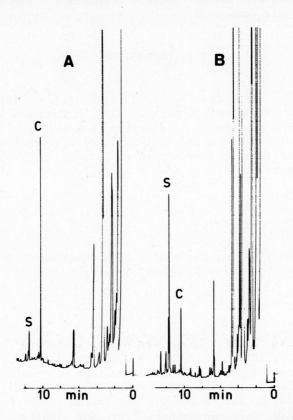

FIGURE 5. Chromatograms from a normal distilled spirit
(A), and one with a severe haze problem (B). C: TMS derivative of cholesterol (internal standard, 1 mg/ℓ), S: TMS
derivative of beta-sitosterol, respectively 0.1 mg/ℓ and
2.1 mg/ℓ in A and B. The chromatographic conditions are
similar to those of ref. 15, but with an initial temperature
of 180°C.

    The principal compounds responsible are beta-sitosterol
and its glucoside, and while the latter is best determined
by HPLC (25), the former can be determined rapidly by a
simple extraction along the lines of the procedure used for
volatile compounds, followed by evaporation and silylation
of the organic phase. The critical threshold is at the mg/l
level, as shown in Figure 5 (an extreme case), and although

beta-sitosterol is not solely responsible, such problems can be avoided by monitoring finished products for abnormal levels of this compound.

IV. CONCLUSIONS

Capillary gas chromatography, using glass or fused silica columns, has an important part to play in the wine and spirit industry. The advantages of increased sensitivity, reduced sample size, and decreased analysis time are especially useful for a small laboratory which is required to carry out routine quantitative analyses on large numbers of samples.

ACKNOWLEDGMENTS

We would like to acknowledge the contribution to much of this work by our late colleague, Pierre De Smedt.

REFERENCES

1. "Journal Officiel de la République Française", 2 octobre 1973, p. 1238.
2. "Official Methods of Analysis of the Association of Official Analytical Chemists" (W. Horwitz, ed.), 13th edition, p. 155, A.O.A.C., Washington (1980).
3. Ingraham, D.F., Shoemaker, C.F., and Jennings, W.G., J. Chromatogr. 239, 39 (1982).
4. Maarse, H., personal communication, 1982.
5. Grob, K., and Grob, G., J. High Resolut. Chromatogr. Chromatogr. Commun. 2, 109 (1979).
6. Bertrand, A., Ann. Fals. Exp. Chim. 67, 253 (1974).
7. Brunelle, R.L., Schoeneman, R.L., and Martin, G.E., J. Assoc. Off. Anal. Chem. 50, 329 (1967).
8. Drawert, F., Lessing V., and Leupold, G., Chromatographia 9, 373 (1976).
9. Drawert, F., and Leupold, G., Chromatographia 9, 397 (1976).
10. Mattick, L.R., Rice, A.C., and Moyer, J.C., Am. J. Enol. Vitic. 32, 111 (1981).
11. Ribereau-Gayon, P., and Bertrand, A., Comptes Rendus Acad. Sci. Paris, 273, 1761 (1971).
12. Ryan, J.J., and Dupont, J.A., J. Agric. Food Chem. 29, 65 (1981).

13. Poole, C.F., and Verzele, M., J. Chromatogr. 150, 439 (1978).
14. Casoli, A., and Colagrande, O., Ind. Bev. 57, 29 (1982).
15. De Smedt, P., Liddle, P.A.P., Cresto, B., and Bossard, A., J. Inst. Brew. 87, 349 (1981).
16. Madani, C., and Chambaz, E.M., Chromatographia 11, 725 (1978).
17. Black, R.A., and Andreasen, A.A., J. Assoc. Off. Anal. Chem. 57, 111 (1974).
18. Nykänen, I., 27th IUPAC Congress, Helsinki, 1979.
19. Tressl, R., Renner, R., and Apetz, M., Z. Lebensm. Unters.- Forsch. 162, 115 (1976).
20. Etievant, P.X., J. Agric. Food Chem. 29, 65 (1981).
21. Lehtonen, M., J. Chromatogr. 202, 413 (1980).
22. Puech, J.L., and Jouret, C., Ann. Fals. Exp. Chim. 75, 81 (1982).
23. Braus, H., Eck, J.W., Mueller, W.M., and Miller, F.D., J. Agric. Food Chem. 5, 458 (1957).
24. Black, A., and Andreasen, A.A., J. Assoc. Off. Anal. Chem. 56, 1357 (1973).
25. Byrne, K.J., Reazin, G.H., and Andreasen, A.A., J. Assoc. Off. Anal. Chem. 64, 181 (1981).

INSTRUMENTAL NEUTRON ACTIVATION ANALYSIS OF ALUMINUM, CALCIUM, MAGNESIUM AND VANADIUM IN GREEK WINES

Apostolos P. Grimanis
Maria Vassilaki-Grimani
George D. Kanias

Radioanalytical Laboratory
Chemistry Division
Nuclear Research Center "Demokritos"
Athens, Greece

I. INTRODUCTION

Greece is a wine-producing country with a great variety of wines.

The chemical analysis of wines is very important for quality control and handling of wines. Several methods are used for the determination of inorganic and organic constituents of musts and wines (1). For the analysis of trace elements in wines, flameless atomic absorption and neutron activation analysis (NAA) are the most sensitive and accurate methods for a great number of trace elements.

NAA is internationally recognized as a well-established analytical method for the determination of minor and trace elements and as such it is a very important peaceful use of research reactors. The method consists in determining the quantity of an element present in a sample by measuring the radioactivity when the sample is bombarded with thermal neutrons in a research reactor. The development of multichannel analyzers and semiconductor detectors for counting gamma-ray activities

Copyright © 1983 by Academic Press, Inc.
All rights of reproduction in any form reserved.
ISBN 0-12-168902-6

and the possibility of gamma-ray spectra processing by computer techniques greatly extended the scope of NAA. In certain cases NAA can be employed successfully as an instrumental, nondestructive multielement method based upon multichannel gamma-ray spectrometry of the neutron activated sample (2). NAA due to its sensitivity, rapidity, precision and accuracy is considered as an ideal method for the simultaneous determination of a large number of trace elements in several materials including foods and beverages.

Several investigators have developed and/or applied NAA methods to the determination of trace elements in wines (1,3-6). In our previous study NAA was applied for the determination of 7 trace elements (As,Br,Cl,Cu,K,Mn and Na) in 48 samples of red and white table wines from different wine production areas of Greece (7). The present study deals with the determination of Al,Ca,Mg and V in Greek wines by Instrumental NAA.

Aluminum is a normal consistuent of wines with concentrations ranging from 0.5-10 mg/L. Aluminum is seldom used in wine processing as it leads to cloudiness. When the Al content in wine is higher than 10 mg/L, owing to possible contamination in storage in Al containers, undesirable changes in clarity, color, odor and taste occur (8). When the wine producer suspects such a contamination, Al determination must be performed.

The calcium and magnesium contents of wines are influenced by soil conditions, use of filter aids, storage in concrete tanks, treatment with fining agents, use of ion-exchange resins, the concentration of ethanol and other constituents such as tartrates and sulfates, the pH and the time and temperature of storage (1). The determination of calcium and magnesium in wines is very important. Calcium combines with tartrates and precipitates as calcium tartrate even months after bottling (9). There are also indications that magnesium may be of importance in tartrate stability and to acid taste (1).

Vanadium is seldom determined in wines due to its low content which is below the sensitivity limit of different analytical methods used for wine analysis. Vanadium, like other trace metals, may affect flavor either by poisoning the yeast enzyme system or by changing the morphology or physiology of the yeasts (10). Recently Myron et al. (11) and Byrne and Kosta (12) have developed flameless atomic absorption spectroscopy and NAA methods respectively for vanadium determination at the nanogram level and applied them successfully to vanadium determination in foods and beverages. Although these methods are very sensitive, accurate and suitable for vanadium determination in foods and beverages they are not instrumental.

The purpose of this study was: a) to develop an Instrumental NAA for the rapid, precise and simultaneous determination of Al, Ca, Mg and V in wines and b) to apply this method to the determination of these elements in 49 samples of experimental and commercial red and white table wines from different wine production areas of Greece.

II. EXPERIMENTAL

Apparatus: A coaxial germanium-lithium drifted Ge(Li) detector (ORTEC Model WIN 15 Series) was used. The output signals were passed into a preamplifier (ORTEC Model 119) and a spectroscopy amplifier (ORTEC Model 571). The Ge(Li) detector system was connected with a 4096 channel pulse height analyzer (Canberra Model Series 80). The analyzer was coupled with a teletype (Digital Model LA 120). The whole system was used for γ-ray spectrometry. The system was adjusted so that every channel corresponded to 1 keV. System resolution was about 2.1 keV and the peak to Compton ratio was about 40:1 for the 1332 keV peak of ^{60}Co. The efficiency of the Ge(Li) detector was about 15%.

Standard aluminum, calcium, magnesium and vanadium solution: A mixed standard solution was prepared to contain 6 µg of aluminum per ml, 4 mg calcium per ml, 200 µg of magnesium per ml and 0.1 µg of vanadium per ml by carefully diluting 1 ml aliquot of the mixed standardized aluminum nitrate, calcium nitrate, magnesium nitrate and ammonium metavanadate solution with three times distilled water to the proper volume.

Standard manganese solution used as a neutron flux monitor: A standard manganese solution was prepared to contain 60 µg of manganese per ml by carefully diluting an aliquot of standardized manganese sulfate solution to the proper volume. This standard solution was used as a neutron flux monitor

Sample collection: Bottles of 24 red and 14 white wines of different harvest time from different wine production areas of Greece were selected for analysis. These wines were experimental ones known to be authentic varietal wines made available for this study by the Wine Institute of Athens, Greece. The red experimental wines were from the following 4 production areas with their variety of grapes given in parenthesis: Crete (Kotsifali-Mandilaria), Halkidiki (Grenach-Limnio), Nemea (Aghiorgitiko), Patras (Mavrodaphni). The white experimental wines were available from the following 3 production areas: Attiki (Sabbatiano), Limnos (Moschato Alexandria), Santorini (Assirtico).

Bottles of 8 red commercial wines from 5 production areas (Attiki, Halkidiki, Naoussa, Nemea, Rhodes) and 4 white commercial wines from 3 production areas (Attiki, Crete, Halkidiki) of known commercial firms were purchased locally and analysed.

Methodology

Irradiation: All neutron irradiations were carried out in the reactor of the Nuclear Research Center "Demokritos" using the pneumatic transfer (rabbit) system.

Five ml from each wine bottle were transferred by pipette into polyethylene tubes. Aliquots of 5 ml of the mixed standard solution were pipetted into identical polyethylene tubes. Aliquots of 0.5 ml of the standard manganese solution used as a neutron flux monitor were pipetted into Clay Adams polyethylene tubes (6×8mm). The tubes were heat-sealed, wrapped in plastic sheets and irradiated for 5 min in a thermal neutron flux of about $2\times10^{12}n\cdot cm^{-2}\cdot sec^{-1}$

The nuclear data for thermal neutron activation of aluminum, calcium, magnesium and vanadium are presented in Table I. Due to the short half life of the induced activities of ^{28}Al, ^{49}Ca, ^{27}Mg and ^{52}V only one wine sample - or the standard sample - and one manganese flux monitor sample were irradiated at a time.

Counting: Each irradiated wine sample was transferred into a 10 ml Erlenmeyer flask and 3 min after the end of irradiation the flask was counted for 3 min on the germanium (lithium drifted) detector. Irradiated standards were also transferred into 10 ml Erlenmayer flasks and measured 3 min after the end of irradiation for 3 min on the same detector. After counting the areas under the photopeak corresponding to the γ-ray energies of ^{27}Mg at 1014 keV, of ^{52}V at 1434 keV, of ^{28}Al at 1778 keV and of ^{49}Ca at 3083 keV of samples and standards were compared.

TABLE I.   Nuclear data for thermal neutron activation of Al, Ca, Mg and V

Stable isotopes	Natural abundance %	Activation cross section barns	Product of thermal neutron irradiation	Half-life min	Principal γ-rays keV
^{27}Al	100	0.21	^{28}Al	2.3	1778
^{48}Ca	0.18	1.1	^{49}Ca	8.8	3083
^{26}Mg	11.17	0.03	^{27}Mg	9.45	844,1014
^{51}V	99.76	4.5	^{52}V	3.76	1434

TABLE II. Concentrations of Al, Ca, Mg and V in experimental red dry table wines from 5 different production areas of Greece.

Production area grape variety	Aluminum (mg/L)	Calcium (mg/L)	Magnesium (mg/L)	Vanadium (µg/L)
Crete				
Kotsifali(3)*				
Minimum	0.84	31	77	6.5
Maximum	1.2	77	122	8.3
Average	(1.1)	(55)	(103)	(6.5)
Std. deviation	0.21	23	23	1.7
Mandilaria(2)*				
Minimum	2.7	89	139	21
Maximum	4.9	131	202	34
Average	(3.8)	(110)	(170)	(27)
Std. deviation	1.5	29	44	4.9
Halkidiki				
Grenach (3)*				
Minimum	0.93	46	101	1.0
Maximum	1.7	48	114	4.0
Average	(1.3)	(47)	(109)	(2)
Std. deviation	0.39	1.0	6.8	1.7
Limnio (3)*				
Minimum	1.0	44	107	3.0
Maximum	2.2	82	118	31
Average	(1.6)	(58)	(112)	(16)
Std. deviation	0.60	21	5.7	14
Nemea				
Aghiorgitiko (8)*				
Minimum	0.88	31	102	1.0
Maximum	3.7	77	152	9.3
Average	(1.9)	(59)	(123)	(8.7)
Std. deviation	0.93	15	19	9.5
Patras (sweet wine)				
Mavrodaphni (4)*				
Minimum	0.84	48	75	2.2
Maximum	3.4	79	143	6.1
Average	(1.7)	(60)	(108)	(4.4)
Std. deviation	1.1	13	35	1.7

*Number of harvest times and samples

Since the wine samples and standards were not irradiated simultaneously, their activities were corrected for flux variation during irradiation. This thermal neutron flux variation during the individual irradiation was estimated through the irradiated manganese flux monitor samples. These samples were also measured for 30 sec in the germanium detector.

Nuclear intereferences: Possible nuclear interferences exist from fast neutron reactions which also produce ^{28}Al from $^{28}Si(n,p)^{28}Al$ and $^{31}P(n,\alpha)^{28}Al$ as well as ^{27}Mg from $^{27}Al(n,p)$ ^{27}Mg and $^{30}Si(n,\alpha)^{27}Mg$ and ^{52}V from $^{52}Cr(n,p)^{52}V$ and $^{55}Mn(n,\alpha)$ ^{52}V. These nuclear intereferences are considered negligible for the analysis of wines, because the fast neutron cross sections of the stable isotopes ^{28}Si, ^{31}P, ^{27}Al, ^{30}Si, ^{52}Cr and ^{55}Mn for the above nuclear reactions are very low, the fast neutron flux of the reactor in the position of the pneumatic system is at least 10 times less than that of the thermal neutron flux and the concentrations of Si, P, Al, Cr and Mn are very low in wines.

Precision and accuracy: The reproducibility of the proposed Instrumental NAA method was checked by analysing for aluminum, calcium, magnesium and vanadium 7 samples of the same wine. The overall relative standard deviation was: for Al 6.9%, for Ca 15%, for Mg 9.6% and for V 12.2%. The accuracy of the proposed INAA method was checked by analysing a standard biological material (Bowen's Kale) for aluminum, calcium, magnesium and vanadium (13). Values found for these trace elements and literature values were in good agreement.

Rapidity: The time required to complete the analysis of a wine sample for aluminum, calcium, magnesium and vanadium by the proposed Instrumental NAA method is about 15 min.

Application of the Instrumental NAA method: The method was successfully applied to the determination of aluminum, calcium, magnesium and vanadium in 49 wine samples from several wine production areas of Greece.

III.   RESULTS AND DISCUSSION

1. Experimental Wines

Table II shows the concentrations of aluminum, calcium, magnesium and vanadium in experimental red table wines from four different wine production areas of Greece (Crete, Halkidiki, Nemea and Patras). The Table also includes the different grape varieties from each wine production area and a number (in parenthesis) which shows the number of harvest times and the samples analysed. Ranges and averages of concentrations as well as standard deviations are also presented in the same Table.

The highest concentrations of aluminum, calcium, magnesium and vanadium in red wines were found in the grape variety Mandilaria from Crete. The other experimental red wines from the 4 wine production areas of Greece have similar contents of aluminum, calcium and magnesium. Great ranges of vanadium concentrations in red experimental wines were found. Such ranges were observed in wines of the same grape variety but of different harvest times (chronological variations might be significant) - with the exception of the two red wines from Crete - as well as among average concentrations of vanadium in red wines from the 6 grape varieties of the 4 wine production areas of Greece. Chronological variations of aluminum were also observed in the red wines Aghiorgitiko from Nemea and Mavrodaphni from Patras.

Table III presents the concentrations of aluminum, calcium, magnesium and vanadium in experimental white wines from three wine production areas of Greece (Attiki, Limnos, Santorini). The Table also includes the grape variety from each wine production area, the harvest times and the samples analyzed given with a number (in parentheses), ranges, averages and standard deviations.

The highest amounts of aluminum and calcium were found in the experimental white wines from Santorini (Assirtico) while those of magnesium and vanadium were observed in the wines from Limnos (Moschato) and Attiki (Sabbatiano), respectively. Similar average concentrations of aluminum, calcium, magnesium and vanadium were found in white experimental wines from the three grape varieties of the three wine production areas of Greece. Great ranges of concentrations of aluminum and vanadium were observed in white wines from Limnos (Moschato) as well as of aluminum from Attiki (Sabbatiano) and of vanadium from Santorini (Assirtico). These ranges are probably due to different harvest times (chronological variations).

2. Commercial wines

The concentrations of aluminum, calcium, magnesium and vanadium in 12 red and white commercial wines of known commercial names from 6 different production areas of Greece are shown in Table IV. Values of elements given for each wine are means of duplicate analyses.

TABLE III.   Concentrations of Al, Ca, Mg and V in experimental white dry table wines from three different wine production areas of Greece.

Production area grape variety	Aluminum (mg/L)	Calcium (mg/L)	Magnesium (mg/L)	Vanadium (µg/L)
Attiki				
Sabbatiano (5)*				
Minimum	1.0	48	82	13
Maximum	4.1	101	106	26
Average	(2.0)	(68)	(91)	(22)
Std. deviation	1.2	22	10	5.6
Limnos				
Moschato (4)*				
Minimum	0.72	26	73	1.7
Maximum	3.2	58	118	47
Average	(1.7)	(41)	(95)	(13)
Std. deviation	1.1	16	20	22
Santorini				
Assirtico (5)*				
Minimum	2.3	54	64	3.2
Maximum	4.6	92	117	66
Average	(2.9)	(79)	(86)	(20)
Std. deviation	0.98	16	20	26

*Number of harvest times and samples

The highest concentrations of aluminum, calcium, magnesium and vanadium were found in red wines from Rhodes. The values of aluminum found in these wines are in the range of the maximum suggested limit for aluminum in wines (1).

The lowest contents of aluminum were found in red and white wines from Halkidiki, while those of magnesium were observed in red and white wines from Attiki. Red wines from Halkidiki have the lowest values of calcium. The lowest concentration of vanadium was observed in the red wine Tsantali from Naoussa.

Calcium and magnesium concentrations do not show noticeable differences between red and white wines or among the different wine production areas with the exception of the red wines from Rhodes. In general great ranges of concentrations of aluminum and vanadium were observed among all the red and white commercial wines analysed.

TABLE IV.   Concentrations of Al, Ca, Mg and V in commercial red and white dry table wines of Greece.

Production area Commercial name	Aluminum (mg/L)	Calcium (mg/L)	Magnesium (mg/L)	Vanadium (µg/L)
Attiki				
Cambas Hymettos (red)	4.6	88	89	19
Halkidiki				
Chateau Carras (red)	1.0	52	100	36
Domain Porto Carras Cote de Meliton (red)	1.2	67	147	31
Naoussa				
Tsantali (red)	2.2	69	163	9.6
Boutari (red)	2.3	83	144	61
Nemea				
Hercules (red)	2.6	90	122	23
Rhodes				
Rhodes (red)	10	139	198	67
Chevalier de Rhodes (red)	6.8	67	173	95
Attiki				
Cambas Hymettos (white)	4.6	74	88	25
Kourtaki (white)	2.9	79	130	38
Crete				
Minos (white)	5.2	98	119	32
Halkidiki				
Domain Porto Carras Cote de Meliton (white)	1.2	99	106	47

3. Summary of analyses

A summary of the results of the analyses of aluminum, calcium, magnesium and vanadium in experimental and commercial
red and white table wines of Greece are presented in Table V.
This table includes number of samples, ranges, averages and
standard deviations. The results are grouped by wine type.

In experimental wines it was found that concentrations of
vanadium are slightly higher in white wines than in red ones.
The opposite is true for magnesium, while similar concentrations of aluminum and calcium were observed both in red and
white wines.

In commercial wines similar concentrations of Ca, Mg and
V were found in red and white wines while slightly higher concentrations of Mg were observed in red wines than in white ones.

TABLE V.   Summary of analyses of Al, Ca, Mg and V in
           experimental and commercial red and white
           table wines of Greece.

Wine type (Number of samples)	Aluminum (mg/L)	Calcium (mg/L)	Magnesium (mg/L)	Vanadium (μg/L)
Experimental red table (23 samples)				
Minimum	0.84	31	75	1.0
Maximum	4.9	131	202	34
Average	(1.9)	(65)	(121)	(11)
Experim. white table (14 samples)				
Minimum	0.72	26	64	1.7
Maximum	4.1	101	118	66
Average	(2.2)	(63)	(91)	(18)
Commercial red table (8 samples)				
Minimum	1.0	52	89	9.6
Maximum	10	139	198	95
Average	(3.8)	(82)	(142)	(43)
Commercial white table (4 samples)				
Minimum	1.2	74	88	25
Maximum	5.2	99	130	47
Average	(3.5)	(87)	(111)	(36)

Commercial red and white wines have generally more aluminum and vanadium and slightly more calcium and magnesium than experimental ones.

Ranges and averages of calcium and magnesium found in this study in Greek experimental and commercial red and white table wines are in line with ranges and means reported for table wines from other countries (1). The ranges and average values for aluminum found in this study in Greek table wines are higher than those reported for California table wines but they are similar with those reported in French and Bulgarian wines (1,6, 14). Values of vanadium in Greek wines found in this study are lower than those reported in literature (8) but similar with those reported for Yugoslavian wines (12).

IV. CONCLUSIONS

A rapid Instrumental NAA method has been developed for the simultaneous determination of aluminum, calcium, magnesium and vanadium in wines.

The method was successfully applied to the determination of these elements in 49 samples of experimental and commercial red and white table wines from different wine production areas of Greece.

Chronological variations of aluminum and vanadium were observed in several experimental red and white Greek wines.

The concentrations of Al, Ca, Mg and V found in Greek wines were in line with those reported for wines from other countries.

ACKNOWLEDGMENTS

The authors wish to thank the Wine Institute of Athens, Greece for giving the experimental wines analysed and especially the Director of this Institute Dr. S. Kourakou-Dragonas for useful suggestions and helpful discussions.

REFERENCES

1. Amerine, M.A. and Ough, C.S.,"Methods of Analysis of Musts and Wines". J. Wiley and Sons, New York, (1980).
2. Amiel, S., "Nondestructive Activation Analysis", Elsevier Scientific Publishing Co., New York, (1981).
3. Grimanis, A.P., Nat. Bur. Stand. (USA) Spec. Publ. 312, Vol. 1, 197, (1969).

4. Beridge, G.I., Macharashvili, G.R. and Mosulishvili, L.M., Radiokhimya, Coden: Radkau Ser. 11, No. 6, 726, (1969).
5. Dedvariani, T.G., Beridge, G.I., Zakharov, E.A., Macharashvili, G.R. and Doneliya, G.I., Soobshch. Akad. Nauk. Gruz, SSR, 1973: Coden SAKNAM Ser. 72, No. 1, 161, (1973).
6. May, S., Leroy, J., Piccot, D. and Pinte, G., J. Radioanal. Chem. 72, 305, (1982).
7. Grimanis, A.P., Vassilaki-Grimani, Maria, Kanias, G.D., in "The Quality of Foods and Beverages" (G. Charalambous, G. Inglett eds) p. 349, Academic Press, New York, (1981).
8. Amerine, M.A., and Joslyn, M.A., "Table wines", University of California Press, Berkeley, Los Angeles, London (1970).
9. Ough, C.S., Crowell, E.A. and Benz, J., Journal of Food Science 47, 825, (1982).
10. Frolov-Bagreev, A.M. and Andreevskaia, E.G., Vinodelie i Vinogradrstvo SSSR 10 (6), 38 (1950).
11. Myron, D.R., Givand, S.H. and Nielsen, F.M., J. Agric. Food Chem. 25, No. 2, 297 (1977).
12. Byrne, A.R. and Kosta, L., "The Science of the Total Environment", 10, 17, (1978).
13. Bowen, H.J.M., At. Energy Rev. 13, 451 (1975).
14. Lechev, V. and Geneva Ts., Lozar, Vinar 24, No. 8, 33, (1975).

APPLICATION OF PATTERN RECOGNITION TECHNIQUES IN THE DIFFERENTIATION OF WINES

J. Schaefer[1], A.C. Tas[1], J. Velisek[2],
H. Maarse[1], M.C. ten Noever de Brauw[1] and P. Slump[1]

1. INTRODUCTION

In the wine trade the authenticity of the product is a very important aspect. On the label detailed information is given on country and district from which the wine originates as well as on the quality rating and the grape species. In practice, this information is not always fully reliable and therefore consumers as well as wine merchants would benefit from a method to control the authenticity. So far, a small (3-5 members) taste panel has excerted the control by assessing the wine. Although these taste panel members have a remarkable ability to "identify" coded wine samples, changes in the wine production are the cause that they cannot possibly always arrive at the correct conclusion. Additional techniques are needed to differentiate between the various wines and to establish the authenticity. Unfortunately, the differences in composition within a number of wine samples of the same type are of the same order as those between samples of different types of wine. Discrimination between types of wine with the aid of one or two parameters is therefore only possible in exceptional cases. However, a versatile tool is available to extract information from the many data obtained by chromatographic analysis of wine samples in the form of multivariate data analysis, also known as pattern recognition. In literature some examples are to be found (1-6). Siegmund and Bächman (4) and Kwan (5) illustrated the possibilities of the technique by determining the origin of wines using the data on the content of

1) Institute CIVO-Analysis TNO, Zeist, The Netherlands
2) Institute of Chemical Technology, Prague, Czechoslovakia

Instrumental Analysis of Foods
Volume 2

trace elements. Noble (6) succeeded in discriminating between three types of wine: White Riesling, Chardonnay and
French Colombard by analyzing the volatile compounds of the
wine. In our investigation we tried to differentiate between
two German white wines from the same district, but made from
different grape species: Müller-Thurgau and Riesling. Wines
of these two grape species were analysed on the concentration of volatile compounds, amino acids and non-volatile
acids. We applied pattern recognition techniques in order to
classify the wines.

II. MATERIALS AND METHODS

A. Wine samples

In an investigation such as this it is essential that
the wine samples are authentic. There should in fact not be
any doubt as to the origin and the grape species from which
they have been made. To be quite certain in this respect the
wine samples were collected with the German manufacturers
themselves. We obtained 20 samples of Rheingauer Müller-Thurgau white wines, and 25 samples of Rheingauer Riesling white wines of 1971-1981, representing different quality classes (table 1). Each sample consisted of two 0,7 l
bottles which were stopped with a cork. The samples were
stored horizontally at a temperature of 5 °C.

B. Analysis

1. Volatile compounds. The volatile compounds in the wine
   samples were analysed in two ways:

a. Low boiling compounds were analysed gas chromatographically by direct injection of the wine. A bottle of wine
   was opened and a sample was immediately injected on system 1. as described in section II.B.5.
b. Higher boiling compounds were separated from the wine by
   extraction; 100 ml of wine was extracted four times in a
   separation funnel with 15 ml pentane/ether 35+65. The
   combined extracts were extracted four times with 10 ml
   5 % NaHCO$_3$ solution to separate the acidic compounds.
   After drying of the extract with Na$_2$SO$_4$ the volume
   was reduced to 0.5 ml by distillation of the solvent at
   atmospheric pressure using a Vigreux column. Then the extract was analysed gaschromatographically on system 2.
   (see section II.B.5).
c. The volatile compounds were identified by a gas
   chromatograph-mass spectrometer-computer combination

TABLE 1. Wine samples, Rheingauer Müller-Thurgau (1-20) and
Riesling (21-45), used in the training set

---

1. Lorcher Kapellenberg, QbA, 1980
2. Johannisberger Erntebringer, QbA, 1981
3. Hattenheimer Schützenhaus, QbA, 1981
4. Kiedricher Sandgrub QbA, 1981
5. Rüdesheimer Magdalenenkreuz, Kabinett, 1978
6. Hattenheimer Heiligenberg, Kabinett, 1978
7. Johannisberger Erntebringer, Kabinett, 1978
8. Hattenheimer Schützenhaus, Kabinett, 1979
9. Lorcher Schlossberg, Kabinett, 1979
10. Lorcher Kapellenberg, Kabinett, 1980
11. Johannisberger Erntebringer, Kabinett, 1980
12. Lorcher Pfaffenwies, Spätlese, 1976
13. Geissenheimer Fuchsberg, Spätlese, 1971
14. Geissenheimer Fuchsberg, Spätlese, 1972
15. Rüdesheimer Magdalenenkreuz, Kabinett, 1973
16. Geissenheimer Fuchsberg, Kabinett, 1973
17. Rüdesheimer Burgweg, QbA, 1974
18. Rüdesheimer Burgweg, Kabinett, 1975
19. Rüdesheimer Magdalenenkreuz, QbA, 1975
20. Rüdesheimer Magdalenenkreuz, Kabinett, 1975
21. Rüdesheimer Berg Rottland, Spätlese, 1971
22. Rüdesheimer Bischofsberg, Auslese trocken, 1976
23. Rüdesheimer Burgweg, QbA, 1979
24. Rüdesheimer Rosengarten, Kabinett, 1975
25. Rüdesheimer Rosengarten, Spätlese, 1979
26. Rüdesheimer Berg Rottland, Spätlese, 1979
27. Rüdesheimer Burgweg, Spätlese, 1976
28. Rüdesheimer Berg Rottland, Auslese, 1976
29. Rüdesheimer Bischofsberg, Auslese trocken, 1976
30. Rüdesheimer Bischofsberg, Auslese, 1976
31. Rüdesheimer Klosterlag, QbA, 1978
32. Rüdesheimer Klosterlag, Kabinett, 1979
33. Rüdesheimer Klosterlag, QbA, 1977
34. Rüdesheimer Rosengarten, Spätlese, 1971
35. Rüdesheimer Berg Rottland, Spätlese, 1975
36. Rüdesheimer Bischofsberg, Auslese, 1976
37. Rüdesheimer Burgweg, QbA, 1980
38. Rüdesheimer Kirchenpfad, QbA, 1980
39. Rüdesheimer Magdalenenkreuz, QbA, 1979
40. Rüdesheimer Magdalenenkreuz, Kabinett, 1979
41. Rüdesheimer Burgweg, Kabinett, 1979
42. Rüdesheimer Magdalenenkreuz, Auslese, 1976
43. Rüdesheimer Burgweg, QbA, 1978
44. Rüdesheimer Magdalenenkreuz, QbA, 1977
45. Rüdesheimer Magdalenenkreuz, QbA, 1973

(GC–MS–COMP). Since it was obvious that the same volatile compounds were present in all of the wine samples, identification was carried out only on two wine samples, nrs. 13 and 42 (see table 1). The same gaschromatographic conditions were applied as described for the quantitative analysis in a. and b. For a description of the GC–MS–COMP system see section 2.2.4.

2. Non-volatile acids

The non-volatile acids were analysed according to a method described by Fantozzi and Bergeret (7) and Fernandez et al. (8).

To 10 ml of wine 50 ml of ethanol 96 % was added in order to fulfill the requirement that the ethanol content of the sample must be not less than 80 %. After standing one hour at 5 °C the sample was filtrated; 10 ml of the filtrate was transferred to a centrifugating tube containing 0,2 g of Celite 545. A saturated solution of neutral lead acetate $(Pb(CH_3COO)_2$ and 30 ml of 85 % ethanol was added. After standing for one hour at room temperature the mixture was centrifugated at 3000 r/min., decanted, washed and centrifugated three times with 10 ml of 85 % ethanol. Then the sediment, containing the led salts of the acids, was dried at 100 °C and transferred to a silylation vessel containing 0,2 g anhydrous $CaSO_4$. The vessel was closed then and subsequently 0,5 ml pyridine, 0,25 ml HMDS and 0,1 ml TMCS were injected through the cap. After standing for one hour at 50 °C the reaction mixture was analysed gas chromatographically on system 3 (see section II.B.5).

3. Amino acids

The amino acids were determined on a Biotronik model LC 6001 E amino acid analyzer equipped with a Spectra Physics model SP 4100 datasystem. The chromatographic conditions destinated for protein analysis were used. By this method asparagine and glutamine are not completely separated from serine and glutamic acid.

The free amino acids were determined by direct injection of the wine in the system. The total amount of amino acids was determined after hydrolysis of the wine with 6N HCl during 24 hours under reflux.

4. Identification

Identification of the volatile compounds was carried out by GC–MS–COMP.

The gas chromatographic conditions were:
Apparatus: Varian 3700
Column: glass, 50 m x 0,5 mm i.d.
Stationary phase: SP 2300
Oven temperature: 5 min. at 50 °C, 4 °C/min. to 200 °C.
Carrier gas: He, 5 ml/min.
The gas chromatograph was coupled to a Finnigan-MAT 112 mass
spectrometer by way of an open split interfase. Every 2,5
sec. a mass spectrum was obtained from the GC effluent and
stored in a Finnigan-MAT SS200 computer system.

5. Gaschromatography

For the analysis of volatile compounds and non-volatile
acids three systems were used:
System 1. Low boiling compounds
Apparatus   : Intersmat 16
Column   : glass, 50 m x 0,3 mm i.d.
Stationary phase: Carbowax 400
Oven temperature: 75 °C
Carrier gas   : He, 3 ml/min.
Detector   : FID, H_2 30 ml/min., air 500 ml/min.
Injected volume : 0,5 ul, splitter 1:10

System 2. Higher boiling compounds
Apparatus   : Packard Becker 433
Column   : glass, 25 m x 0,3 mm i.d.
Stationary phase: CP Wax 51
Oven temperature: 3 min. at 35 °C, 8 °C/min. to 200 °C
Carrier gas   : He, 3 ml/min.
Detector   : FID, H_2 30 ml/min., air 500 ml/min.
Injected volume : 0,5 ul

System 3. Non-volatile acids
Apparatus   : Intersmat 16
Column   : glass, 25 m x 0,3 mm i.d.
Stationary phase: CP Sil 5
Oven temperature: 3 min. at 100 °C, 4 °C/min. to 200 °C
Carrier gas   : He, 3 ml/min.
Detector   : FID, H_2 30 ml/min., air 500 ml/min.
Injected volume : 0,5 ul

C. Computerized Pattern Recognition

1. Introduction

At the present time modern analytical instruments produce large amounts of data, often obtained in the form of

chromatograms and spectra or listed in tables as numerical intensity or area values.

The analytical chemist often finds it impossible to make an optimal use of the total amount of information provided. Multivariate data analysis in the form of pattern recognition (PARC) has been found a versatile tool for extraction of information from such data sets. The necessity for applying PARC does not exist only in the case that large data sets have to be elaborated. PARC can also be beneficially used for smaller data sets of a more complex structure. When, for example, measurements are carried out on complex samples such as food products, to determine properties like the quality or the origin of the samples, it is not clear which variables or combinations of variables are related to these properties. In such cases application of PARC will reveal such relationships, if present (9). It is quite obvious therefore that PARC has been applied in the analysis of several food products (10, 11, 12).

By using PARC a geometric description of the measured objects can be given. When an object is described by m parameters, it can be considered as a point in an m-dimensional space, the mesured parameters being the coordinates of each point in m-space. All the objects, measured on the <u>same</u> variables, occupy places in the same m-space. Using such a geometrical description, the data can be analyzed by different PARC-methods, some of which will be described in this paper.

2. Data preprocessing and feature selection

Variables were first sequenced according to the magnitude of the ratio of their interclass variability and intraclass variability (various weight). Subsequently, the data were autoscaled, converting each variable into a feature with a mean of zero and a variance of one over the whole data set. For a PDP 11/34 computer system the data set (45 objects, 115 parameters) is too large to be handled as a whole by most of the subroutines used. Therefore, the data set was divided into three subsets of about the same size.
The first set contained the first 40 selected features, the second set the next 40 features, and so on.

All three subsets were screened by principal components (PC) analysis (see section II.C.3).
Class discrimination power of the variables was used for selection, so it is to be expected that features carrying relevant information will mainly be concentrated in the first subset. However, there is no guarantee that the most optimal combination of features is transferred to this sub-

set. Feature selection was limited to prevent that features discriminating only by chance will dominate too much.

3. Methods

Data were analyzed using the pattern recognition program ARTHUR, implemented on a PDP 11/34 DEC computer system.

a. Display methods

For a visual examination, the objects can be projected from m-space to the two dimensional space (13). In this way groups of objects (clusters) and outliers can be visualized.

The two display methods most frequently used are non-linear mapping (NLM) and principal components (PC) projection.
NLM tries to conserve the interobject distances during projection from m-space in two-space. With PC plots (also named eigenvector plots or Karhunen-Loeve plots) objects are projected in the plane of two principal components, which are constructed as linear combinations of the original variables. These new variables are uncorrelated (orthogonal) and give the best approximation in a limited number of variables of the variance of the original data. The use of the PC projection has the advantage that several windows (= projections) can be used to gain an impression of m-space. Moreover, the contribution of the original variables to the principal components is known.
In this study PC projection was used for display and for interpretation of the first principal components which account for a major part of the variance.

b. K Nearest Neighbour Classification (KNN)

Similar objects can be expected to occupy positions close together in m-space. Therefore the simple Eucledian distance (14) is a good measure of similarity. KNN is considered as a standard method of PARC which classifies an object according to the majority of its k-nearest neighbours in m-space, using a distance matrix of objects (15). The method also shows the position of objects in m-space relatively to each other.
In our study the method was used for both classification and screening of the m-space.

Even when a small number of objects (d) is involved, many variables (m) can be used for classification (d/m < 1 is allowed) (16). The classification performance can be improved, however, by applying an appropriate feature selection to remove noisy variables (17).

c. SIMCA analysis

In the SIMCA analysis each class is represented by a disjointed principal component (PC) model. Around the model a confidence interval is constructed. In this way a hyperbox is obtained for each class. This PC model describes the systematic variance within a class. For a more detailed description see Wold and Blomquist (18, 19).
The method enables the detection of outliers, the determination of interclass distances and the occurrence of class overlap. Features can be selected according to their participation in the class model (modelling power) and their influence on the class separation (discriminating power). In this way noisy variables can be removed and a reduced data set is obtained with an improved classification performance.
SIMCA analysis of the wine data will be carried out at the Pharmaceutical Institute of the Free University, Brussels by dr M. Derde and prof D.L. Massart and will be subject of a future publication.

Summary of the characteristics of the techniques described in this section; PC analysis does not make use of knowledge of class membership and can be considered as a method of unsupervised learning. KNN is a technique to be used in a supervised learning mode for classification, as well as for estimation of the position of the objects in m-space in a unsupervised way.

III. RESULTS AND DISCUSSION

A. Instrumental analysis

A number of 26 volatile compounds could be identified with the aid of GC-MS-COMP while 11 remained unknown by lack of reference compounds. In table 2 the results of the identification and the quantitative determination by GC are compiled. Only a small number of the non-volatile acids could be identified. They are given in table 3 just as the results of the quantitative determination of those compounds. The results of the quantitative determination of aminoacids – free and after hydrolysis – are given in table 4. Two peaks not belonging to protein amino acids were found in all the chromatograms. One of them obviously belonged to ethanolamine, the other one remained unknown.

TABLE 2. Volatile compounds in Müller-Thurgau and Riesling wines. Mean and range of concentrations

		Müller-Thurgau		Riesling	
nr.	compound	mean µg/1	range µg/1	mean µg/1	range µg/1
1		516	62- 3332	180	83- 648
2	acetaldehyde	8076	2585-15083	7284	4454-11329
3	acetone	1851	230- 4284	2062	437- 5016
4	ethylacetate	5163	2634- 7775	6191	3233-12130
5		2728	750- 6013	2780	831- 7863
6	methanol	8099	5777-12250	9742	6698-13653
7	propanol	5292	1929- 9894	5634	3046-10838
8	isobutanol	17840	7370-37310	15900	6820-22080
9		258	138- 373	249	143- 308
10	2-methylbutanol	8266	5544-12840	9362	6718-14779
11	3-methylbutanol	32228	22380-58660	33438	24820-52330
12		2220	395- 3807	1963	693- 4557
13		934	413- 1808	662	100- 1157
14	n-hexanol	7438	1690-21310	9594	3650-42100
15		4363	500- 9356	3720	486-11752
16		10471	1500-23290	16100	2380-40000
17		30772	12890-43930	34926	15610-53030
18		532	323- 763	566	204- 1054
19		3736	1355- 8315	4355	1070- 9010
20		28	14- 101	20	12- 51
21	isobutylacetate	165	18- 508	73	5- 401
22	n-butanol	71	43- 159	61	26- 94
23		25	10- 55	18	10- 50
24	isoamylacetate	334	56- 1235	411	47- 1605
25	ethylacetate	838	300- 1308	635	315- 1033
26	isobutyric acid	26	9- 70	14	8- 23
27	2,3-butanediol	132	66- 256	118	69- 172
28		13	8- 19	17	11- 50
29	hexanoic acid	26	10- 71	21	11- 38
30	3-methylthiopropanol	377	27- 1476	641	88- 1508
31		26	10- 76	29	12- 68
32	diethylsuccinate	71	8- 255	41	8- 140
33	octanoicacid	72	27- 137	120	12- 386
34	2-phenylethanol	11378	4075-38000	8943	4350-20575
35	2,4-hexadienoicacid	1339	90- 4370	1935	371- 4718
36	N-isoamylacetamide	12	4- 26	16	8- 29
37	diethylmaleate	224	75- 503	188	75- 412

TABLE 3. Non-volatile acid in Müller-Thurgau and Riesling
wines. Mean and range of concentrations

		Müller-Thurgau		Riesling	
nr.	compound	mean μg/l	range μg/l	mean μg/l	range μg/l
1		259	130- 405	243	121- 385
2	succinic acid	275	153- 479	346	179- 463
3		9	6- 18	9	4- 15
4		8	4- 14	9	5- 12
5		15	2- 40	19	5- 35
6		12	4- 22	14	2- 33
7	malic acid	2278	729-4529	2255	451-3704
8		86	41- 141	93	48- 159
9		19	2- 44	30	4- 77
10	tartaric acid	390	193- 710	404	230- 954
11	citric acid	55	10- 94	59	11- 95
12		39	7- 93	37	3- 124
13		124	5- 335	83	6- 178
14		12	1- 21	10	2- 37
15		78	9- 155	69	2- 207
16		42	22- 65	53	19- 121
17	ferulic acid	184	51- 449	360	31-1346
18		50	13- 115	64	24- 121
19		55	10- 100	73	7- 195
20	caffeic acid	384	221- 764	299	15- 446

TABLE 4. Amino acids, free and after hydrolysis, in Müller-Thurgau and Riesling wines. Mean and range
of concentration

nr. compound	Müller-Thurgau		Riesling	
	mean μmol/l	range μmol/l	mean μmol/l	range μmol/l
1 aspartic acid - free	433	186- 841	447	216- 720
2 aspartic acid - hyd.	800	355-1243	980	573-1480
3 threonine - free	320	76- 720	410	107- 537
4 threonine - hyd.	571	321-1056	679	485-1035
5 serine - free	398	130-1381	377	220- 677
6 serine - hyd.	703	419-1603	782	536-1200
7 glutamic acid - free	606	252-1453	667	317-1170
8 glutamic acid - hyd.	2516	1227-6030	2800	888-6300
9 proline - free	4913	2084-6910	4220	2460-6800
10 proline - hyd.	4462	1970-6815	3726	2140-5610
11 glycine - free	277	135- 473	281	173- 356
12 glycine - hyd.	810	462-1126	970	648-1590
13 alanine - free	1197	378-3850	967	478-1720
14 alanine - hyd.	1639	575-4330	1432	719-2570
15 valine - free	314	145- 761	316	205- 554
16 valine - hyd.	530	120- 986	624	372-1150
17 methionine - free	64	23- 138	90	18- 156
18 methionine - hyd.	64	17- 175	77	20- 215
19 isoleucine - free	167	61- 362	171	100- 312
20 isoleucine - hyd.	337	143- 544	402	218- 829
21 leucine - free	314	110- 546	335	181- 698
22 leucine - hyd.	452	168- 706	537	228-1155
23 tyrosine - free	118	43- 258	129	65- 186
24 tyrosine - hyd.	160	27- 349	189	88- 256
25 phenylalanine - free	207	78- 430	207	98- 309
26 phenylalanine - hyd.	280	91- 502	312	179- 582
27 γ-aminobutyric acid - free	794	203-1272	717	132-1350
28 γ-aminobutyric acid - hyd.	782	166-1300	678	155-1170
29 histidine - free	164	84- 393	148	93- 190
30 histidine - hyd.	210	61- 471	223	160- 317
31 ornithine - free	190	19- 808	163	2- 760
32 ornithine - hyd.	304	29-1100	246	18-1075
33 lysine - free	233	45- 331	297	131- 593
34 lysine - hyd.	337	82- 526	488	232- 996
35 arginine - free	2477	632-6240	1930	345-4210
36 arginine - hyd.	2585	664-6570	2070	411-4540
37 ethanolamine - free	203	88- 253	210	130- 262
38 ethanolamine - hyd.	268	219- 361	320	247- 441
39	153	54- 343	259	64- 966

The results stated in tables 2, 3 and 4 show that discrimination between the two types of wine cannot be effected
by one single component. For some compounds the mean value
may be different for the two wines, but the range is high,
so there is a huge overlap.

B. Data analysis

1. Description of the data matrix

Müller-Thurgau wines (20) and Riesling wines (25) were
measured on 19 free and 20 bonded amino acids, 37 volatile
compounds and 20 non-volatile acids. Moreover the quotients
of 19 free and bonded amino acids were calculated and included in the data analyses. Hence a total of 115 parameters
was obtained.

2. Pattern recognition

a. Principal component analysis

PC Analysis of the first subset
    The first subset of 40 features was analyzed by PC. The
objects were plotted in the plane of PC 1 and PC 2, coded as
PC 12 (Fig. 1), PC 1 and 3 (Fig. 2) and PC 2 and 3 (Fig. 3).
From the PC 12 plot it can be concluded that the greater
part of the wines can be differentiated according to their
class. A striking phenomenon is a subgroup being clearly
distinguishable on the left side of the plot (Fig. 4). All
members of this group belong to the so-called "Auslese"
wines, which are produced from special, selected grapes.
Being a subgroup of Riesling wines (category 2) they occupy
a distinct position in the plot.

As PC 12 and PC 13 show, the first PC renders the best
differentiation. The main contribution to the first PC was
formed by a number of acidic compounds, like ferulic acid,
caprylic acid and succinic acid.

Inspection of the measured GC data revealed that the
mean concentrations of these compounds are higher for "Auslese" wines and Riesling wines, compared to the Müller-Thurgau wines. The same trend in the data was observed with
3-methylthio-1-propanol. On the other hand, the concentration of ethyl lactate was lower in these wines.
In addition, the ratio of free and bonded aspartic acid and
glycine (Q-Asp and Q-Gly) loaded to a substancial extent on
this PC.

The major contribution to PC 2 is given by 14 amino acid loadings (bonded and quotient), which appear to be negatively correlated to the subsequent 8 GC peaks (volatile compounds and acids). The first three PC's account for circa 50 % of the variance of the original variables.

PC analysis of the second subset

PC analysis was applied to the second subset of 40 features (feature 41-80). The objects were plotted again in the plane of the first two PC's (Fig. 5).

As is shown in the plot, the differentiation between the two main categories has been disappeared almost entirely. Only the subgroup of "Auslese" wines shows a distinct cluster. It is clear that this subset still contains some useful information as regards classification.

PC analysis of the third subset

The third subset of 35 features was analyzed in the same way. The PC 12 plot (Fig. 6) shows that it is impossible to differentiate between the two categories with this set of features. However a subgroup of five out of six "Auslese" wines are found close together in the projection.

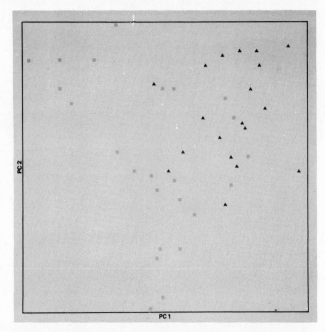

FIGURE 1. Plot of PC 1 and 2

▲ Müller-Thurgau   ■ Riesling   ● Auslese

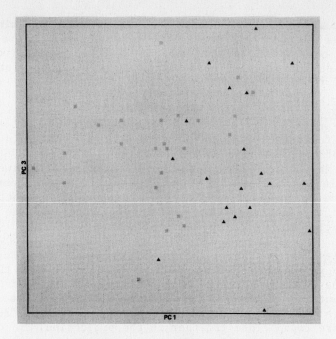

FIGURE 2. Plot of PC 1 and 3

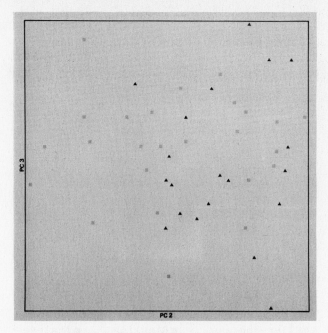

FIGURE 3. Plot of PC 2 and 3

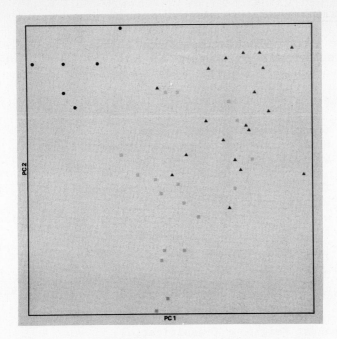

FIGURE 4. Plot of PC 1 and 2 (subgroup detection)

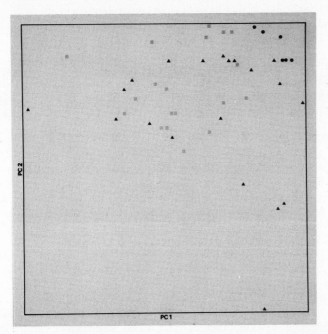

FIGURE 5. Plot of PC 1 and 2 (second subset of variables)

Some variable/variable plots

The trends which were visible in the PC analysis of the first subset can be studied in more detail by plotting of some two-variable plots.

Our observations have been laid down as follows in:

Fig. 7 (ferulic acid/caprylic acid):
   Higher concentrations for "Auslese" wines (except wine 42 R22)

Fig. 8 (ferulic acid/3-methylthio-1-propanol):
   Higher concentrations for "Auslese" wines (except wine 42 R22 which shows a lower concentration acid, but a high methylthiopropanol content)

Fig. 9 (ferulic acid/aspartic acid):
   Higher acid concentrations and medium aspartic acid concentrations for "Auslese" wines (wine 42 R22 shows a low aspartic acid concentration)

b. K Nearest Neighbour Classification

The first subset of 40 features was used for KNN classification. The classification results for 1NN, 3NN and 4NN, and the distances to the three nearest objects are presented in table 5.

In accordance with the results of the PC analysis the distance matrix shows that "Auslese" wines occupy a rather distinct position in m-space. Predominantly, the individual members have these subgroup members in their vicinity.

IV. CONCLUSION

The data analytical methods chosen are expected to give a reliable picture of our data set. A further improvement of the class distinction will be carried out by defining PC models for each class and subgroup, using the SIMCA method.

V. ACKNOWLEDGEMENT

We thank mr. J. Bouwman for implementing the Arthur package on the PDP 11/34 computer and Ir J.T.N.M. Thissen for stimulating discussions.
We thank mr. H. Laber from Asbach & Co, Rüdesheim am Rhein for the collection of the wine samples.

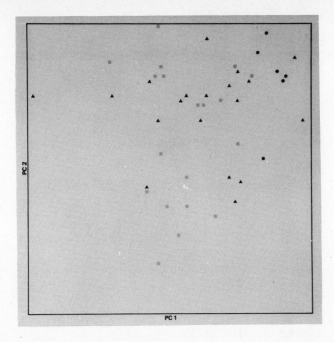

FIGURE 6. Plot of PC 1 and 2 (third subset of variables)

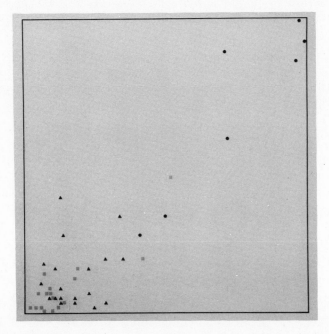

FIGURE 7. Plot of ferulic acid (hor) and caprylic acid (vert)

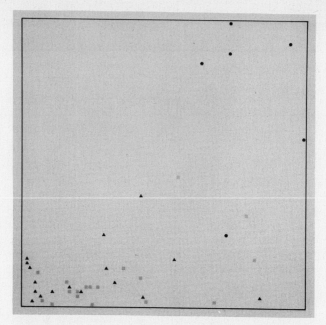

FIGURE 8. Plot of ferulic acid (vert) and
3-methylthio-1-propanol (hor)

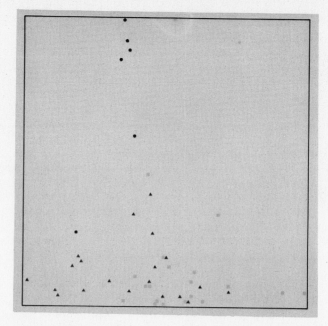

FIGURE 9. Plot of ferulic acid (vert) and aspartic acid (hor)

TABLE 5. KNN classification results

object	classification				nearest neighbours					
	cat.	1NN	3NN	4NN	obj.	dist.	obj.	dist.	obj.	dist.
1	1	1	1	1	2	0.88	4	0.94	9	0.98
2	1	1	1	1	4	0.85	1	0.88	41	0.97
3	1	1	1	1	4	1.05	9	1.16	12	1.21
4	1	2	2	2	41	0.83	40	0.84	2	0.85
5	1	2	2	2	45	0.91	18	0.98	43	0.99
6	1	1	1	1	9	0.76	20	0.84	44	0.84
7	1	1	1	1	11	0.96	17	0.99	9	1.03
8	1	1	1	1	17	0.72	9	0.81	14	0.83
9	1	1	1	1	6	0.76	44	0.79	12	0.80
10	1	1	1	1	4	1.19	39	1.23	8	1.26
11	1	1	1	1	17	0.85	7	0.96	8	1.00
12	1	1	1	1	9	0.80	44	0.93	6	0.97
13	1	2	2	2	34	0.96	21	1.02	14	1.04
14	1	1	1	1	9	0.82	17	0.82	44	0.83
15	1	1	1	1	9	1.20	43	1.20	45	1.24
16	1	1	1	1	8	1.01	14	1.04	12	1.04
17	1	1	1	1	8	0.72	20	0.81	9	0.81
18	1	1	2	2	20	0.63	24	0.69	44	0.77
19	1	2	1	1	27	0.76	20	0.82	18	0.83
20	1	1	1	1	18	0.63	44	0.74	17	0.81
21	2	2	1	2	34	0.81	14	0.93	20	0.95
22	2	2	2	2	28	0.98	29	0.98	36	0.99
23	2	2	2	2	32	0.84	39	0.85	25	0.89
24	2	2	1	1	35	0.65	18	0.69	20	0.88
25	2	2	2	2	26	0.76	32	0.87	35	0.89

TABLE 5. (continued)

object	classification				nearest neighbours					
	cat.	1NN	3NN	4NN	obj.	dist.	obj.	dist.	obj.	dist.
26	2	2	2	2	25	0.76	35	0.91	8	0.93
27	2	1	1	1	19	0.76	18	0.79	35	0.83
28	2	2	2	2	36	0.70	22	0.98	29	1.07
29	2	2	2	2	30	0.91	22	0.98	27	1.01
30	2	2	2	2	29	0.91	27	0.98	36	1.07
31	2	2	2	2	32	0.90	23	0.96	39	1.00
32	2	2	2	2	23	0.84	39	0.86	25	0.87
33	2	2	2	2	35	0.87	19	0.90	39	0.95
34	2	2	2	2	21	0.81	13	0.96	14	0.99
35	2	2	2	2	24	0.65	18	0.79	27	0.83
36	2	2	2	2	28	0.70	22	0.99	27	1.04
37	2	2	2	2	23	0.96	35	1.00	24	1.05
38	2	2	2	2	32	1.03	39	1.10	37	1.12
39	2	2	2	2	23	0.85	32	0.86	17	0.86
40	2	2	2	2	41	0.73	44	0.83	4	0.84
41	2	2	2	2	40	0.73	44	0.78	4	0.83
42	2	2	2	2	45	1.01	22	1.12	30	1.15
43	2	2	2	2	44	0.83	5	0.99	40	1.00
44	2	1	1	1	20	0.74	18	0.77	41	0.78
45	2	1	1	1	5	0.91	44	0.96	18	0.99

% correct classif.

class. 1		80	75	75
class. 2		88	76	84

VI. REFERENCES

1. Kwan, W.O. and Kowalski, B.R., J. Food Sci. 43 (1978) 1320
2. Rapp, A., Hastrich, H., Engel, L. and Knipser, W., Bull. O.I.V. 53 (1980) 91
3. Symonds, P. and Cantagrel, R., Ann. Fals. Exp. Chim. 75 (1982) 63
4. Siegmund, H. and Bächmann, K., Z. Lebensm. Unters. u. Forsch. 166 (1978) 298
5. Kwan, W.O., Kowalski, B.R. and Skogerboe, R.K., J. Agric. Food Sci. 27 (1979) 1321
6. Noble, A.C., Flath, R.A. and Forrey, R.R., J. Agric. Food Sci. 28 (1980) 346
7. Fantozzi, P. and Bergeret, J., Ind. Agric. Alim. 90 (1973) 731
8. Fernandez-Flores, E., Kline, D.A. and Johnson, A.R., J. Assoc. Offic. Anal. Chem. 53 (1970) 17
9. Kowalski, B.R. and Bender, C.F., J. Am. Chem. Soc., 94 (1972) 5632
10. Smeyers-Verbeke, J., Massart, D.L., and Coomans, D.J., J. Assoc. Off. Anal. Chem. 60 (1977) 1382
11. Greef, J. van der, Tas, A.C., Bouwman, J., Noever de Brauw, M.C. ten and Schreurs, W.H.P., Anal. Chim. Acta, in press
12. Leegwater, D.C. and Leegwater, J.A., J. Sci. Food Agric. 32 (1981) 1115
13. Kowalski, B.R. and Bender, C.F., J. Am. Chem. Soc., 95 (1973) 686
14. Massart, D.L., Dijkstra, A. and Kaufman, L., Evaluation and optimization of laboratory methods and analytical procedures. A survey of statistical and mathematical techniques, Elsevier, Amsterdam, 1978
15. Varmuza, K., Pattern Recognition in Chemistry, Springer-Verlag, 1980, page 63
16. Wold, S., Proceedings of the IUFOST symposium on Food Research and Data Analysis, September 1982, Oslo, Norway
17. Coomans, D. et al., Anal. Chim. Acta, 133 (1981) 241
18. Wold, S., Pattern Recognition, 8 (1976) 127
19. Blomquist, G., et al., J. of Chromatogr., 173 (1979) 7

A STUDY OF NITROGEN FERTILIZATION AND FRUIT MATURITY AS AN APPROACH FOR OBTAINING THE ANALYTICAL PROFILES OF WINES AND WINE GRAPES

María Carmen Polo
Marta Herraiz
María Dolores Cabezudo

Instituto de Fermentaciones
Industriales, C.S.I.C.
Madrid (Spain)

INTRODUCTION

A certain differentiating role has been attributed to the amino acid composition in the various grape varieties(1-3) . It is also known that nitrogen constituents in their different forms (amonium nitrogen, amino nitrogen and peptidic nitrogen) and in the different parts of the plant (root, grape leaf petioles and grapes) depend on fertilization practices as well as on the the growing stage, when the samples have been collected (4-7). Are classical statements: the positive relationship between the total nitrogen and the amino nitrogen of the grapes, the higher levels of proline (the major amino acid in ripe grapes)compared to arginine (the major amino acid in unripe grapes) and the use of arginine as an indicator of the plant nitrogen requirements.

A degree of optimum maturity, common to all varieties of grapes, does not exist . For example, grapes for sparkling wine making are harvested earlier than grapes for white and red table wines. This is favourable to find out differences between varieties. However, ripeness criteria change according to the geographical areas for the same variety . Therefore, the lack of uniformity of the samples of an identical variety group reduces the initial advantage.

357

Some difficulties arise in the moment of planning the experimental tests adressed to find the chemical constituents which better show the differences between grape varieties. The final options are the following: either to get the analytical profile of each crop variety, using samples from plants cultivated under identical conditions, or to get this using grapes from different geographical areas, where viticultural practices and climate are not the same.

The first alternative is more interesting from the botanical than from the enological point of view. It might occur that some grapes samples belonging to one of the varieties for which the analytical profile has been defined does not fit into it, for many reasons: because they had been fertilized with more or less intensity or because quantities of delivered nitrogen being identical, the fallen rains - in abundant or scarcely - or the soil structure, made nitrogen assimilable in more or less proportion.

The second alternative is more realistic. The representative standard samples which belong to each group, as they come from different soils, climates and fertilization practices (within certain limits) will show the differences which strictly depend on the variety as well as those due to the other factors. However, an excessive hetereogeneity in samples of the same group should be avoided, as it would mask the differences between groups. The maturity grade should be considered, as much to group samples with an equivalent grade, as to point out the differences between groups, when harvesting practices are different.

All this is of double importance because the grape must is the raw material for white wine making, and red musts plus skins is the substrate which gives the red table wines. For this reason new causes of hetereogeneity should be added. They are derived from yeast metabolism, from the fermentation methods and from the other practices up until bottling. Beforehand, it can be affirmed that to build the chemical profile of any wine type, the samples available will vary within wide limits.

This investigation was made primarily to establish the heterogeneity admissible among the samples in order to select the analytical variables which show the most relevant aspects of the ampelographic varieties from which they proceed. The samples used for the tests are from nine Spanish varieties of Vitis vinifera.

MATERIAL AND METHODS

Viticultural

The fruit used was obtained from grapevines belonging to nine varieties, as follows:Xarel-lo, Macabeo and Parellada (wine grapes for sparkling wine making), Airén, Jaén and Pardillo (for white table wines) and Garnacha, Tempranillo and Bobal (for red table wines). Generally, the NPK fertilizing was done according to traditional methods, alternating in some cases the fertilization at the bottom with the foliar fertilization the year after. Samples of plots without nitrogen contribution and with a strong fertilization were also studied.

Harvest

Clusters of nearly all varieties were harvested at 15 day interval to create samples with a different grade of maturity. For the election of the various geographical harvesting areas, those with endemic varieties have been taken into consideration: Cataluña (varieties for sparkling wines), Madrid and La Mancha for the white wine groups, Toledo and Zaragoza for the Garnacha samples, La Rioja for Tempranillo grapes, and Levante for the Bobal ones. We had also available samples of the same grape varieties cultivated in other areas.

Samples

Grape musts from the above varieties and wines made from the same grapes. Wines have been made: a)with only one selected yeast, Sacch.cerevisiae , b) on pilot plant scale with the local yeasts of La Mancha and Levante, and c) industrially, with spontaneous yeasts of Levante and Cataluña.

Methods of analysis

Each of the musts was identified through 35 analytical variables which included: conventional analysis; free amino acids (quantitative HPLC analysis: reversed phase of the dansyl derivatives, polarity gradient methano/phosphate,fluorescens detector)(8); and minor volatile substances, extract by Freon 11 and separated by means of a GC packed column (Carbowax 20M+ Igepal Co880; 20:80; 6.8m x 2.2 mm;temperature program:80-180 ºC, 2º/min).

The same approach was used for wines, which have been identified through 64 analytical variables on the basis of more complex chromatograms and because of the major volatiles

(9)(acetaldehyde, diethyl acetal, esters from ethyl formate up
to isopentyl acetate, alcohols from methanol up to 3-methyl-1-
-butanol and 2-phenyl ethanol). Must and wine organic extracts
were also injected in a Superox 0.1 capillary column (16m x 0.2
mm int.diam.; temperature programm: 80ºC 10min, 80-184ºC 2º/min
split ratio: 1/100), which efficiency equals 3000 N_{eff}/m(k'=10,
80ºC) and lacks absorbancy for any of the Grob's test (10).

Data Treatment

Stepwise Discriminant Analysis by means of the BMDP7M
Programm (11).

RESULTS AND DISCUSSION

To calculate analytical profiles with a wide scope of
application depends on a careful selection of the samples to
constitute the standard populations. It was judged convenient
to establish certain conditions to the standard samples, in
order to maintain a balance between the uniformity due to the
common origin of samples of the same group and the heterogeneity
observed. In this respect, a previous study of the obtained
analytical data has been carried out through. The matter was to
verify until which extend, an inadequate N fertilization of the
crops or an inconvenient maturity of the grapes -lack or
excess --- would produce anomalous samples. In such a case,some
limits to certain analytical variables will be established in
order to choose the samples within those limits as standard
samples and to exclude those which do not fulfil the conditions.

Grapes: Total Nitrogen vs. Amino Nitrogen

The nitrogen contents depend on the N fertilization intensity
There is a quite satisfactory correlation between total and
amino nitrogen, without distinction of the grape's variety
-white or red-, among the nine varieties studied, as it is
shown in Fig.1 , and even independently of the maturity grade,
and for a wide range of total nitrogen levels:

$$\text{Total N (mg/l)} = 174.39 + 1.51 \text{ Amino N (mg/l)}$$
$$(n=48) \quad (r^2=0.842)$$

Grapes: Proline (PRO) and Arginine (ARG)
vs. Fruit Maturity

The scattered values quoted in the literature concerning these two amino acids, and our own experience, as well as the dependence of their concentration on the plant vegetative stage makes it necessary to establish such a correlation in our own numerical terms.

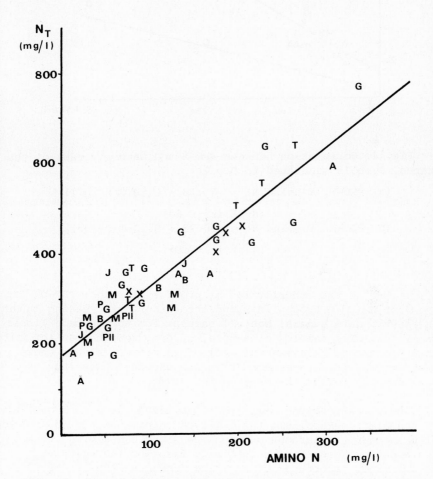

Fig. 1. Correlation between the total nitrogen and amino nitrogen contents in grape musts. Sample varieties: A= Airén, J= Jaén, P= Pardillo, X= Xarel-lo, M= Macabeo, Pll= Parellada, G= Garnacha, T= Tempranillo and B= Bobal.

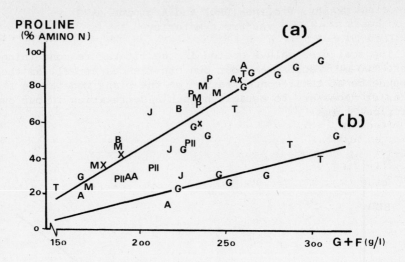

Fig. 2. Correlation between PRO-N and maturity rate of the grapes. Symbols as stated in Fig.1

(a) PRO-N (% amino N) = -69.23 + 0.58 G+F (g/l)

$$(n=37) (r^2=0.83)$$

(b) PRO-N (% amino N) = -50.73 + 0.32 G+F (g/l)

$$(n=9) (r^2=0.81)$$

In Fig. 2 the contribution of proline nitrogen (PRO-N) to the must amino nitrogen is shown according to increasing content in the two main sugars. Not all samples fit to the same linear model (a). Nine samples of a total of 46 offer the particularity that the gradual increase of proline is slower (b) and that even waiting for grapes acquire an unsuspected maturity , proline will not be the major amino acid.

Due to the proline-arginine relationship as a consequence of the plant metabolism, the observed phenomenon should coincide with another similar one of opposite signe concerning the arginine contents. Fig. 3 shows it clearly. The relative always minor importance of arginine nitrogen (ARG-N) as maturity process advances does not appear in the same way in the nine samples which proved to be an exception for PRO-N content. These anomalous samples came from old plants and had been strongly fertilized. Nevertheless, other nine samples from younger plants, which received similar treatment, showed the expected proline and arginine nitrogen percentages according to their sugar contents; for which reason, strong fertilization

seems not to be the final reason. We are in favour of the
hypothesis that an excessive quantity of nitrogen fertilizer
applied to old plants is the cause of the anomalous bahaviour.

There are two situations, as in Fig. 4. When grapes come
from correctly treated crops (Situation I), to an average of
230 g of glucose+fructose per liter of must,correspond 63 and
27% of the amino nitrogen, to PRO-N and ARG-N, respectively.
However, in anomalous samples (Situation II) these levels are
changed and the PRO-N is 25% of the amino nitrogen and ARG-N
is 67%. In both cases, the remaining amino acid nitrogen is
similar (10% and 8%) and very inferior to the contribution
of these two amino acids.

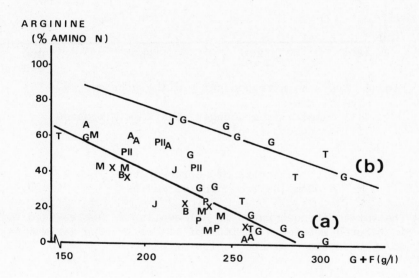

Fig. 3. Correlation between ARG-N of the musts and the
maturity rate of the grapes. Symbols as stated in Fig. 1

(a) ARG-N (% amino N) = 132.24 - 0.46 G+F (g/1)

$(n = 37)$ $(r^2 = 0.74)$

(b) ARG-N (% amino N) = 141.10 - 0.32 G+F (g/1)

$(n = 9)$ $(r^2 = 0.80)$

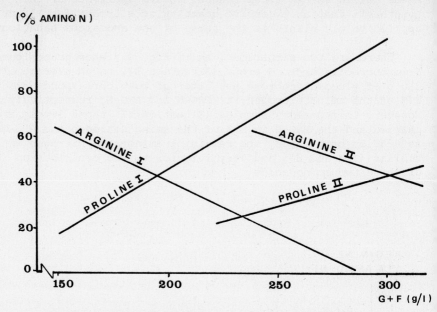

Fig. 4. Arginine nitrogen and proline nitrogen vs.
maturity rate of grapes.
      (I) Samples from younger plants well treated
    (II) Anomalous samples

Grapes: Selection of Representative
Samples from nine varieties

The total nitrogen rates in all available samples used to
be distributed between a minimum of 115 mg/l and a maximum of
645 mg/l, with the characteristics of Table 1.

Table 1. Distribution of the Total Nitrogen[*] rates

Normal Distribution	Grapes for:		
	Sparkling Wines	White Table Wines	Red Table Wines
$\bar{x}$	306.9	281.3	389.4
s	97.5	93.7	123.8
$\bar{x} \pm s$	209–404	187–375	265–513

    * mg/l

This made us adopt the approach general that grapes used as
representative of variety groups belonging to thse three
classes should contain total nitrogen levels between the range
$\overline{X} \pm s$, and come from not very old plants.

Concerning maturity grade, estimated through the glucose+
fructose concentrations, experimental results are distributed
according to the characteristics summarized in Table 2.

Table 2. Distribution of the Glucose + Fructose*
rates

Normal Distribution	Grapes for:		
	Sparkling Wines	White Table Wines	Red Table Wines
$\overline{X}$	189.6	222.3	234.5
s	18.2	28.9	39.9
$\overline{X} \pm s$	171 - 207	193 - 251	194 - 274

* (g/l)

To the requirements already established, should be added
the condition that the standard samples of each variety group
have the glucose + fructose concentration within the limits
$\overline{X} \pm s$ stated in Table 2 for their corresponding class.

Wines. Nitrogen Compounds and Sugar Content
for the Selection of Representative Samples.

Wines were made using grape musts which met the requirements
of the previous paragraphs. A correlation was found between the
total and the amino nitrogen of the resulting wines, as folows:

$$Total\ N\ (mg/l) = 45.6 + 1.64\ Amino\ N\ (mg/l)$$
$$(n = 35) \qquad (r^2 = 0.88)$$

The slope is not very much different from that found in
musts (Fig. 1) and this correlation can be useful to determine
the nitrogen balance of unknown samples. Wines from very unripe
grapes do not fit this linear relationship, nor do wines from
those musts which had been considered anomalous before.

Amino acids are an important substrate for the yeasts to
accomplish the alcoholic fermentation. Independently of Proline,
the remaining amino acids drop practically to zero during the
first stages. When fermentation is carried out on red grape must
the important contribution of the amino acids of the skins
(proline, alpha and gamma mino butyric acids, arginine and in
lesser quantity, isoleucine) is noticiable. Musts are enrichened
in these amino acids up to 2% of ethanol; then, a drastic
decrease takes place, reaching a minimum for 8-9%. The remaining
amino acids (ASP,GLU,THR,GLY,alphaALA and VAL decrease gradually
from the beginning and acquire very low concentrations after
3-4% of ethanol with slower decrease of SER.

Showing the concentration of proline of 23 white wines and
18 red wines versus that of the corresponding musts we come to
the following results:

$$\text{WHITE SAMPLES} \quad : \quad \text{PRO (mg/l)}_{\text{wine}} = 0.993 \text{ PRO (mg/l)}_{\text{must}}$$

$$\text{RED SAMPLES} \quad : \quad \text{PRO (mg/l)}_{\text{wine}} = 0.918 \text{ PRO (mg/l)}_{\text{must}} + 339.6$$

Both slopes are not different (statistically speaking)from
1. That is why PRO concentration is the same in both types of
samples (musts and wines). The ordinate 339.6 shows the PRO from
the skins, added at the usual rate in Spain. Musts from unripe
grapes or from mature ones but excessively fertilized are again
exceptions. In the first case, the quantity of proline in the
wine is higher than that in the must, and in the second case it
is lower.

Due to the fact that all studied varieties are used in dry
wine making, the wine standard samples should be restricted to
those which have a sugar concentration lower than 1 g/l
(according to the Spanish Dry Wine Regulations) as well as to
fulfil the requirements derived from this discussion.

Grapes and Wines: Election of the Initial
Set of Analytical Variables

The overall wine quality relies upon aroma intensity and
flavor, and both sensorial qualities allow the tasters to
recognize the grape varieties. Therefore, there is no doubt that
a great deal of chemical information on the aroma constituents
is very valuable. For this reason,must and wine organic extracts
were injected in the Superox capillary column in order to
explore differences between groups of samples.Chromatograms of

the three types of wines (red, white and base wines for sparkling ones) are shown in Figs. 5a,5b and 5c as examples . Differences between peak heights can easily be observed, but lack of accuracy has been detected in a few peak area measurements. It occurs in spite of the fact that this is an interesting column: high efficiency and fairly good tailing behaviour, besides the good reproductibility of the retention parameters and appropriate film thickness. It can be attributed to the split sampling technique used and consequently the injection system should be improved before making use of all these peaks in the statistical data treatment. Shomburg's research (12) is promising as he trys to combine the advantages of split sampling and modern on-column (13) and cold injection techniques respectively. On the other hand. The Programmed Temperature Vaporizer injection technique (14) makes it possible to obtain quantitative results considered excellent. All these procedures will be evaluated as there is no universal injection system for these complex mixtures.

 GC packed columns produce not such good separations as capillary columns do, but they are less troublesome for the quantitative analysis. The main features of the wine chromatograms obtained with the packed column already mentioned are shown in Fig. 6 where peaks used in computation have been numbered. Musts chromatograms only show the peaks 1,3,7,8,9,13, 19 and 22, most of them in higher proportions than in wine samples.

 This GC information, joined to the HPLC analysis of 14 free amino acids and to data derived from the conventional analysis, constitutes a panoramic picture of must and wine composition . It has been considered "a priori" suitable to draw out the analytical varia-les which contribute the most to differentiate the nine variety groups.

Stepwise Discriminant Analysis Application
for Selecting Analytical Variables.

 The classical method of Linear Discriminant Analysis (LDA) has been used for different purposes, in spite of its limitations. This statistical procedure cannot be strictly applied to those data sets whose number of variables greatly exceeds the number ob objects. Besides, one assumes that no other classes can be present than those represented in the training set. That is why a minimum of information is needed in order to avoid the risk involved if an unknown sample does not belong to any of the classes taken into account; LDA will

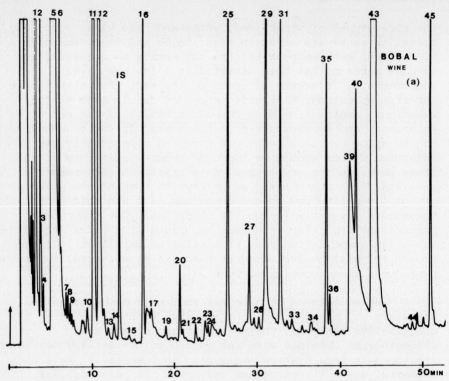

Fig. 5a. Chromatogram of the organic extract of a Bobal
wine with the Superox 0.1 capillary column. Identified peaks:
(1) hexanal, (5) iso amyl alcohols, (6) ethyl caproate, (7)
iso amyl butirate, (8) hexyl acetate, (11) ethyl lactate, (12)
1-hexanol, (IS) internal standard = methyl caprylate, (16)
ethyl caprylate, (17) 1-heptanol,(20) benzaldehyde, (27) ethyl
caprate, (29) diethyl succinate, (33) propyl caprate, (35)
1-phenyl ethanol, (36) 2-phenyl acetate, (40) benzyl alcohol,
(43) 2-phenyl ethanol, (44) lauryl alcohol.

necessarily assign it to one of the classes represented in
the initial data matrix. The BMDP7M program performs the LDA
in such a way that variables are chosen in a stepwise manner.
At each step the variable that adds most to the separation of
the groups is entered.

    At present, chemometrics researches (15, 16) are in favour
of Pattern Recognition methods by means of Principal Components
models which do not have the LDA objections. Programs are
performed in order to define different "class models" through
the training data set, and new objects can be classified
according to the extent to which they fit them. Hence, the

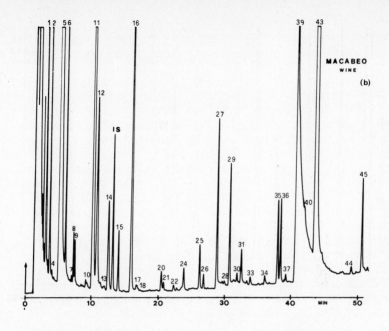

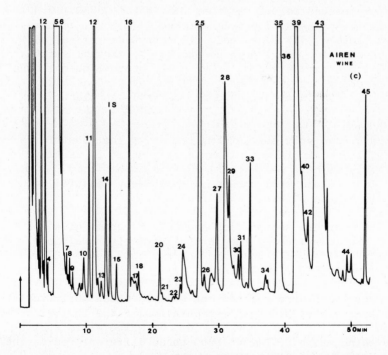

Figs. 5b, 5c. Chromatograms of the organic extracts of two variety white wines. Peak numbers as stated in Fig. 5a

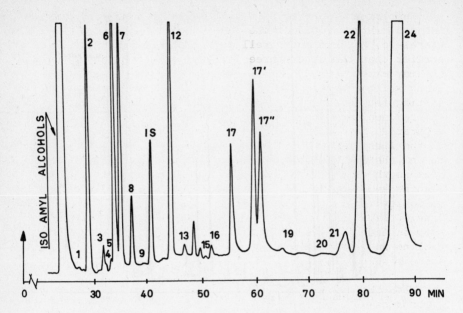

Fig. 6.Chromatogram of the organic extract of a white wine
with the packed column Carbowax 2OM+Igepal Co88O (20:80).
Identified peaks: (2) ethyl caproate, (3) hexyl acetate, (6)
ethyl lactate, (7) 1-hexanol, (12) ethyl caprylate, (15)
1-octanol, (16) ethyl pelargonate, (17') diethyl succinate,
(17") ethyl caprate, (2O) ethyl undecanoate+1 phenyl ethanol ,
(21) 2 phenyl ethyl acetate, (22) benzyl alcohol, (24) 2 phenyl
ethanol. (IS) internal standard = methyl caprylate.

analysis also provides information about the possible existence
of outliers. Expert experience says that it is usually
desirable to have ten variables or so to get good classification
but, if they are well chosen and precisely measured, excellent
results can be obtained with fewer.

    Both Pattern Recognition and BMDP programs give information
about the relevance of the initial variables available. We
justify the use of the second one for this purpose, as it is
less computer time consuming.

    Four tests were carried out to find which analytical
variables contribute to the separation of the following groups:
a) musts from six white varieties, b) white wines from the
preceeding musts, c) musts from three different red varieties
and, d) red wines from those musts.

Results which refer to the first and second case are in Table 3. Three groups of the must samples belonged to varieties (Xarel-lo, Macabeo and Parellada) which are generally collected earlier than the other three (Airén,Jaén and Pardillo). Thus, one may expect that were will exist to types of variables among which separate the six groups of musts: those which distinguish between two wide classes (less mature grapes and mature ones) and someone else which divide the classes already separated, into smaller variety groups. Baumé degree belongs to the first type, and the two GC peaks 8 and 22, as well as pH and the ratio Glucose/Fructose, to the second one. The reasons why the four amino acids (proline, arginine, alpha alanine and gamma amino butyric acid) appear as discriminant variables might be: either because they rely upon the maturity rate or because they are a consequence of the differences between each grapevine metabolism. It seems probably that proline and arginine are due to the first reason. Nevertheless, these results will be again discussed below.

When these musts become wines, amino acids have no further importance anymore except for the small role played by hydroxyproline. However, one HPLC peak (Rr Norvaline=0.76 , tentatively identified as an amine) is the second variable chosen. It could be a consequence of decarboxylation of some amino acid. The other substances, developed during alcoholic fermentation, contribute decisively to point out the differences between these white wine varieties. The GC peak 8 and pH have been selected for white musts and white wines discrimination , and it should be underlined.

Similar arguments relate to musts and red table wines from the varieties Garnacha, Tempranillo and Bobal. The analytical variables differentiating between populations are shown in Table 4 which shows as well those minor constituents (GC peaks 19,1 and 7) which allow the must groups to be differentiated (together with other variables). Among them, malic acid and dry extract refer to the maturity rate, which is reasonable since the Glucos+Fructose concentration,stated in Table 2 $(\bar{x} + s)$ for these red grapes, ranges between a 80 g/liter interval. This interval is the same as that one established for the group of white wine grapes plus sparkling wine grapes in the same table. However, the only amino acid chosen for this differentiation was arginine. Therefore, regarding must results in Tables 3 and 4, we can assume that alpha alanine and gamma amino butyric acid contents are significantly different in white grape musts and they are not very much different in red musts. pH and the GC peak 8 are again necessary in order to separate the three must families.

Table 3. Analytical Variables Selected by Means of the
BMDP7M Program Allowing a 100% Correct Classification
                of the Standard Samples
Samples from: Xarel-lo, Macabeo, Parellada, Airén ,
                 Jaén and Pardillo

	M u s t s		W i n e s
Step	Variables	Step	Variables
1	GC peak 8*	1	2 phenyl ethanol
2	arginine	2	HPLC peak (Rr=0.76)**
3	alpha alanine	3	GC peak 17*
4	GC peak 22*	4	total acidity
5	Glucose/Fructose	5	GC peak 8*
6	pH	6	colour intensity
7	proline	7	diethyl acetal
8	gamma amino butyric ac	8	methanol
9	Baumé degree	9	iso butanol
		10	hydroxyproline
		11	pH

(*) See Fig. 6   (**) Rr to Norvaline)

Table 4. Analytical Variables Selected by Means of the
BMDP7M Program Allowing a 100% Correct Classification
                of the Standard Samples
Samples from: Garnacha, Tempranillo and Bobal

	M u s t s		W i n e s
Step	Variables	Step	Variables
1	dry extract	1	GC peak 8*
2	GC peak 19*	2	3 methyl 1 butanol
3	GC peak 8*	3	HPLC peak (Rr=0.18)
4	malic acid		
5	GC peak 1*		
6	pH		
7	GC peak 7*		
8	arginine		

(*) See Feg. 6   (**) Rr to Norvaline

Three analytical variables are enough to distinguish these three types of red table wines. Once more, the GC peak 8 is among them and the relevant role played by the 3 methyl 1 butanol and an unknown HPLC peak (Table 4) has been made manifest.

CONCLUDING REMARKS

The search for the must and wine anlytical profiles implies the investigation of the training data set structure. Obviously, a careful selection of samples and variables through which they are identified is needed. Otherwise, bias in these two aspects might cause the research to fail, since statistics cannot replace the basic knowledge of the problem to be solved. Two main points mark the strategy to be followed:a) standard samples should be previously evaluated for a minimum of requirements, derived from the more common agricultural practices, and b) the initial set of variables should contain those able to give a coarse separation of the samples and another able to divide into as many groups as desired, the large classes shaped before.

It is feasible to establish beforehand the requirements which must samples should fulfil on the basis of the optimal or traditional plant treatments. Nevertheless, not very many requirements can be precisely establish for wine samples , because many biochemical aspects of must fermentation remain obscure. The pragmatism of restrictin the wine samples to those coming from selected musts, fermented under optimal techniques, avoids possible risks.

BIBLIOGRAPHY

1. Kliewer,W. M., J. Food Sci.34, 274 (1969)
2. Kliever,W. M., J. Food Sci.35, 17 (1970)
3. Polo, M. C. and LLaguno, C., Conn.Vigne Vin,8, 81 (1974)
4. Ough, C. and Amerine M.A., Am. J. Enol.Vit.,17,163(1966)
5. Ough, C., Vitis, 7, 321 (1968)
6. Ough, C., Am. J. Enol.Vit., 20,213 (1969)
7. Bell, A.A., Ough, C.S. and Kliewer, W.M., Am. J. Enol.Vit. 30, 124 (1979)
8. Martin, P., Suarez, A., Polo, C., Cabezudo, M.D. and Dabrio, M.V., Anal.Bromatol (Spain), 32, 289 (1980)
9. Cabezudo, M.D., Gorostiza, E.F., Herraiz, M., Fdez-Biarge, J., García-Dominguez, J.A. and Molera, M.J., J. Chromatog. Sci. 16, 61 (1978)

10. Grob, K. Jr.,Grob, G. and Grob, K. J. Chromatog.156,1(1978)
11. BMDP7M Program. Health Sciences Computing Facility. University of California. Los Angeles.Calif.
12. Schomburg, G., Husmann, H., and Schulz, F., J. of HRC and CC 5, 565 (1982)
13. Grob, K., and Grob, K. Jr., J. Chromatog. 151, 311 (1978)
14. Poy, F., Visani, S., and Terrosi, F., J. Chromatog. 217, 81 (1981)
15. Wold, S. and Research Group for Chemometrics. Symposium on Applied Statistics. Compenhagen. January 22, 1981.

ANALYSIS OF THE FLAVORS IN AGED SAKE

Toshiteru Ohba
Makoto Sato
Kojiro Takahashi
Makoto Tadenuma

National Research Institute of Brewing, Japan

I. INTRODUCTION

A. Production Method of Sake

The outline of sake making is shown in Fig. 1.
Polished rice.   Polished grains of nonglutious rice are
used as raw material. The rice suited specially for sake making
has many important characteristics-extra-large size of grains,
low in protein, elastic softness that moderate the dissolution
in the main mash, etc. Raw rice is polished in the rice mill,
polishing ratio reaches 70%, cleaned by washing machines and
then soaked in cold water. This soaked rice is steamed in the
continuous rice-steaming machines and then cooled.
Water.   Pure and natural water is essential to sake making. An adequate amount of minerals such as K, P, Na, and Ca
is significantly required at the early stage of fermentation.
Koji(Culture of the mold on steamed rice).   Fractions
of steamed rice are mixed with spores of mold, Aspergillus
oryzae, and they are kept in an automatically controlled humidity chamber (temperature about 35°C) for about 40 hours.
This turns the steamed rice into Koji (moldy rice).
Moto (Seed mash).   Koji is mixed with water, steamed
rice and sake yeast (Saccharomyces cerevisiae) in the pre-
fermentation tank. This seed mash is a concentrated culture
of pure and healthy living cells of sake yeast.
Moromi (Main mash).   Each day a new mixture of steamed
rice, water, Koji, and moto is put into the fermentation tanks.
The fermentation process is starting at a low temperature (10

375
Copyright © 1983 by Academic Press, Inc.
ISBN 0-12-168902-6

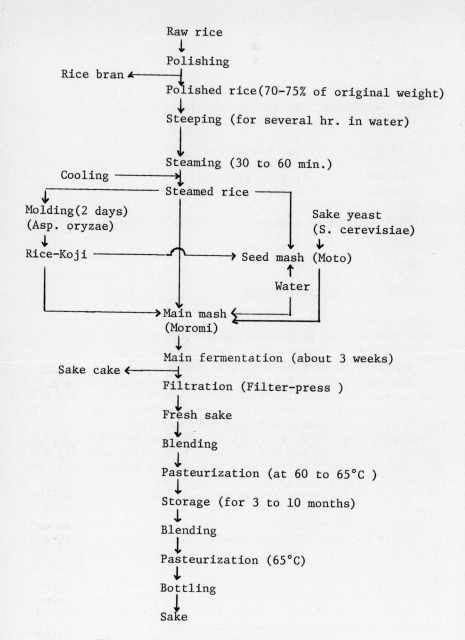

Fig. 1. Process chart of sake making.

°C). On the third and fourth days, additional volumes of steamed rice, water, and Koji are added to the tank in order to
secure gradual growth of yeast in the mash. During the fermentation of main mash, the rice starch is dissolved and saccha-
rified by amylase of Koji, and this converted mixture is fermented into by the action of sake yeast. The both processes,
saccharification and alcohol fermentation, advance simultaneously in a well-balanced manner. This is the most unique complicated method called "parallel fermentation".

Filtration of mash. When the fermentation of the mash is
finished, it is put into the filter-press.

Sake-cake (Unfiltered cake). The mash residue remains on
the filtering cloth as cake.

Fresh sake and filtration. As fresh sake thus obtained
still contains yeast and turbid materials, it has to be filtered once more through a cotton filter-press to obtain clear
sake.

Pasteurization and storage of sake. Fresh sake is pasteurized at the temperature of 65°C and is transferred to large storage tanks. After three to ten months aging(storage)
under carefully controlled temperature and humidity, sake is
brought to the very peak of mellow maturity. During the storage
period, samples are tested by specialists at frequent intervals
for quality control to ensure the flavor and taste.

B. Sake Production in Japan

The main alcoholic beverages consumed in Japan are sake,
beer, and whisky. While beer consumption grew rapidly in a
recent decade and amounted to about 4,521 thousand Kl in 1980,
the growth rate of sake consumption has been retarded, and about 1,473 thousand Kl were consumed in 1980. However, the
consumption of sake still keeps the first rank if the volumes
are calculated in terms of pure alcohol: alcohol contents are
about 4% in beer and 16% in sake. The amount of rice used for
sake making was nearly 560 thousand tons in 1980. The numbers
of sake factories in 1980 were 2,947.

Sake is not only consumed in Japan, but also exported
worldwide in annually increasing amounts. In TABLE I, the
records of export in 2 recent years are shown.

C. Pasteurization and Storage

The fresh sake after the filtration of main mash is blended to give the final products desired and then heated to 60°C
-65°C by passing through a tube-type heat exchanger made of

TABLE I. Export of Sake in Recent Years (Kl)

Country	1980	1981
U.S.A.	1,533	1,657 (50.6 %)
Formosa	309	411 (12.6)
Canada	308	284 (8.7)
W. Germany	172	181 (5.5)
Hong Kong	121	133 (4.1)
England	104	103 (3.1)
Greece	1	1
Others	479	502 (15.3)
Total	3,027	3,272 (100.0)

tinned copper, aluminium, or stainless steel. Recently, plate-type heat exchangers have become available.

In this pasteurization, the enzymes, α-amylase and protease, etc., are inactivated and the so-called "hiochi bacteria", which spoil sake during storage, are killed.

Immediately after pasteurization, the hot sake is transfered to the sealed vessels for storage.

The maturation of pasteurized sake during storage is probably due to chemical reaction, oxidative reaction, and physicochemical changes. Usually the color of sake becomes deeper as the maturation proceeds. The maturation and the coloration are accelerated at higher temperature.

Usually, the rate of the changes of color and flavors are rapid compared with the rate of the changes of taste.

Therefore, the temperature of storage room should preferably be kept under 20°C, considering a balance between the maturation velocity and the time of bottling.

The browning of sake during storage depends chiefly on amino-carbonyl reaction and melanoidin formation. The contribution of melanoidin to the color of sake is about 40-60%.

Sake is very sensitive to sunlight. Bottled sake increases its color intensely if exposed to sunlight during its transportation or window display. This is accompanied by the deterioration of flavors.

Also, the coarse taste of fresh sake disappeares during aging, it gradually grows smooth and harmonious. But the over-aging carries out the changes of the taste. The taste turns to the bitterness. The bitter compounds which increase with aging, are L-prolyl-L-leucine anhydride, methyl thioadenosine, tetrahydro-harman-3-carboxylic acid, and harman etc. (1-4).

II. FLAVORS OF AGED SAKE

A. Introduction

There is remarkable difference of flavor between fresh sake and aged sake. The flavor of fresh sake is fragrant and delicious odor derived mainly from esters and higher alcohols. On the contrary, the flavor of aged sake is quiet and mellow. When the aging proceeds more over, the flavor grows resemble to the odor of Raochu (alcoholic beverage in China).

B. 3-Hydroxy-4,5-dimethyl-2(5H)-furanone, a Burnt Flavoring Compound Isolated from Aged Sake

A burnt flavoring compound, which imparts to aged sake its characteristic and dominant flavor, was isolated by Diaion HP-20, silicic acid and Dowex 1-X8(CH_3COO^-) column chromatography, and chloroform extraction. Based on gas chromatography, UV, and GC-MS spectral data, it was identified as 3-hydroxy-4,5-dimethyl-2(5H)-furanone and its structure was also confirmed by synthesis. It was suggested that compound was formed by the condensation of α-ketobutyric acid with acetaldehyde which occured from degradation of threonine (5).

1. Materials and Methods

Sample sake.   Sake was aged as follows: fresh sake (180 liters) brewed in this laboratory was warmed for 1 month at about 60°C in order to hasten the aging. Storage for 1 month under these conditions is equal to aging for about 2 years at 30°C.

Isolation of burnt flavoring compound (BFC) from sake. Aged sake (180 liters) was concentrated to half its volume, in oder to remove ethanol, in vacuo at 40°C. The concentrate was passed through a column of Diaion HP-20 (10 liters), and the column was washed with distilled water (20 liters). BFC was eluted with 60% ethanol (50 liters) the eluate was concentrated to a volume of 3 liters. The concentrate was extracted 3 times with 3 volumes of chloroform at pH 2. The extract was evaporated in vacuo at 40°C. The residue (ca. 200 ml) was adjusted to pH 9 with 1 N sodium hydroxide and extracted twice with equal parts by volume of chloroform. BFC was not extracted at pH 9. The aqueous layer was re-extracted 4 times with equal parts by volume of chloroform at pH 7.5. The chloroform extract containing BFC was evaporated in vacuo at 40°C. The concentrate was applied to a silicic acid column (silicic acid

obtained from Malinkrodt Chemical Works was packed with n-hexane in a 2 cm od. by 7 cm column). The column was washed with
50 ml of n-hexane followed by 50 ml of 15% diethyl ether in
n-hexane. BFC was then selectively eluted with 300 ml of 70/30
n-hexane-diethyl ether mixture. The eluate was concentrated
to a volume of 2 ml and neutralized with conc. anmmonia, and
the solution was applied to a column of Dowex 1-X8 (CH_3COO^-,
200-400 mesh, 50 ml). The elution was carried out with 250 ml
of 20% ethanol followed by 250 ml of 0.005 M acetic acid.
The latter fraction was extracted 3 times with 100 ml of chloroform at pH 2. The chloroform extract was evaporated in vacuo
at 40°C. The oily fraction obtained had a typical burnt flavor
and showed one peak on gas chromatogram.

Synthesis of 3-hydroxy-4,5-dimethyl-2(5H)-furanone(HDMF).
The synthesis of HDMF was performed according to the methods
described by Schinz et al. and Sulser et al. (6-8).
Yield: 4.85g (18.5% yield from ⍺-ethoxalyl propionate ethyl
ether). p-Nitrobenzoate derivative: mp 99°C. Anal. Found: C,
56.35, H, 3.97; N, 4.84. Calcd. for $C_{13}H_{11}O_6N$: C, 56.34, H, 4.
06, N, 5.08%.

Gas chromatography.   A Hitachi model K-53 gas chromatograph equipped with flame ionization detector was used. The
column was stainless steel (1m x 3mm) packed with $DEGS-H_3PO_4$
(5-1%) on 60-80 mesh Chromosorb W (AW), and it was held isoth-
ermally at 150°C or 173°C. Nitrogen was used as carrier gas at
flow rate of 60 ml/min. For the determination of BFC and synthesized samples, ethyl stearate was used as internal standard.

Gas chromatography-mass (GC-MS) spectrometry.   GC-MS
spectra were measured with a Jasco-Finningan model 1015-D mass
spectrometer operated at an ionization voltage of ca. 70 eV.

Sensory evaluation of HDMF.   HDMF was dissolved in distilled water. A series of solutions of increasing concentration,
each twice as strong as the preceding one, was prepared (concentrations of sample : 15.26-3,905.5 ppb, 9 steps). In the
ascendind series method, the flavor sample was sniffed, and
tasted by dropping the solution on the tongue, and the concentrations at which odor was detected were noted. Panel members
were 8 well-trained members of our labolatory.

2. Results and Discussion

a. Properties of BFC

The gas chromatogram of the concentrate extracted with
chloroform at pH 7.5 is shown in Fig. 2.

The general properties of BFC were compared with those of
HDMF. The UV absorption spectra of BFC in ethanol solution and
alkaline solution are shown in Fig. 3. The UV absorption peak
was 232 nm, and the bathochromic shift (9) proposed for an

α-hydroxy substituent in an acylic conjugated ketone is 35 nm;
the calculated value of 267 nm is thus in fair agreement with
our value of 268 nm.

The MS spectrum of BFC was shown in Fig. 4. The UV and
MS spectral data of HDMF were completely identical with those
of BFC isolated from aged sake.

On the basis of the above results, BFC isolated from
aged sake was identified as 3-hydroxy-4,5-dimethyl-2(5H)-
furanone. Furthermore, NMR and IR spectral data and the results of elemental analysis of p-nitrobenzoate of HDMF were

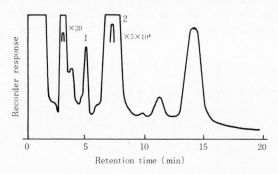

Fig. 2. Gas chromatogram of the flavor concentrate.

Flavor concentrate was the fraction extracted with
chloroform at pH 7.5 in the isolation procedure of
BFC from sake.
Peak No. 1, BFC; 2, monoethyl succinate.

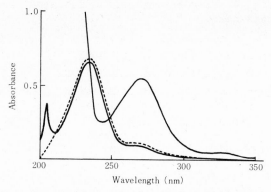

Fig. 3. UV absorption spectra of BFC isolated
from aged sake.

----, in acidic ethanol solution (pH 1);━━━, in
ethanol solution (pH 4.4); ──────, in alkaline
solution (pH 10).

completely identical with those of HDMF.

HDMF has been synthesised by Schinz et al. (6) as an
α-keto-γ-lactone homologue, but had not previously been isolated as a natural product.

    b. Determination of the odor unit for HDMF and distribu-
       tion in other fermented foodstuffs
    The threshold values for HDMF were determined by the sen-
sory evaluation test mentioned above, HDMF content in aged
sake was determined by GLC. HDMF was perceived as burnt flavor
by all panel members. The odor units obtained from the ratio
of concentration in sake to threshold value are shown in TABLE
II. HDMF contributes to the flavor of aged sake.

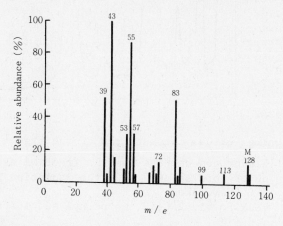

Fig. 4. MS spectrum of BFC isolated from
aged sake.

TABLE II. Odor Units of HDMF

	Odor threshold value (a) (ppb) (A)	Concentration in aged sake (ppb) (B)	Odor unit (B/A)
I (b)	5804		0.02-0.07
		140-430	
II (c)	76.3		1.84-5.64

(a). Sample was dissolved in distilled water.
(b). Odor threshold value obtained by sniffing the
     sample solution.
(c). Odor threshold value obtained by tasting the
     sample solution.

In other fermented foodstuffs, HDMF was found in Raochu (made in Japan, more than 30 ppb) and Soy sauce (more than 50 ppb) by GLC and GC-MS. Recently, HDMF was isolated from Sherry and cane molasses (10,11). It was reported that HDMF was the key compound of sugary flavor.

c. Formation and degradation mechanism of HDMF

Formation and degradation mechanism of HDMF were proposed as shown in Fig. 5.

A 50 mM solution (50 ml) of each of 20 amino acids (Lys. HCl, His·HCl, Arg, Asp, Thr, Ser, Glu, Pro, Ala, Cys, Val, Met, Ile, Leu, Tyr, Phe, Try, CySH·HCl, Orn·HCl (all L-configuration) and Gly) and α-aminobutyric acid was heated for 10 hr at 110 °C under acidic conditions (pH 1). After extracting with 30 ml of chloroform, each sample was subjected to GLC analysis. The burnt flavoring compound was formed only from threonine.

Acetaldehyde, glycine, α-ketobutyric acid (α-KBA) and others have been reported as acid- degradation products of threonine (12-17), and we also detected these compounds by TLC and GLC. Roedel et al. reported and we also confirmed that HDMF was formed from the reaction of α-KBA with acetaldehyde (18).

From these results, it was concluded that HDMF was formed by the condensation of α-KBA with acetaldehyde which were produced by acid-degradation of threonine. This conclusion may also be supported by the fact that the threonine content of aged sake is about 40% of that of fresf sake.

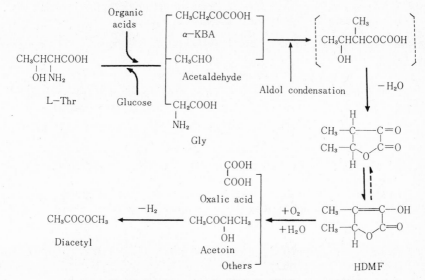

Fig. 5. Formation and degradation of HDMF.

The effects of organic acids on the formation of HDMF were not clarified, but the yield of HDMF from threonine with organic acid was almost proportional to the amount of glycine which was simultaneously produced from threonine. Therefore, it was thought that the organic acid perhaps accelerated the degradation of threonine.

The molar ratio of starting materials has an important effect on the yield of HDMF, and the highest yield was obtained from equal molar proportions of threonine and organic acid. In proportion to the decrease in the ratio of organic acid to threonine, the formation of 3-hydroxy-4-methyl-5-ethyl-2(5H)-furanone(HMEF), which has been isolated from vegetable-protein hydrolysates by Sulser et al. (19), increased. While that of HDMF decreased.

HMEF was not found in aged sake, and the reason was thought to be the higher concentration of organic acids than threonine in sake (concentration in sake: Thr, 2.7 mM; malic acid, 3.1 mM; lactic acd, 6.7 mM; succinic acid, 6.9 mM).

HDMF is a very unstable compound and the pure substance spontaneously decomposed within 2-4 weeks when kept in a closed flask at room temperature.

According to Schinz et al. (6), enolized α-keto-γ-lactones are decomposed by moisture and oxygen to give oxalic acid and ketols. The decomposition products of HDMF detected by GLC were oxalic acid, acetoin, and diacetyl.

It is interesting to study whether other α-keto-γ-lactones, homologous with HDMF, might also be found in sake. Homologous compounds were synthesized from α-KBA with some aldehydes under the same conditions, but compounds corresponding to the synthesized samples were not found in sake.

C. Volatile Sulphur Compound

Hydrogen sulfide (H_2S), methyl mwrcaptan (CH_3SH), dimethyl sulfide (DMS), and dimethyl disulfide (DMDS) were detected

TABLE III. Threshold Values of Volatile Sulphur Compounds in Sake.

Compound	Threshold value (ppb) (A)	Content (ppb)(B)	Odor unit (B/A)
MeSH	4	0-14	~3.5
DMS	10	0-10	1.0
DMDS	6.5	0- 7	1.1

with FPD-GC by the head space method in sake making process. The main sources of sulphur compounds in brewing are thought to be the chemical degradation of sulphur-containing amino acids, and the metabolism of microorganisms.

DMDS was not found in fresh sake, but found in aged sake. The content of DMDS is about 7 ppb. As the threshold value of DMDS is 6.5 ppb, it is thought that DMDS contributes to the flavors of aged sake. Also, methyl mercaptan was increased by the pasteurization. It was elucidated that DMDS and methyl mercaptan were formed from methionine or cysteine by the degradation during pasteurization and storage of sake. The threshold values of sulphur compounds are shown in TABLE III (20).

Although the sulphur compounds present in a small quantity in sake, they are disliked as the unfavorable attributes on account of the low threshold values.

D. Other Flavor Compounds Isolated from Aged Sake

In addition to sulphur compounds and HDMF, many compounds were isolated only from aged sake. The compounds isolated from sake are shown in TABLE IV.

1. Materials and Methods

Sample sake.   Aged sake stored for six years at room temperature was analyzed.

Fractionation of flavor compounds.   Aged sake (5 liters) was diluted to twofold with water. The diluted sake was passed through a column of Amberlite XAD-2 (1 liter), and the column was washed with distilled water (10 liters). The flavor compounds were eluted with 95% ethanol (3 liters) followed by 3 liters of water and the eluate was concentrated to a volume of 2 liters. The concentrate was extracted 3 times with 2 volumes of dichloromethane at pH 7. The extract was evaporated in vacuo at 35°C. The residue (ca. 10 ml) was adjusted to pH 10 with 1 N sodium hydroxide and extracted 3 times with 2 volumes of dichloromethane (Basic extracted fraction). The aqueous layer was adjusted to pH 2 with 1 N hydrochrolic acid and extracted 3 times with 2 volumes of dichloromethane (Acidic extracted fraction).

Basic and acidic extracted fraction were concentrated to ca. 1 ml in vacuo at 35°C. Each concentrate was applied to silicagel column (Silicagel was packed with n-hexane in a 2.5 od. by 10 cm column), respectively. The column was washed with 100 ml n-hexane. The typical flavor of aged sake was recognized in the fraction eluted with 50 ml of 70/30 n-hexane-ethyl acetate mixture at acidic extracted fraction. This fraction

TABLE IV. Flavor Compounds Detected from Sake.

Alcohol	methanol, ethanol, n-propanol, iso-buthanol, n-buthanol, iso-amylalcohol, n-amylalcohol, active-amylalcohol, n-hexanol, β-phenetyl alcohol, tryptohol,
Ester	ethyl ester of C_1-C_{14} straight-chain aliphatic acid, propyl acetate, buthyl acetate, iso-butyl acetate, n-amyl acetate, iso-amyl acetate, β-phenethyl acetate, ethyl iso-valerate, ethyl iso-caproate, ethyl pyruvate, ethyl keto-iso-valerate, ethyl keto-iso-caproate, ethyl oxy-iso-valerate, ethyl oxy-iso-caproate, ethyl p-oxycinnamic acid ethyl ester, ethyl lactate, * ethyl phenylacetate, * ethyl vanillate, * ferulic acid ethyl ester, *ethyl p-oxybenzoate
Acid	C_1-C_{16} straight-chain aliphatic acid, iso-butyric acid, iso-valeric acid, iso-caproic acid, iso-capric acid, iso-caprylic acid, oleic acid, pyruvic acid, keto-valeric acid, phenyl pyruvic acid, benzoic acid, ferulic acid, indole acetic acid, protocatechuic acid, gallic acid, *coumalic acid,*p-oxy-cinnnamic acid,*p-hydroxy-phenyl acetic acid, *phenyl acetic acid, * vanillic acid
Carbonyl compound	formaldehyde, acetaldehyde, propionaldehyde, iso-valeraldehyde, caproic aldehyde, diacethyl, acetoin, acetone,*benzaldehyde,* cinnamic aldehyde, * fulfural, * p-oxybenzaldehyde, * vanillin, * phenyl acetaldehyde
Amine	ethanol amine, iso-butyl amine, putrescine cadaverine, phenethyl amine
Sulphur compound	hydrogen sulfide,methyl sulfide, dimethylsulfide , * dimethyl disulfide
Others	*3-hydroxy-4,5-dimethyl-2(5H)-furanone,* maltol

* : Detected only from aged sake.

was concentrated to ca. 1 ml in vacuo at 35°C and submitted to
GC-MS.

GC-MS spectrometry.   GC-MS spectra were measured with a
Shimazu model LKB-9000 mass spectrometer operated at an ionization voltage of 40 eV. The column was grass column (1 m x 3 mm)
packed with 3% Silicone OV-17 on 80-100 mesh Chromosorb W HP,
and operated at 70°C to 160°C (programmed at 5°C/min).

2. Results and Discussion

The gas chromatogram of the fraction eluted with 70/30
n-hexane-ethyl acetate mixture at acidic extracted fraction
of aged sake is shown in Fig. 6. Phenyl acetic acid, p-hydroxy
benzaldehyde, vanillin, and ethyl vanillate were isolated only
from aged sake.

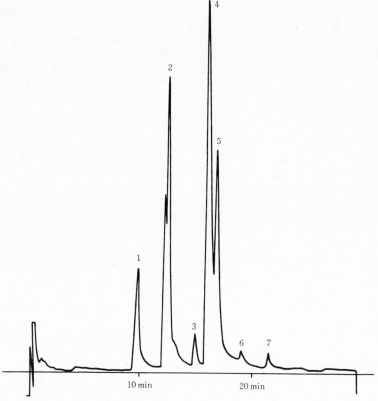

Fig. 6. Gas chromatogram of the flavor concentrate.

Peak No. 1, benzoic acid; 2, phenyl acetic acid; 3,
unknown; 4, p-hydroxybenzaldehyde; 5, vanillin; 6,unknown;
7, ethyl vanillate.

3. Formation of Phenyl Acetic Acid

Authors isolated phenyl acetic acid from sake for the first time. The scheme of the formation of phenyl acetic acid is considered as shown in Fig. 7. Phenyl acetic acid was formed from phenylalanine by the Strecker degradation in the process of amino-carbonyl reaction.

E. Conclusion

The flavor caused by aging of sake is usually called "hineka" that is disliked as an unfavorable attribute. The following compound were identified in aged sake by FPD-GLC and GC-MS: methyl mercaptan, methyl sulfide, dimethyl disulfide, 3-hydroxy-4,5-dimethyl-2(5H)-furanone (HDMF), furfural, benzaldehyde, phenyl acetic acid, p-hydroxybenzaldehyde, vanillin, and ethyl vanillate.

Dimethyl disulfide was not found in fresh sake, but was formed from methionine or cysteine during storage of sake.

HDMF was perceived as burnt flavor by all panel members.

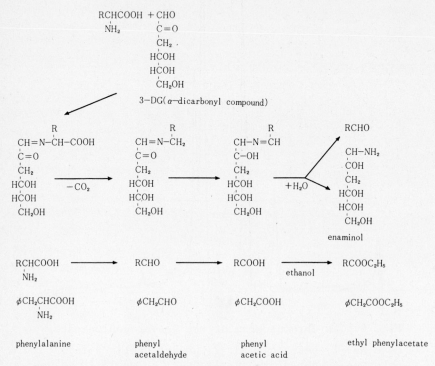

Fig. 7. Mechanism of formation of phenyl acetic acid.

Among the flavors of aged sake, burnt flavor was the most characteristic one. Odor threshold value of HDMF by tasting was 76 ppb. The content of HDMF in aged sake was 140-430 ppb. It was suggested that HDMF was formed by the condensation of α-ketobutyric acid with acetaldehyde which occured from degradation of threonine. In other foodstuffs, HDMF was found in Raochu, Sherry, Soysause, and molasses.

Vanillin was made from ferulic acid in sake making. The origin of ferulic acid in sake was lipid-phenol and sugar-phenol in rice grain. They were decomposed by Aspergillus oryzae to form ferulic acid (21).

It was considered that phenyl acetic acid was formed phenylalanine by the Strecker degradation.

It is concluded that the flavor of aged sake is a complex flavor composed of the above compounds.

III. AN OFF-FLAVOR COMPOUND, DIMETHYL SULFIDE, IN SAKE BREWED WITH OLD RICE

When old rice stored for one or more years at room temperature was used for sake making as the raw material, an unfavorable smell, the so-called "komaishu", in pasteurized and stored sake has been recognized by sensory test.

A. Identification of Dimethyl Sulfide in Sake

The unfavorable off-flavor compounds were removed by passage of nitrogen gas through the warm sake at 40°C and trapped in a bottle cooled with liquid nitrogen or in mercury chloride solution. The collected off-flavor compound were analyzed by a gas chromatograph equipped with a flame photometric detector. Based on mp and IR spectral data of the mercury derivative, the unfavorable smell compound was identified as dimethyl sulfide (DMS). DMS contents in sake made from old rice were about 0.4-44.2 ppb (average 9.6 ppb), while contents of it in sake made from new rice were almost trace amounts.

A significant correlation was obtained between DMS content and sensory strength of the off-flavor in sake at the 1% degree of significance. It was possible to reproduce the off-flavor by the addition of DMS to sake made from new rice. Therefore, it is concluded that DMS was the main compound which caused the off-flavor in sake made from old rice. But there were small contents of DMS and weak off-flavor in sake made from new rice (22).

B. Quantitative Analytical Methods of DMS

The quantitative analytical methods of DMS in sake and DMS precursor in rice were shown in Fig. 8. and Fig. 9, respectively. DMS precursor contents were estimated as the amounts of DMS. DMS precursor contents were measured by heating the sample in a sealed vial at 120°C for 20 min in 1 N NaOH (heat alkali method). DMS evolved was determined by gas chromatography with flame phtometric detector and diisopropyl sulfide as internal standard was used (23).

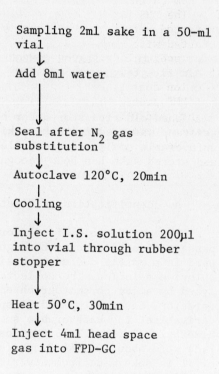

Sampling 50-100mg rice
powder in a 50-ml vial
↓
Add 5ml water
↓
Add 5ml 1 N NaOH
↓
Seal after N$_2$ gas
substitution
↓
Autoclave 120°C, 20min
↓
Cooling
↓
Inject I.S.(1.63ppm
diisopropyl sulfide
in EtOH) 200 µl into
stopper
↓
Heat 50°C, 30min
↓
Inject 4ml head space
gas into FPD-GC

Fig. 8. Analytical method
of DMS precursor in
rice.

Sampling 2ml sake in a 50-ml
vial
↓
Add 8ml water
↓
Seal after N$_2$ gas
substitution
↓
Autoclave 120°C, 20min
↓
Cooling
↓
Inject I.S. solution 200µl
into vial through rubber
stopper
↓
Heat 50°C, 30min
↓
Inject 4ml head space
gas into FPD-GC

Fig. 9. Analytical method
of DMS and DMS
precursor in sake.

C. DMS Precursor in Sake

It has been found in many foods and beverages that DMS is evolved on heat treatment and that one of the precursors is S-methylmethionine sulfonium salt. Likewise, it was elucidated that DMS precursor in fresh sake brewed with old rice was S-methylmethionine (24).

D. DMS Precursor in Rice

The identity of DMS precursors in rice was investigated. Production of DMS from new unpolished rice was estimated at 0.3 ppb, while from 80% polished rice it could not be detected. DMS production from old unpolished rice stored at room temperature and from 50% polished rice were measured at 8 and 0.6 ppm as shown in Fig. 10, respectively (25).

The DMS precursor in new and old rice bran could not be extracted with fat-extracting solvents, but from new rice bran defatted with methanol it could easily be extracted with water, sodium chloride or dilute acid solution. DMS precursor in old rice bran was extracted with alkaline solution or SDS solution containing 2-mercaptoethanol. These results indicate that DMS precursor in new rice bran is one of low molecular weight, and DMS precursor in old rice bran is present in protein, especially glutelin.

The DMS precursor from defatted new rice bran was purifi-

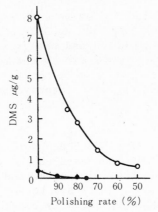

Fig. 10. Changes in DMS precursor contents with polishing.

DMS precursor contents in polished rice were determined by the heat-alkali method shown in Fig.8.
○ old rice, ● new rice

ed by Bio-Gel P-2 chromatography and ion exchange chromatographies on Dowex 50-X4(H), Amberlite IRA 411(OH), and IRC 50(H). The DMS precursor obtained from new rice bran was judged to be S-methylmethionine sulfonium by amino acid analysis (25).

DMS precursor in old rice was extracted successively with H_2O, 0.5 M NaCl,70% EtOH, and 0.05 N NaOH. Most of the precursor was found to be present in the glutelin fraction by Sephadex 4B column chromatography of the SDS-2-mercaptoethanol extract of the crude fraction.

Crude protein treated with α-amylase from Bacillus subtilis was hydrolyzed with HCl. DMS precursor in HCl hydrolyzate was purified repeatedly by Bio-Gel P-2, Dowex 50-X4 (H), Amberlite IRA 411 (OH), and IRC-50 (H) column chromatography and crystallized as its picric acid salt. The amino acid composition of purified DMS precursor, and the melting point and IR of its picrate were identical with those of authentic methyl methionine sulfonium chloride and its picrate. On the basis of the above results, it was concluded that DMS precursor in old rice was glutelin containing S-methyl methionine (26).

E. Cause of Formation of DMS Precursor in Old Rice

When sake brewed from new rice is pasteurized and stored, the unfavorable smell does not occur. Therefore, it was investigated to find the cause of formation of DMS precursor in old rice.

Old rice grains stored for fifteen years without treatment with sterilizers or insecticides were found to contain a small amount of DMS precursor (100-500 µg DMS/Kg dry wt.). However, twenty-five rice samples which had been fumigated and then stored at room temperature for two to three in the government storehouse, contained a considerable amount of DMS precursor (7,400-32,000 µg DMS/Kg dry wt.). Storage temperature (-3 to 30°C) had little effect on the content of DMS precursor in the rice grains.

Brown rice samples were fumigated with methyl bromide and phostoxine and analyzed for DMS precursor; fumigation with methyl bromide remarkably increased the content of DMS precursor in the grains, but that with postoxine had little effect (TABLE V). Fumigation of wheat, soy beans, casein, and methionine with methyl bromide resulted in remarkably increase DMS precursor in each case. Polishing brown rice grains decreased the content of DMS precursor, but grains fumigated with methyl bromide were found to contain DMS precursor in the inner parts.

These results indecate that the remarkable increase in the content of DMS precursor in old rice grains is mainly

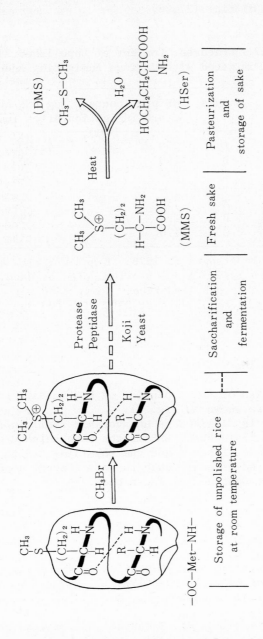

Fig. 11. Mechanism of evolution of DMS in sake
brewed from old rice.

ascribable to fumigation with methyl bromide, which is usually
used for sterilization of the grains and as an insecticide
during storage of rice (27,28).

TABLE V. DMS Precursor Content of Unpolished Rice
Fumigated with Various Fumigating Agents.

Fumigating agent	Sample	DMS pre. content (μg DMS/Kg dry wt)	
		Before fumigation,	After fumigation
Methyl bromide	A	250	41,000
	B	190	38,000
	C	170	33,000
Phostoxine	A	250	270
	B	190	190
	C	170	170

A: Rice sample cropped in 1978. B and C : rice sample
in cropped in 1979. A and B : fumigated as unpolished rice.
C : fumigated as unhulled rice. Fumigation was carried out
in a desiccator for 5 days at 25°C with 100 ppm of fumigating agent.

F. Conclusion

The mechanism of the evolution of DMS in sake brewed
from old rice was shown in Fig. 11.
Though DMS is disliked as unfavorable smell in sake, it
is favorable smell in beer. DMS is the main sulphur compound
through processing of beer, and the remaining DMS in finished
beer is considered to contribute to beer flavor, especially
to the flavor of Continental lager beer.

REFERENCES

1. Takahashi,K., Tadenuma, M., Kitamoto,K., and Sato, S.,
   Agric.Biol.Chem. 38:927 (1974).
2. Takase, S., and Murakami, H., Agric.Biol.Chem. 30:869
   (1966).
3. Sato, s., Tadenuma, M., Takahashi, K., and Nakamura, K.,
   Nippon Jozo Kyokai Zasshi, 70:821 (1975).
4. Tadenuma,M., Takahashi, K., and Sato, S., ibid., 70:585
   (1975).

5. Takahashi, K., Tadenuma, M., and Sato, S., Agric.Biol.
   Chem. 40:325 (1976).
6. Schinz, H., and Hinder, M., Helv.Chem.Acta. 30:1349 (1947)
7. Schinz, H., and Rossi, A., ibid. 31:1953 (1948).
8. Sulser, H., Habegger, M., und Büchi, W., Z.Lebensm-
   Untersuch.-Forch. 148:215 (1972).
9. Jaffe, H.H., and Orchin, M., " Theory and Application of
   Ultraviolet Spectroscopy", p.215. John Wiley and Sons,
   New York, 1962.
10. Dobois, P., Rigaud, J., and Dekimpe, J., Lebenm.-Wiss.
    u.-Technol. 9:366 (1976).
11. Takimoto, T., Kobayashi, A., and Yamanishi, T., Pro. Japan
    Acad.Ser.B, 56:457 (1980).
12. Braunstein, A. E., and Vilenkina, G.Y., Dokl.Akad.Nauk
    SSSR, 66:243(1962).
13. Hino, T., Kimituka, A., Ito, K., Ogasawara, T., and
    Takenishi, T., Nippon Nogeikagaku Kaishi, 36:314 (1962).
14. Kato, S., Kurata, T., Ishitsuka, R., and Fujimaki, M.,
    Agric.Biol.Chem. 34:1826 (1970).
15. Lien, O.G., and Greenberg, D.M., J.Biol.Chem. 200:367
    (1953).
16. Wieland, Th., and Wiegandt, H., Angew.Chem. 67:399 (1955).
17. Brockmann, H., und Franck, B., Naturwissenshaften, 42:180
    (1955).
18. Roedel, W., and Hempel, U., Nahrung, 18:133 (1974).
19. Sulser, H., Depizzol, H., and Wibüchi, J.Food.Sci.
    32:611 (1967).
20. Sato, S., Tadenuma, M., Takahashi, K., and Koike, K.,
    Nippon Jozo Kyokai Zasshi, 70:588 (1975).
21. Yoshizawa,K., Komatsu, S., Takahashi, I., and Otsuka, K.,
    Agric.Biol.Chem. 34:170 (1970).
22. Takahashi, k., Ohba, T., Takagi, M., Sato, S., and Numba,
    Y., Hakkokogaku Kaishi, 57:148 (1979).
23. Kitamoto, K., Ohba, T., and Namba, Y., Nippon Jozo Kyokai
    Zasshi, 74:677 (1979).
24. Kitamoto, K., Ohba, T., and Namba, Y., Hakkokogaku Kaishi,
    60: in press (1982).
25. Namba, y., Ohba, T., Kitamoto, K., Hirasawa, A., and
    Karahashi, S., ibid. 60:27 (1982).
26. Namba, Y., Ohba,T., Kitamoto, K., and Hirasawa, A.,
    ibid. 60:35 (1982).
27. Kitamoto, K., Ohba, T., and Namba, Y., Nippon Jozo Kyokai
    Zasshi, 76:491 (1981).
28. Maegawa, K., Yokoyama, T., Tabata, N., Shinke, R., and
    Nishina, H., Hakkokogaku Kaishi, 59:483(1981).

THE DETERMINATION OF VOLATILE PHENOLS IN RUM AND BRANDY BY GC AND LC

Matti Lehtonen[1] and Pekka Lehtonen[2]

[1]The Laboratory of Rajamäki Factories of the State Alcohol Monopoly (Alko), SF-05200 Rajamäki, Finland

[2]The Research Laboratories of the State Alcohol Monopoly (Alko), SF-00101 Helsinki 10, Finland

I. INTRODUCTION

Phenols contribute to the aroma of many foodstuffs and alcoholic beverages. Smoked and roasted products in particular contain large quantities of phenols (1,2). Phenols also occur in wines (3) and in matured spirits (4-6). They give their own characteristic quality to the aroma of alcoholic beverages. For example, at least part of the perceived smokey aroma in Scotch whiskies derives from phenols (7).

Until recently gas chromatography has been the usual way of analyzing phenols. Now, however, liquid chromatography has attained equal status as an alternative method. The chromatography can be performed either with the free phenols or on derivatives, although neither gas nor liquid chromatography is able to resolve various isomers when free phenols are used. Consequently, it is common to prepare suitable derivatives before the chromatography is carried out. A large number of derivatives are available for use in gas chromatography. Phenols readily form trifluoro- and chloroacetates (8,9). Their use, however, is restricted by the ease with which they are hydrolyzed in aqueous solution. In contrast, α-bromo-2,3,4,5,6-pentafluorotoluene and 1-fluoro-2,4-dinitrobenzene form derivatives with phenols that are stable in water (10,11). Moreover, employing these derivatives makes it possible to use the specific electron capture detector in the gas chromatographic analysis. This detector is far more sensitive than the original flame ionisation detector (12).

Free phenols can also be analyzed by liquid chromatography using any of three different detectors: a spectrophotometer, an electrochemical or a fluorescence detector. Each has its advantages and limitations. For example, an electrochemical detector is more specific and sensitive than a spectrophoto-

Instrumental Analysis of Foods
Volume 2

397

meter detector, but it is not possible to use gradient
elution with it. Masoud and Cha (13) have compared the
detection responses of some phenols with these three
detectors. The environmental phenols were analyzed by Shoup
and Mayer (14) and phenolics in commercial beverages by Roston
and Kissinger (15) using electrochemical detection. In trace
analysis of phenols the electrochemical detector is the most
suitable because it is the most sensitive.

The aim of this work was to apply gas and liquid chromatography to the analysis of phenols and to investigate the
contribution phenols make to the aroma of rums and brandies.

II. EXPERIMENTAL

A. Gas Chromatographic Analysis.

Before the gas chromatographic determination the phenols
were converted to 2,4-dinitrophenyl ethers by the nucleophilic
Schotten-Bauman substitution reaction (12). The derivatives
were extracted with n-hexane, dried over sodium sulphate, and
analyzed on a Hewlett Packard 57030A gas chromatograph fitted
with a ^{63}Ni electron capture detector and a HP 18740B inlet
splitter. A Hewlett Packard SP-2100 fused silica column,
50 m x 0.3 mm, was used for the analysis. Internal standard,
3,4-dimethylphenol, was added to the samples before
derivatization.

B. Liquid Chromatographic Analysis.

The liquid chromatograph used was a Hewlett Packard model
1084B HPLC equipped with a Hewlett Packard 79875A variable
wavelength detector or a Bioanalytical systems LC-4A
amperometric detector with a glassy carbon electrode TL-5A.
Absorption spectra of the phenols were recorded in the UV
region 200-300 nm on a Kontron Uvikon Spectrophotometer.

In the analysis with a spectrophotometer detector eluent
was pumped at a flow rate of 1 ml/min and the amount injected
was 10 μl. A stainless steel column (25 cm x 4 mm) packed with
Spherisorb S5 ODS2 (pore size 5 μm, Phase sep., Queensferry,
Great Britain) was used. The separation of phenols was
achieved in 17 minutes with a gradient elution from 45 % to
55 % (v/v) of methanol in water. The wavelength used for the
detection was 275 nm.

With electrochemical detection the measurements were
performed at ambient temperature. A chromatographic column
(20 cm x 4 mm) packed with μBondapak C_{18} (pore size 10 μm,

Waters Ass., Milford, USA) was used. The eluent was 50 % (v/v) methanol in water containing 0.1 M sodium perchlorate and 0.005 M trisodium citrate. The pH of eluent was adjusted to 5.0 with acetic acid. The electrode potential was 0.9 V.

III. RESULTS AND DISCUSSION

A. Phenols in Rum and Brandy.

Gas chromatograms of the phenolic fractions isolated from rums and brandies are presented in Figure 1. It is evident that the beverages contain largely the same phenolic components. A major difference is that rums alone contain dihydroeugenol. Some of the phenolic compounds detected have not previously been reported in these beverages. In brandies these new compounds are o-, m-, and p-cresols and thymol; p-methylguaiacol is new in cognac. Nishimura and Masuda (4) have identified p-methylguaiacol in the phenolic fraction from whisky and Timmer et al. (5) from rum. Thymol has been reported as an aroma component in tea (16). The appearance of dihydroeugenol only in rums shows that, although it structurally resembles eugenol, its origin is not the oak lignin of the maturing casks. Possible sources are the sugar cane molasses or juice used as raw material or bacterial fermentation.

B. The Content of Phenols in Rums and Brandies.

The average concentrations of phenolic compounds in brandies and rums are shown in Table 1. There are some variations from drink to drink, with the brandies generally containing lower concentrations, although the content of the major component in the phenolic fractions, p-ethylguaiacol, can rise to more than 0.5 ppm in some brandies. A second important component in brandies, eugenol, can reach almost 0.3 ppm.

The phenolic content of rums differs markedly from that of brandies. The concentration of guaiacol, p-ethylphenol, p-ethylguaiacol and eugenol is much higher, and dihydroeugenol is found only in dark rums. There is generally more phenol in rums too, but here the difference is less obvious. As a consequence of the methods of production, white rums contain far less phenolic compounds than dark rums. Nevertheless, the concentration of guaiacol, p-ethylphenol, p-ethylguaiacol and eugenol in white rums exceeds that in brandies.

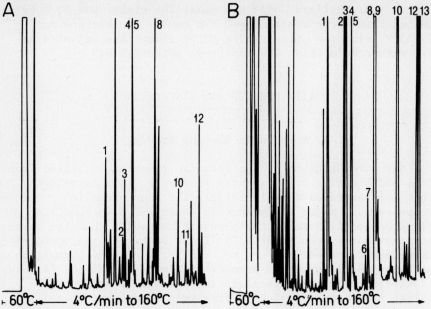

Fig. 1. Gas chromatograms of the phenols in brandy (A) and rum (B) recorded on OV-101 fused silica capillary column. 1. Phenol, 2. o-Cresol, 3. Guaiacol, 4. m-Cresol, 5. p-Cresol, 6. o-Ethylphenol, 7. 2,4-Dimethylphenol, 8. p-Ethylphenol, 9. p-Methylguaiacol, 10. p-Ethylguaiacol, 11. Thymol, 12. Eugenol, 13. Dihydroeugenol.

Table 1. The concentration ranges of phenols found in cognacs, brandies and rums (ppm). Number of brands: cognac 10, brandy 11, and rum 18.

Phenol	Cognac	Brandy	Rum	
			Dark	White
Phenol	0.01-0.06	0.01-0.07	0.03-0.26	tr[a]-0.11
o-Cresol	nd[b]-0.02	nd-0.02	tr-0.10	nd-0.02
m-Cresol	nd-0.02	nd-0.01	tr-0.05	nd-0.02
p-Cresol	tr	tr	tr-0.20	nd-0.02
Guaiacol	nd-0.04	tr-0.06	0.09-0.83	nd-0.13
p-Ethylphenol	0.03-0.06	0.01-0.16	0.05-1.96	nd-0.88
p-Ethylguaiacol	0.20-0.66	0.01-0.20	0.09-1.83	nd-0.18
Eugenol	tr-0.15	tr-0.27	tr-1.36	nd-0.18
Dihydroeugenol	-	-	tr-0.84	-

[a]tr = trace, 0.001 < tr < 0.01 ppm
[b]nd = none detected, < 0.001 ppm

C. The Difference between the Phenolic Contents of Brandies and Rums.

An attempt was made to classify the different beverages using the results from discriminant analysis. Data from the two variable discrimination function are depicted in Figure 2. The difference in the location of the centroids is extremely significant. The figure includes results from the analysis of whisky samples (17).

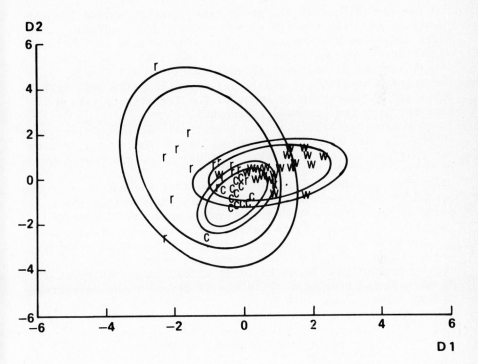

Fig. 2. The classification by discriminant analysis made on the basis of the concentration of phenols in whisky (w), rum (r), cognac and brandy (c). The inner ellipses contain 95 % of the samples and the outer 99 %. The coordinates of the centroids are for whisky (0.724, 0.241), rum (-1.058, 0.512) and brandy (-0.231, -0.819). Total discriminating power of variables, $F_{16, \; 124} = 5.89$, is significant at the 1 % level.

The most discriminatory phenolic compounds are o-cresol, p-ethylphenol and p-ethylguaiacol. Whiskies and rums can most easily be distinguished from the other beverages, while the difference between cognacs and other brandies is less clear. The differences between the various beverages in their contents of phenolic compounds are too small to allow their accurate classification into separate classes.

D. The Reproducibility and Sensitivity of the Gas and Liquid Chromatographic Methods.

The suitability of the gas chromatographic method for the quantitative determination of phenols was assessed using both model compounds and genuine samples. The average coefficient of variation for the model compounds was 5 %, while with actual samples it was only slightly worse at 6 % (Table 2). The average sensitivity was 10 pg, which compares well with other gas chromatographic methods. The sensitivity varied somewhat from one compound to another.

The reproducibility of the liquid chromatographic analysis of phenols with the UV detector was investigated by injecting the standard mixture five times and using the gradient elution shown in Figure 3. The standard deviations of peak height and peak area measurements (Table 2) show that the accuracy of the method based on peak heights is clearly greater than that based on peak areas. The average value of a coefficient of variation for peak heights is 2.5 % and for peak areas 5 %.

The sensitivities in the liquid chromatographic methods are variable. Using the electrochemical detector the sensitivity (signal:noise ratio 3) of the method is about 20 pg of injected compound. With the UV detector the sensitivity is much lower, 2 ng.

E. Liquid Chromatography of Phenols.

The reversed-phase liquid chromatographic properties of ten different phenols found in distilled alcoholic beverages were examined using both a spectrophotometer and an electrochemical detector. These phenols, arranged by their retention times on gradient elution, are listed in Table 3, which also contains the absorption maximum wavelengths in 50-50 (v/v) water-methanol and in methanol. The absorption maximum wavelenghts are essentially the same for each compound in both solvents.

Table 2. The recovery of phenols in the GC and HPLC analysis (mg/1).

| | GC | | | | HPLC | | |
| | Rum | | Cognac | | | | |
Phenol	Added	Found	Added	Found	Added	Found[a]	Found[b]
Phenol	0.13	0.12	0.13	0.12	3.0	2.89	2.91
o-Cresol	0.08	0.08	0.08	0.09			
m-Cresol	0.16	0.16	0.16	0.14	4.5	4.51	4.44
p-Cresol	0.06	0.06	0.06	0.06			
Guaiacol	0.13	0.14	0.13	0.12	3.0	2.92	2.91
p-Ethylphenol	0.10	0.11	0.10	0.10	3.0	2.86	2.98
p-Ethylguaiacol	0.26	0.27	0.26	0.26	3.0	3.04	2.90
Eugenol	0.22	0.23	0.22	0.22	3.0	2.84	2.82
Dihydroeugenol	0.22	0.22	–	–	3.0	2.94	3.22

[a]from the peak heights
[b]from the peak areas

Table 3. Retention time (RT) using a gradient from 45 % to 55 % of methanol in water for 17 minutes and the wavelength at absorption maximum (λ_{max}) in 50-50 water-methanol and in pure methanol for some phenols.

No	Compound	RT (min)	λ_{max} (nm, 50-50 water-methanol)	λ_{max} (nm, methanol)
1.	Phenol	3.94	271	272
2.	Guaiacol	4.40	274	276
3.	m-Cresol	5.47	272	274
4.	p-Cresol	5.49	279	279
5.	o-Cresol	5.62	272	273
6.	3,4-Dimethylphenol	7.44	278	279
7.	p-Ethylphenol	8.12	277	278
8.	p-Ethylguaiacol	8.98	280	281
9.	Eugenol	10.31	280	281
10.	Dihydroeugenol	13.90	280	281

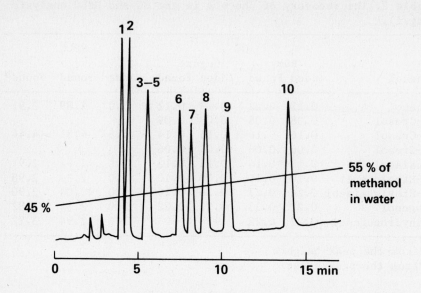

Fig. 3.   A chromatogram of a standard mixture of phenols
obtained with the spectrophotometer detector. The gradient
used is presented in the figure. The flow rate was 1 ml/min
and the injected amount of the cresols (compounds 3–5) was
50 ng of each and the others 100 ng of each. The detector was
operating at 275 nm and 0.05 a.u.f.s. See Table 3 for peak
identifications.

All phenols except the cresols were easily separated using a
simple gradient (Figure 3).

In reversed-phase liquid chromatography the most polar
compounds elute first. This is seen for phenols in Fig. 3,
where the number of substituted carbon atoms seems to
determine the elution order. When there are the same number of
carbon atoms in the substituent groups the elution order is as
follows. Methyl substitution causes longer retention time than
methoxy substitution (compounds 3–5 and 2, respectively), p-
ethyl substitution causes a longer retention time than
dimethyl substitution (compounds 7 and 6, respectively), and
phenol substituted with a saturated propyl group has a longer
retention time than phenol substituted with a propenyl group
(compounds 10 and 9, respectively).

Two different eluents were prepared and tested for the
work with the electrochemical detector. The first one
contained 0.1 M sodium perchlorate and 0.005 M sodium citrate
in 50–50 (v/v) water-methanol, the pH being adjusted to 5.0

with acetic acid. The second contained 0.05 M sodium metaperiodate in 50–50 (v/v) water–methanol. The first one gave a better response and therefore, was selected for further work.

To determine the optimal electrode potential for phenols the detection responses for phenol, p-cresol and p-ethylguaiacol were measured in the range 0.5 V – 1.1 V at 0.05 V intervals. When the area of the chromatographic peak no longer increased it was concluded that the compound is totally oxidized and it is of no use to use a higher potential because a lower electrode potential is better for the baseline. The optimum electrode potential was 0.95 V for phenol, 0.85 V for p-cresol and 0.7 V for p-ethylguaiacol. An electrode potential of 0.9 V was selected as a compromise for further work.

The retention times together with the response factors relative to phenol were measured and are listed in Table 4. The values are averages of three independent injections. The injected amount was 5 ng of each compound. A chromatogram of a standard mixture containing 500 pg of some phenols is presented in Figure 4.

Table 4. Retention time (RT) and molar relative response factor (RRT) relative to phenol obtained with the electrochemical detector. The response factors are calculated from the peak areas. See text for chromatographic conditions.

No	Compound	RT (min)	RRT
1.	Phenol	3.09	1.00
2.	Guaiacol	3.34	1.33
3.	m-Cresol	4.57	1.09
4.	p-Cresol	4.59	1.22
5.	o-Cresol	4.83	1.02
6.	3,4-Dimethylphenol	6.96	1.10
7.	p-Ethylphenol	8.04	1.05
8.	p-Ethylguaiacol	8.54	1.12
9.	Eugenol	10.45	1.26
10.	Dihydroeugenol	16.50	1.06

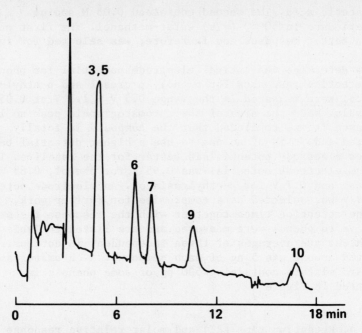

Fig. 4. A chromatogram of some phenols obtained with the
electrochemical detector. Injected amount of each phenol was
500 pg. The eluent was a 50-50 (v/v) mixture of water-methanol
containing 0.1 M sodium perchlorate and 0.005 M trisodium
citrate, adjusted to pH 5.0 with acetic acid. The electrode
potential was 0.9 V. See Table 3 for peak identifications.

IV CONCLUSION

Rums and brandies contain variable quantities of volatile
phenols. Rums can clearly be differentiated on the basis of
their phenolic contents, and dark rums are unique in
containing dihydroeugenol.

The gas and liquid chromatographic methods described are
suitable for the quantitative determination of phenols. An
advantage of using the 2,4-dinitrophenyl ether derivatives for
the gas chromatography is that it enables the resolution of
certain isomers that cannot be separated when the free
compounds are analyzed by either gas or liquid chromatography.
The sensitivity of the liquid chromatographic method with an
electrochemical detector was almost as high as that obtained
by gas chromatography with an electron capture detector. The
use of a UV detector in liquid chromatography, however,
resulted in a much lower sensitivity.

References

1. Luten, J. B., Ritskes, J. M., and Weseman, J. M., Z. Lebensm. Unters. Forsch. 168, 289 (1979).
2. Tressl, R., Grünewald, K. G., Köppler, H., and Silwar, R., Z. Lebensm. Unters. Forsch. 167, 108 (1978).
3. Etievant, P. X., J. Agric. Food Chem., 29, 65 (1981).
4. Nishimura, K., and Masuda, M., J. Food Sci. 36, 819 (1971).
5. Timmer, R., ter Heide, R., Wobben, H. J., and de Valois, P. J., J. Food Sci. 36, 462 (1971).
6. Lehtonen, M., in "The Chromatographic Determination of Some Trace Organic Compounds in Alcoholic Beverages", Dissertation p. 25, Helsingin Yliopiston Monistuspalvelu, Helsinki, (1982).
7. Swan, J. S., and Burtles, S. M., Chemical Society Reviews 7, 201 (1978).
8. Shulgin, A. T., Anal. Chem. 36, 920 (1964).
9. Argauer, R. J., Anal. Chem. 40, 122 (1968).
10. Kawahara, F. K., Anal. Chem. 40, 1009 (1968).
11. Cohen, I. C., Norcup, J., Ruzicka, J. H. A., and Wheals, B.B., J. Chromatogr. 44, 251 (1969).
12. Lehtonen, M., J. Chromatogr. 202, 413 (1980).
13. Masoud, A. N., and Cha, Y. N., J. High Resolut. Chromatogr. Chromatogr. Commun. 5, 299 (1982).
14. Shoup, R. E., and Mayer, G. S., Anal. Chem. 54, 1164 (1982).
15. Roston, D. A., and Kissinger, P. T., Anal. Chem. 53, 1695 (1981).
16. Renold, W., Näf-Muller, R., Keller, U., Willhalm, B., and Ohloff, G., Helv. Chim. Acta 57, 1301 (1974).
17. Lehtonen, M., J. Assoc. Off. Anal. Chem. 66, (1983) in press.

APPLICATION OF GC AND HPLC IN THE PREDICTIVE QUALITY
EVALUATION OF WINES

Gianfrancesco Montedoro
Mario Bertuccioli

Istituto di Industrie Agrarie
Dipartimento di Scienze e Tecnologie Alimentari
Università degli Studi
Via S. Costanzo
06100 - Perugia - Italy

The originals techniques were developeds for isolation, i
dentification and quantitative evaluation of volatiles and non
volatiles components. The first group were estimated both by e
richment of "headspace" vapors on porous polymer (Tenax GC
traps) their desplacement and introduction through a closed sy
stem to the GC capillary column and direct injection of the
sample on the GC column.

The second group were evaluate both by HPLC (organic a-
cids) and selective fractionation by Methilcellulose (MC) and
successive colorimetric analysis (polyphenols).

Some wines samples were submitted on the sensorial analysis ("panel test"). Sensory assessment combined with quantitative istrumental analysis has been used to describe the quality of wine's aroma.

I. INTRODUCTION

The quality of foods constitutes an element of fundamental importance as far as their appetency is concerned thus
their acceptability by the consumer and, more generally speaking, as index of marketable and nutritional evaluation of the
same one. The latter may be considered from different points

of view: a) hedonistic (acceptability and typicality); b) ge-
nuineness one ("original" natural composition); c) hygienic-sa
nitary one (introduction or formation of components unrelated
with the natural composition of food such as to modify its nutritional and dietetic properties including its toxicity).

In the actual state all these different qualitative a-
spects are of difficult analytical evaluation because numerous
components take part in them, not only difficult to be evaluated qualitatively, but also to be quantified as most of them a
re present within the thresholds of concentration inferior to
mg/l. (1).

Among the many different aspects the one which is taken
into serious consideration be researchers is the hedonistic o-
ne.

In this sense the quality definitions of a wine may be nu
merous.

The meaning intended to be given the term is manifold and
in any case it always underlines either the criterion of choice or the presence of particular properties and typicality
and the possess of a certain superiority or excellence. (2).

In this field the most valid definitions concerne some em
blematic aspects such as: "the quality of a wine is the whole
of the properties which make it acceptable or desirable". This
goes on well with the assumption that a wine is part of the
class of food in which "the flavour is the fundamental factor
of choice and to which the consumers attribute an affective
and symbolic factor a gratifying power". There is a series of
factors which contribute to form this characteristic: the climate and soil and, in this ambit, the vine and what man by selection can get from the content of the different constitutive
parts of the berry, operating through the harvesting wine-ma-
king, storage and ageing.

All these factors influence the formation and concentration of the "primary" original compounds and of those "induced" (derivatized) by the technological processes themselves.

The wine flavour, like the beer, cyder and other the fermented beverages, in general, results of being of great comple
xity. (3).

In the grapes have already been found out about 400 volatile constituents in a single chromatographic operation in gaseous phase on capillary columns; of these, 250 have undoubte-

dly been identified. To this complexity of composition is to
be added a no better complexity of origin.

The flavour of wines is in fact the final product of a
long biological biochemical and technological sequence,and not
only the product of grapes fermentation. The compounds groups
which can result being present in the productive wine sequence, can be classified as follows: "primary" original flavours
(81 hydrocarbons, 31 alcohol, 40 esters, 48 free fat acids, 28
aldehydes and acetates, 18 ketones, various terpenic compounds
to bring back to the linanol, geraniol, nerol, terpineol, citronellol and Ho-trienol);"induced" original flavours (methyl
alcohol, galacturonic acid, neutral sugars, trans-2-hexenal,
cis-3-hexenal, trans-2-hexenol, ethyl-vinil-ketone, 2-pentil-
furan); flavours produced by fermentation (superior alcohols,
ethyl-acetate and ethyl-lactate, ethyl-capronate, ethyl-capri-
late, ethyl-caprinate, ethyl-laurate and volatile aliphatic
fat acids); flavours following fermentation (aldehydes, acetates, lactons caming from different parts and by products of
the naftalene and furan).

It is the case to remember once more that more than 500
compounds have been identified, characterizing the aromatic
profile and the visual aspect of many food. Notwithstanding
this, the evaluation of the hedonistic appreciation, doesn't
leave the sensory analysis that is however the unique means of
the wine qualitative appreciation.

Some experimental attempts have been started in order to
objectivate some sensory characteristics and that is to
say, the smell, the taste, the softness, the typicality, etc.
with the instrumental investigation (gas chromatography, mass
spectometry, NMR). (4).

Very recent studies have given the possibility to remark
that there is a relation between the origin of wine and the vo
latile fraction of its constituents and furthermore, even if
in a more difficult way, that is possible to get to its qualitative evaluation, by making use of the discriminant analysis
and the multiple regression analysis in the elaboration of the
data achieved by the quantitative instrumental analysis and
sensorial assessment. (5).

From the above said a fundamental consideration comes out
and, that is to say, that to the usual analytical determinations relative to the alcoholic grade, volatile and total aci-

dity, dry extract and ashes is to be placed side by side, the
evaluation of the volatile substances (alcohols, aldehydes, e
sters, lactons, ethers, terpenes, acetates) and the non volatile ones (organic acids, polyphenols, sugars, peptides) all
responsible for the aromatic profils.

While all methods can give precise results about the con
centration,only some permit to correlate the instrumental and
sensorial analysis, as they are non sample's destructive and
so in a condition to evaluate contestually the real concentra
tion of the responsible constituents and the sensory   cha
racteristics.

On the other hand during the last period of time many re
searchers have shown a great interest in getting the "panel
test" ready on scientific basis according to the new acquisi-
tions on these compounds in relation with the concentration
level of the aromatic compounds, with its physicochemical sta
te and with the interaction between this and the other consti
tuents of the medium (sinergy and obliteration). (6, 7).

It is indeed on the previous statement that our Institute has made up some methods of colorimetric and chromatographic analyses, by which has also been possible not only to i-
solate, separate, identify and dose numerous compounds, of
which many aromatic, but also to quantify the intervention of
these constituents on some sensory quality of fermented drinks
in general and of wine in particular.

II. ANALYTICAL PROCEDURES

A. Methylcellulose Precipitation of Phenols

Complex polyphenolics occur in most foods and beverages
of plant origin. They are responsible for taste sensation,
such as "body" and "bitter" and "astringency" other that for
colours raging.

They can be found in grapes and wines under different
forms, the physicochemical state and sensory   properties
very different themselves with molecular dimensions varying
from 1 to more than 10 elementary molecules. (8, 9, 10).

Always for sensory   purposes generally such compounds,
in function of the molecular weight, are classified into four
different groups: inferior to 300, corresponding to the mono

meric forms of flavans, flavonols and anthocyanins, which a-
re responsible for the colour and "bitter taste"; between 500
and 1500, corresponding to fractions constituted of 2 to 5 e-
lementary molecules and they are generally responsible for the
colour, "body" and "astringent" tastes; between 1500 and 5000,
corresponding to polymers constituted of 6 to 10 elementary mo
lecules, having "astringent" sensations, "body" responsability
and reactivity with proteins, polysaccharides and other compounds, to form pseudo stable solutions; superior to 5000 corresponding to highly polymerized forms of 10 monomers,presents
in the colloidal state.

This last three groups of compounds are the most important, because they constitute, in the case of red wines, 70%-
90% of the total polyphebolics complex.

The phenolic constituents, which partecipates to the for-
mations of the colour of red wines, are the anthocyanins (3
and 3,5 glucosides and acylated glucosides of the delphinidin,
petunidin, malvidin, cyanidin, peonidin) and condensed tannin
(3-flavonols, 3.4-flavandiols and anthocyanins).

Diverse factors regulate this ultimate characteristic,
such as the pH, the rH, the tenor in sulphurous acid and acetaldehyde, the ageing stage of the wine in itself. The anthocyanins are responsible for the bright red colour (absorption
at 520 nm.) while tannins are responsible for the orange red
colour (absorption at 420 nm.).

Many attempts have been carried out in order to reach a
correct evaluation of the two phenolic fractions (monomers and
polymers) and its correlations on the sensory quality of the
wine.

On our side we have developed an analytical procedure
which gave reproducible value on a series of wines with a variability of about 1%. The precipitation of phenols (tannins)
were carried out using the methylcellulose, that differently
from the other precipitants (cinconin , polyvinilpyrrolidino-
ne, etc.) (11) doesen't make any change of the physicochemical
conditions of the supernatant.

This possibility permits to make other assays, including
the one of the monomeric and polymeric anthocyanins the first
group both in colourless and coloured forms.

A summarizing scheme of the separation of different pheno
lics constituents is reported in Figure 1.

M U S T O R W I N E
(Total phenols)

Colorimetric Assay	Methylcellulose precipitation: Supernatant (monomeric, dimeric, trimeric forms)	Methylcellulose precipitation: Precipitate (oligomeric, polymeric forms)
Folin-Ciocalteu	- Phenols (acids, alcohols, flavonoids)	Tannins phenols (condensed, hydrolizable)
Vanillin	- Catechine - Epicatechins	
CH_3CHO	- Colourless Anthocyanins (bisulphite derivative)	
$Na_2S_2O_5$ 0.003 M (1)	- Coloured Anthocyanins (monomeric forms)	
$Na_2S_2O_5$ 0.008 M (2)	- Total Anthocyanins	
But.OH-HCl	- Leucoanthocyanidins or Proanthocyanidins	

Formaldheyde precipitation

	Supernatant	Precipitate
Folin-Ciocalteu	Phenolic acids (Phenols alcohols)	Flavonoids

Figure 1. Methylcellulose fractionation and polyphenolics evaluation

A - SAMPLE PREPARATION
Isolation and concentration of components

a) DISTILLATION
at atmospheric pressure
at a reduced pressure
steam distillation
fractional distillation

b) CONCENTRATION
by solvent extraction
by freezing (cryo-concentration)
by adsorption

B - COMPONENTS SEPARATION
Chromatographic procedures (GLC, HLC, etc.)

C - COMPONENTS IDENTIFICATION
by chemical methods
by physical methods (IR, Vis, UV, NMR, MS)

D - SENSORIAL IDENTIFICATION
olfactive, at the GLC column output
by determination of the perception threshold
by difference tests (comparison, substraction)

by descriptive tests (typical character, quality)

E - QUANTITATIVE ANALYSIS

Figure 2. The analytical steps in the study
of the "aromatic qualities"

B. Chromatographic analysis

The volatile compounds analysis goes on through a series
of phases, the details of which are schown in Figure 2.

The one relative to the extraction and concentration of
the sample, is substantially variable, using, as scown, diffe
rent principles and operative techniques. (12, 13).

These processes may strongly condition the analytical re
sult. As in fact reported in Figure 3 and, each technique leads
to a different result, even if starting from the same sample.

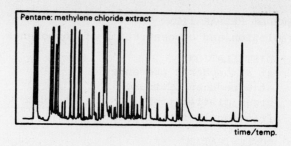

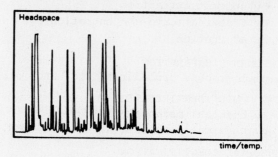

Figure 3. Chromatograms of Tawny port. (7)

Headspace: as evinced by the same pictures, the headspace tec‐
nique differentiate completely from the others as it keeps in
to account the partition of volatile compounds in the middle,
although it is a function of its physico-chemical conditions.
The applying interest of this technique as already mentioned
lies derives for the absence of external interventions (mani‐
pulations) which could modify the actual composition of the
actual material in question. That permit to make at the instru
mental analysis and at the same time and simultaneously at the
sensorial appreciation under conditions similar to those of ta
sting.

    Some methodologic difficulties, linked to the compounds
transfer which are present in the headspace, at the gaschroma‐
tographic column, by making use of capillary columns, have
been overcome, modifying in part the traditional system made
up at the time. The made up operative scheme is reported in Fi
gure 4. The modalities are as follovs:
Step A, Pre-column purge: Gas stream elimination of excess i‐
    nert components (water, ethyl alcohol, a fraction of the
    higher alcohols, etc.), taking into account the nature of

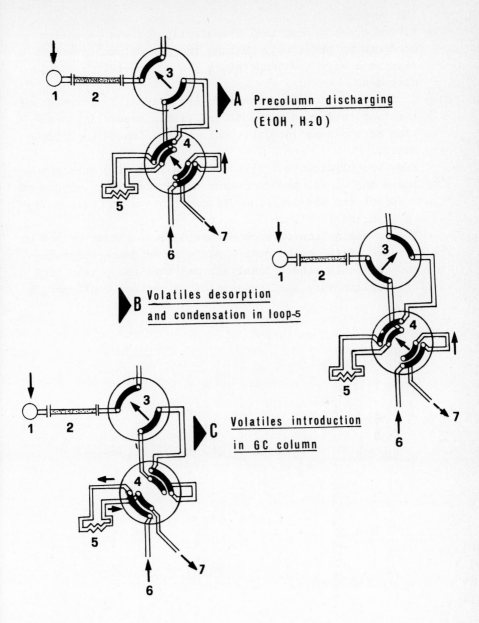

Figure 4. Introduction of volatile components into the capilla ry column. 1 - Auxiliary flow regulator; 2 - pre-column; 3 - four-way valve; 4 - eight-way valve; 5 - loop; 6 - carrier gas input; 7 - GC capillary column.

the support.

Step B, Compounds desorption: Transferring of the adsorbed products by previously heating the pre-column in a loop kept at a very low temperature (-90°C or -170°C, liquid nitrogen).

Step C, Compounds introduction into the column: Evaporation of the loop condensed products and instantaneous introduction of the same into the column (the "on column" proce dure).

Some exemplifications elaborated in analysing corresponding musts and wines, produced from two different cultivars of grapes (Grechetto and Trebbiano toscano, grown in Central Italy) are reported in Figure 5.

The investigation of such chromatograms, characterised by the presence of 150 peaks puts clearly in evidence the resolving power of this kind of analysis and thus the possibility to differenciate very similar samples. A further confirmation

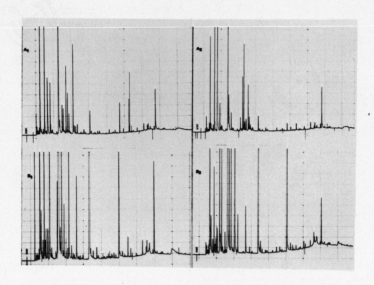

Figure 5. Chromatograms of musts (A_1 and A_2) and of wines (B_1 and B_2) from Trebbiano (1) and Grechetto (2) grapes grown in Middle Italy.

is given by the chromatograms schown in Figure 6 on of them
(the first) referred to a wine and the other (the second) to a
model system from whose comparison puts into evidence significant areas of the superior aromagram linked to the compounds
taken into consideration.

Direct injection GC of the sample: various substances, such as
methanol, higher   alcohols, acetaldehyde, ethyl-acetate,ethyl
-lactate, acetoine, 2.3 butanediol and glycerol , because pre
sent in the wine with a greater concentration, may be directly
dosed using diverse procedures: gas chromatographic, chemical,
physico-chemical and enzymatic ones. (15).

A methodology recently made up by us, has permitted the
simultaneous and direct gas chromatographic phase, constituted by graphitized coal. (16). The operative conditions were
the following: a Varian gas chromatograph, model 3700 equipped
with flame ionization detector was used. Column was 6 ft x 1/4
inch OD glass tubing packed with 80/100 mesh Carbopack C coated with 0.2% w/w Carbowax 1500, a product of Supelco, USA.

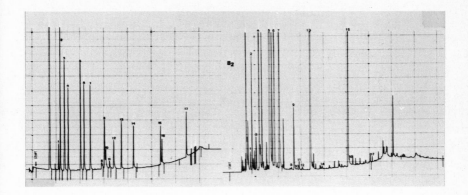

Figure 6. Headspace chromatograms of a wine and of a mixture
of standard compounds, obtained employing a 100 m capillary co
lumn. Peak: 1 = isobutyl acetate; 2 = ethyl butyrate; 3 = hexenal; 4 = isopentyl acetate; 5 = 3-methyl-1-butanol + trans-2
-hexenal; 6 = ethyl capronate; 7 = hexyl acetate; 8 = ethyl lac
tate; 9 = n-hexanol; 10 = cis-3-hexenol; 11 = trans-3-hexenol;
12 = trans-2-hexenol; 13 = ethyl caprylate; 14 = benzaldehyde;
15 = ethyl caprinate; 16 = diethyl succinate; 17 = 2-phenethyl
alcohol.

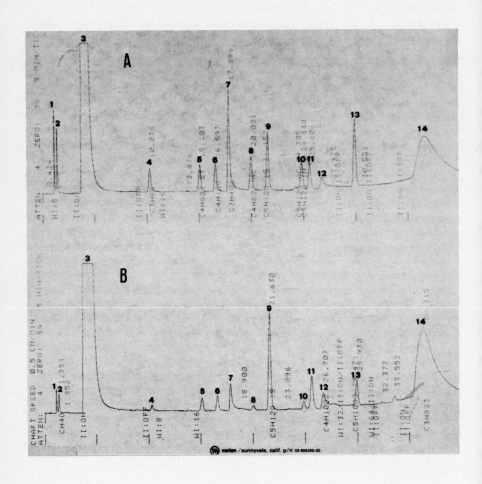

Figure 7. Gas chromatographic traces of standard solution (A) and wine sample (B) on glass column coated with Carbopack C. Peak 1 = acetaldehyde; 2 = methanol; 3 = ethanol; 4 = n propanol; 5 = ethyl acetate; 6 = 2-methyl-1-propanol; 7 = acetic acid; 8 = acetoin; 9 = 3-methyl-2-butanol (IS); 10 = 2-methyl-1-butanol; 11 = 3-methyl-1-butanol; 12 = 2,3 butanediol; 13 = e-thyl lactate; 14 = glycerol.

Table 1. Relative response factors (RRF) and percent variation coefficients (% V.C.) (16)

COMPONENT	RANGE (mg/l)	RRF (1)	% V.C.
ACETALDEHYDE	35.5 - 141.8	2.374726	7.33
METHANOL	39.5 - 158.2	1.737868	4.46
n PROPANOL	20.1 - 80.4	1.178418	2.59
ETHYL ACETATE	90.2 - 360.8	1.688615	1.61
2-METHYL-1-PROPANOL	60.4 - 241.8	0.951351	1.74
ACETOIN	30.0 - 120.0	2.158312	3.95
2-METHYL-1-BUTANOL	40.8 - 163.2	0.866537	1.94
3-METHYL-1-BUTANOL	81.3 - 325.2	0.815265	1.65
2,3 BUTANEDIOL	639.0 - 2130.0	2.009391	4.52
ETHYL LACTATE	130.9 - 520.0	2.221371	3.69
GLYCEROL	3201.0 - 10670.0	4.304329	8.83

(1) refer to 3-METHYL-2-BUTANOL (IS)

Table 2. Ranges and averages values (mg/l) of volatile compounds
of white and red italian wines (16)

COMPONENT	WHITE WINES		RED WINES	
	Range	Average (1)	Range	Average (1)
ACETALDEHYDE	28.2 – 176.9	75.9 ± (27.4)	38.4 – 213.7	103.0 ± (56.3)
METHANOL	27.8 – 114.2	48.0 ± (16.6)	109.9 – 208.7	154.5 ± (22.3)
n PROPANOL	3.4 – 36.2	20.0 ± (6.9)	26.8 – 52.5	35.6 ± (6.8)
ETHYL ACETATE	25.6 – 118.2	47.6 ± (17.3)	34.7 – 186.5	109.6 ± (33.4)
2-METHYL-1-PROPANOL	23.7 – 113.4	56.9 ± (18.4)	50.7 – 164.2	100.8 ± (30.4)
ACETOIN	5.3 – 42.4	14.4 ± (12.4)	14.9 – 101.9	25.7 ± (20.2)
2-METHYL-1-BUTANOL	16.7 – 66.0	38.8 ± (13.7)	32.0 – 128.2	62.8 ± (21.9)
3-METHYL-1-BUTANOL	80.8 – 251.4	159.5 ± (44.3)	95.0 – 450.9	238.3 ± (86.4)
2,3 BUTANEDIOL	30.4 – 321.7	146.3 ± (65.5)	171.9 – 804.1	406.8 ± (182.2)
ETHYL LACTATE	2.7 – 305.6	94.1 ± (78.6)	120.6 – 682.2	297.0 ± (126.0)
GLYCEROL	4970.0 – 10915.1	7807.3 ± (1645.6)	5803.2 – 20041.2	10638.2 ± (4888.8)

(1) Number in parentheses are standard deviations.

Gas chromatography operating conditions were: 0.15 µl sam ples of standard solutions and of wines were injected in the gas chromatographic column manteined at 35°C for 2 min., then programmed to 165°C at 4°C/min. with a helium gas flow at 20 ml/min. The injectior and detector were 250°C.

The technique employed an internal standard: 3-methyl-2-butanol.

The evaluation of the relative response factors (RRF) and the quantitative analyses of compounds were carried out using a Varian integrator mod. VISTA 401.

The exemplification of a gas chromatogram is schown in Fi gure 7. The results achieved, examining about 47 dry white wi nes and 35 red wines of different years, produced from diffe- rent cultivars of grapes coming from diverse wine-cellars and regions, are reported in Table 1 and Table 2. As it is eviden ced by the statistical data, the coefficient of the percenta- ge variation (1.61-8.83) and of the relative response factors (0,815265-4,304329), demonstrate that the variability of the method. In fact the quantitative results obtained are in accor dance with the data reported in the bibliography and obtained by making use of several analytical techniques.

Extraction and GC assay: the volatile aliphatic fatty acids,as well know, contribute positively to the freshness note of the wine. (17, 18). It has also been demonstrated that the said compounds partecipate by esterification with ethyl alcohol to the formation of the corresponding esters. In fact to a con centration increase of these, generally corresponds a decrease of the first. (19). From that the necessity to evaluate these basic compounds, the origin of which is of strictly fermentati ve nature even if it is to ascribe to the linoleic acid, that for the its biological oxidization (yeasts), gives thus origin to these compounds.

Furthermore it is to be added that the free volatile fat ty acids presents a characteristic aromatic note; the percep- tion thresholds of these are variable and linked to the mole- cular weight (20). The concentration found in wines even if they don't form part of the sensory perception, may any- way consent in time a forecast of the sensorial quality of its (21, 22). On this background a gas chromatographic method has been made up.

The procedure is the following: to 500 ml.
of wine are added 50 gr. of sodium chloride and submitted to
extraction with a mixture pentane/methylene chloride in a
relation 2 to 1 for three times; adding previously to the system enantic acid as standard. The extract obtained this way
is concentrated in vacuum and dried for passage on anydrous so
dium sulphate, brought back to a volume of 10 ml., utilizing
only pentane. A modification of this procedure is practised
for red wines. In this case the added sample of the standard
is alkalinized till pH 8 with the adding of solvent and submit
ted to extraction. In this first phase all the organic substan
ces that interfere in the dosage, are eliminated. The residue
is reacidified at pH 2 with hydrochloric acid and extracted a
gain for three times with the same solvent, concentrated,dried,
brought back to volume and submitted to analysis.

The gas chromatographic conditions are (23): glass column: 6 ft x 1,4 inch OD, packed with 10% of S.P.-216 P.S.,
100-120 mesh, Supelco; initial temperature 130°C, then program
med to 180°C, at 10°C/min.; transport gas helium (or air) 20
ml/min.; detector and injector temperature 250°C.

A exemplifying gas chromatogram is shown in Figure 8.

The concentration interval, the coefficients of percenta
ge variations and the relative response factors, are reported
in Table 3, while the analytical data of many wines submitted
to sensorial analysis and rating according to the score card,
in three qualitatively decreasing groups, are reported in Table 4. From the results of the whole investigation we can af-
firm that the method, for accuracy and reproducibility, may
result credible and valid. The sensory positive influence exerted by these compounds, on the white wine, in clearly
evidenced by the comparison between the concentrations of fat
ty acids and the classification of wines, which resulting from
"panel test". The partecipation of these compounds in red wines is different because the influence exerted by all the o-
ther constituents, in particular, those that came to be in the
ageing course.

High liquid chromatography: the change of acidity can be considered an important sensory modification of this kind of
drinks. In fact the malic acid is a bicarboxilic compound,that
is to say, it has got two functions in free state.

It follows that its gustative perception shows itself

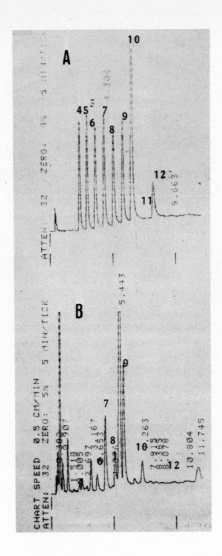

Figure 8. Gas chromatographic traces of standard solution (A) and extract wine sample (B) on glass column coated with SP-216 -PS (10%): Peak 4 = butyric acid; 5 = valeric acid; 6 = caproic acid; 7 = heptanoic acid; 8 = caprilic acid; 9 = pelargonic acid; 10 = caprinic acid; 11 = undecanoic acid; 12 = lauric acid.

Table 3. Relative response factors (RRF) and percent variation coefficients (% V.C.) of fatty acids chromatographic ana
lysis in model sistems.

	RANGE (mg/l)		RRF[1]	% V.C.
BUTYRIC ACID	0.24	1.21	0.84167	7.1
VALERIC ACID	0.25	1.26	1.31267	7.8
CAPROIC ACID	0.21	1.02	1.20967	11.4
HEPTANOIC ACID	0.25	1.24	1.11266	5.7
CAPRILIC ACID	0.23	1.17	1.12539	4.8
PELARGONIC ACID	0.24	1.20	1.01330	2.9
CAPRINIC ACID	0.23	1.14	1.04160	4.6
LAURIC ACID	0.25	1.26	0.99467	6.8

(1) Refer to Undecanoic acid (I.S.)

Table 4. Concentrations (mg/l) of fatty acids in wines
submitted at the "panel test"

WHITE WINES

	Caproic acid	Caprilic acid	Caprinic acid
Group 1	0.50	1.44	1.25
Group 2	0.46	1.25	1.08
Group 3	0.42	1.08	0.27

RED WINES

	Caproic acid	Caprilic acid	Caprinic acid
Group 1	0.08	0.09	0.06
Group 2	0.12	0.15	0.17
Group 3	0.03	0.03	0.03

Group 1: 75.0 - 100.0 marks; Group 2: 62.5 - 75.0 marks;
Group 3: 50.0 - 62.5 marks.

with a striking sour and unripe sense, which confers to the
product a marked sense of unreached ripening.

   In the wines, rich of phenolic constituents, particularly of the tannic fraction, the presence of this acid, brings
a perceptive exaltation of this last component which transfor
ms itself in an exaltation of a rough, coarse astringent and
bitter character. On the other hand, the disappearing of malic
acid and the accumulation of lactic acid (by of the malolactic
fermentation) by effect of the different structural characteristics and sensory of the last one, allows, according to its
concentration, an physico-chemical achivied ripening, just as
it can be found in wines and sparkling wines aged for a long
time. Even if diverse enzymatic methods are at hand to dose
the organic acids separately (tartaric acid, malic acid,citric
acid and lactic acid),for the reasons already mentioned it has
been necessary to procede to one of their comprehensive and si
multaneous dosage. The method followed has been the high liquid chromatographic, using a ion exchange column type AX10.
The operative conditions were the following: flux 2ml/min.;sol
vent A: 0,025 M KH_2PO_4 (pH 4,5); solvent B: 0,25 M (pH 4,5);
gradient 0-70% in 24 min. and successively 70-100%in 5 min.;de
tection wavelength 210 nm (0,05 ABS).A chromatogram standard
is shown in Figure 9,while the analytical data relative to a
series of basic wines and Italian and French sparkling wines o
btained by the Charmat and Champenois method are reported in
Table 5.

III. PREDICTIVE EVALUATION OF WINES QUALITY

A. Instrumental and sensory analysis

   As ready referred,the aroma of alcoholic beverages is always a combination of various differents volatiles and non volatiles components (nuances of colour,odour,taste and fluidi-
ty). Therefore ,47 dry wines,obtained from different cultivars
of grapes,vintages,regions and wine cellars,were submitted to
a "panel test" (10 judges) and subsequantly to chemical,physi-
co-chemical and chromatography analysis,according to the techniques before described.
   The "panel test" was carried out, according to the card
devised by the "Associazione Enotecnici Italiani". The examination, was carried out the following succession: sighting,

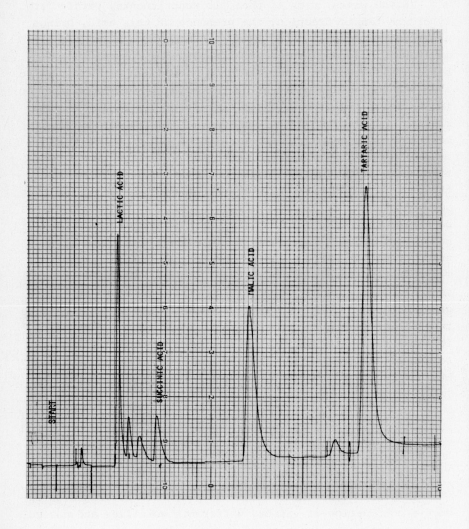

Figure 9. Liquid chromatography chromatogram of sinthetic organic acid mixture. Gradient run: as mentioned under Experimental.

Table 5. Concentrations (gr/l) of organic acids in wines and sparkling wines

CONCENTRATION (gr/l)

WINES [1]

	1	2	3	4	5	6	7	8
Lactic acid	2.49	1.90	0.02	0.02	–	0.02	0.01	0.02
Malic acid	1.85	0.758	5.59	6.70	5.90	5.09	6.46	4.15
Succynic acid	0.63	0.37	–	–	–	–	–	–
Tartaric acid	2.12	1.70	2.66	2.47	2.47	3.30	2.21	3.71

SPARKLING WINES

| | | | C h a r m a t | | | | | Champenois | |
	A[2]	B[2]	C	D	E	F	G	H	I
Lactic acid	1.96	2.14	1.30	4.68	4.33	4.24	1.93	4.31	4.94
Malic acid	1.14	1.16	2.92	0.7	1.04	1.41	1.71	2.46	0.49
Succynic acid	0.63	0.41	0.70	0.40	0.65	0.83	0.64	0.43	1.41
Tartaric acid	2.12	1.63	0.92	1.12	1.96	1.41	0.78	0.58	3.92

(1) 1: Trebbiano; 2: White Pinot; 3: Grolleau; 4: Chanin; 5: Cabernet Franc.; 6: Aligoté; 7: Ugni-blanc; 8: Chardonnay.
(2) Sparkling wine of Trebbiano.

smelling, tasting/taste-smell, assigning,a total of 100 marks, subdidived respectively in 20, 32, 48. All the wines were clas sified in three score classes:

Group 1 = excellent   (75.0-100.0 marks)
Group 2 = good   (62.5-75.0 marks)
Group 3 = below good above fair (50.0-62.5 marks)

B. Statistical methods

The statistical analysis was carried out performing a stepwise discriminant analysis (S.D.A.) and a multiple regres sion analysis (M.R.A.).

Through the S.D.A. was attempts to classify the wines in to same above three groups by means of a set of rank instrumen tal variables selected according to their capacity to separa te one group from others.

By the   M.R.A. was attempts to determine with a good degree of approximation the most significant and responsible compounds of the sensory   note.

The analytical parameters used, the corresponding avera ge values and their standard deviation are reported in Table 6. In Table 7 are reported the discriminating variables stati stically elaborated both by combination among some of them (SC) and by considering them singularly and totally at the sa me time. In the Table 8 at last, are reported the markings found out at the "panel test" (evaluated) and those achieved by means of statistical elaboration of the instrumental analy tic data (estimated). From the joined examination of the three tables it is possible to draw some fundamental considerations: a) the analytical results, put in relation with the three groups of wine, as they resulted during the tasting, are all well discriminated; b) the numerous variables considered both singularly and combined among themselves, show to have a clear discriminating significance for sensory purpores (C_9, M, C_6, etc.); c) all the markings found at the tasting are well cor related with those statistically elaborated except for very ra re exceptions.

IV. CONCLUSION

The combination of the instrumental data with the senso-

Table 6. Evaluated and estimated scores
(total score, set SE) (5)

GROUP 1				GROUP 2				GROUP 3			
Wine	Observed	Predicted	Residual	Wine	Observed	Predicted	Residual	Wine	Observed	Predicted	Residual
2	75.8	77.8	−2.0	1	72.0	68.5	+ 3.5	5	61.2	61.8	−0.6
3	77.9	79.9	−2.0	4	64.5	67.5	−3.0	11	62.2	57.3	+4.9
6	79.6	83.0	−3.4	9	74.8	74.7	+0.1	13	58.4	60.3	−1.9
7	79.2	78.4	+0.8	10	64.3	61.4	+2.9	16	52.6	55.3	−2.7
8	86.1	86.3	−0.2	12	72.1	69.9	+2.2	18	61.6	61.3	+0.3
15	77.9	78.7	−0.8	14	65.6	67.0	−1.4	22	58.1	58.8	−0.7
20	78.4	77.9	+0.5	17	66.4	65.9	+05	24	50.3	49.7	+0.6
21	86.1	86.3	−0.2	19	64.0	68.8	−4.8	27	59.6	61.3	−1.7
23	76.8	75.9	+0.9	29	62.6	63.6	−1.0	30	60.8	59.4	+1.4
25	80.2	80.8	−0.6	34	69.2	68.2	+1.0	31	60.0	59.4	+0.6
26	83.2	83.8	−0.6	35	63.5	62.9	+0.6	33	62.1	60.3	+1.9
28	78.5	77.9	+0.6	39	63.9	66.9	−3.0	36	53.2	54.3	−1.1
32	86.1	85.1	+1.0	41	63.6	64.1	−0.5	45	53.8	55.8	−1.0
37	78.0	79.4	−1.4	43	73.2	71.5	+1.7				
38	82.6	79.3	+3.3	46	65.0	64.6	+0.4				
40	79.2	79.2	=0.0	47	72.9	72.1	+0.8				
42	77.0	77.3	−0.3								
44	91.0	90.7	+0.3								

Table 7. Discriminant variables of sets SC and SE in decreasing order of importance for white wines into three groups (5).

TOTAL SCORE

VARIABLE ENTERED		F - VALUE	
SC	SE	SC	SE
C9	C9	8.44**	18.84**
E	C6	8.14**	14.97**
C	C2	7.98**	14.39**
M	F8	6.07**	12.69**
C10	N	5.74**	8.83**
T	P	5.73**	7.80**
C4	A4	5.20**	6.96**
I	A7	4.28**	6.69**
R	K	4.26**	6.31**
H	O	3.85**	5.94**
V	A6	3.38**	5.86**
C5	H	2.67*	5.75**
L	A	2.65*	5.57**
C12	C10	2.54*	5.04**
C3	F10	2.42*	4.76**
P	E	2.21*	4.42**
X	C	2.14*	4.30**
O	F	2.13*	3.94**
A	A3	2.00	3.79**
C2	V	1.98	3.74**
C7	C13	1.92	3.24**
Y	A5	1.75	2.83**
C1	I	1.52	1.62
K	S	1.46	1.56
F10		1.12	
C6		1.04	
F8		1.00	

ODOUR SCORE

VARIABLE INTERED		F - VALUE	
SC	SE	SC	SE
C4	A13	18.85**	55.05**
I	A4	8.52**	36.42**
C12	I	7.03**	35.47**
Q	X	6.74**	31.68**
C2	F6	6.70**	30.02**
C10	B	5.56**	27.57**
D	C	5.36**	26.37**
N	A5	5.23**	24.68**
B	V	5.18**	24.18**
C9	Y	5.17**	23.28**
H	F	5.15**	23.01**
Y	P	4.25**	22.98**
C7	K	4.09**	21.30**
C	C5	3.42**	19.89**
V	A	3.01**	18.53**
A	N	2.99**	18.11**
E	Q	2.60*	16.41**
C14	R	2.39*	16.02**
P	C2	2.01	15.58**
F6	O	1.77	11.44**
Z	E	1.34	11.33**
R	A8	1.29	11.14**
C1	A3	1.19	11.03**
	F8		10.01**
	C12		9.14**
	C7		8.55**
	C4		8.11**
	C11		5.06**
	C1		4.88**
	C6		2.83*
	C9		2.79*
	C13		2.48*
	A7		1.64

TASTE SCORE

VARIABLE ENTERED		F - VALUE	
SC	SE	SC	SE
M	I	12.12**	24.00**
C6	A6	10.52**	23.60**
C4	M	9.51**	20.73**
X	A	9.30**	16.25**
T	C14	8.76**	13.27**
L	A3	7.52**	12.80**
H	Q	7.30**	11.91**
F8	A7	7.07**	10.90**
Y	N	6.01**	10.62**
R	C9	5.57**	7.56**
N	C12	5.45**	7.55**
C8	A2	5.27**	6.00**
C14	F10	5.02**	5.69**
F6	C8	4.99**	5.44**
C7	Y	4.86**	4.70**
C9	A11	4.72**	3.48**
S	C2	4.45**	3.45**
I	Z	4.36**	3.31**
C11	A14	2.25*	3.00**
G	C3	1.62	2.63*
F	F8	1.59	2.24*
Q	O	1.45	2.21*
O	F6	1.31	1.32
C10		1.29	

** = SIGNIFICANT AT P 0.01
* = SIGNIFICANT AT P 0.05

Table 8   GLC and analytical data average of
three white wine groups (5).

Component	Variable	GROUPS		
		1	2	3
		MEAN +/− STANDARD DEVIATION		
ANALYTICAL				
PH	C1	3.303 +/− 0.162	3.157 +/− 0.072	3.243 +/− 0.101
TOTAL ACIDITY (g/l)	C2	6.221 +/− 0.637	6.931 +/− 0.620	6.460 +/− 0.492
VOLATILE ACIDITY (g/l)	C3	0.428 +/− 0.162	0.437 +/− 0.177	0.371 +/− 0.143
FREE SO_2 (mg/l)	C4	15.089 +/− 7.987	9.587 +/− 5.457	8.861 +/− 7.408
TOTAL SO_2 (mg/l)	C5	112.589 +/− 27.614	118.025 +/− 29.642	114.738 +/− 33.595
COLOR ($D.O_{420}$ x 1000)	C6	81.167 +/− 25.933	83.562 +/− 26.389	107.153 +/− 56.237
TOTAL PHENOL (mg/l)	C7	207.155 +/− 45.483	242.187 +/− 59.814	253.538 +/− 127.415
ETHANOL (VOL. %)	C8	11.893 +/− 0.487	11.910 +/− 0.557	11.968 +/− 0.848
EXTRACT (g/l)	C9	20.822 +/− 1.791	20.518 +/− 2.130	20.277 +/− 1.791
ASH (g/l)	C10	1.707 +/− 0.418	1.496 +/− 0.277	1.703 +/− 0.222
ALKALINITY ASH (meq/l)	C11	17.500 +/− 3.204	16.437 +/− 2.738	17.769 +/− 2.619
MALIC ACID (g/l)	C12	1.770 +/− 1.037	2.345 +/− 0.748	1.657 +/− 1.076
RH	C13	10.926 +/− 4.339	12.456 +/− 3.482	13.069 +/− 0.929
CATECHINS (mg/l)	C14	32.944 +/− 18.675	52.562 +/− 40.740	52.154 +/− 30.973
GAS-CHROMATOGRAPHIC [1]				
i-BUTYL ACETATE *	A	0.566 +/− 0.237	0.489 +/− 0.359	0.421 +/− 0.147
ETHYL BUTIRATE *	B	1.950 +/− 0.718	1.829 +/− 0.842	1.666 +/− 0.748
i-AMYL ACETATE *	C	7.728 +/− 4.205	4.213 +/− 1.398	5.383 +/− 3.592
ETHYL CAPROATE *	D	4.557 +/− 1.868	4.354 +/− 2.054	4.109 +/− 1.970
HEXYL ACETATE *	E	0.624 +/− 0.471	0.342 +/− 0.166	0.443 +/− 0.364
n-HEXANOL *	F	1.237 +/− 0.707	1.226 +/− 0.613	1.258 +/− 0.617
ETHYL CAPRYLATE *	G	5.801 +/− 3.390	4.810 +/− 2.556	4.834 +/− 3.615
ETHYL CAPRATE *	H	1.119 +/− 0.889	0.976 +/− 0.647	0.986 +/− 0.735
DIETHYL SUCCINATE *	I	0.101 +/− 0.118	0.103 +/− 0.112	0.079 +/− 0.034
2-PHENETHYL ACETATE *	K	0.124 +/− 0.082	0.114 +/− 0.108	0.096 +/− 0.057
ETHYL LAURATE *	L	0.128 +/− 0.089	0.118 +/− 0.110	0.121 +/− 0.072
2-PHENETHYL ALCOHOL *	M	0.377 +/− 0.244	0.299 +/− 0.198	0.357 +/− 0.188
ACETALDEHYDE **	N	0.071 +/− 0.019	0.075 +/− 0.025	0.090 +/− 0.040
METHANOL **	O	0.097 +/− 0.119	0.090 +/− 0.129	0.063 +/− 0.015
n-PROPANOL **	P	0.047 +/− 0.014	0.039 +/− 0.016	0.042 +/− 0.012
i-BUTANOL **	Q	0.141 +/− 0.056	0.157 +/− 0.046	0.139 +/− 0.037
ACETOIN **	R	0.022 +/− 0.012	0.014 +/− 0.019	0.011 +/− 0.005
2-METHYL-1-BUTANOL **	S	0.095 +/− 0.038	0.117 +/− 0.036	0.120 +/− 0.040
3-METHYL-1-BUTANOL **	T	0.416 +/− 0.117	0.529 +/− 0.135	0.523 +/− 0.124
2.3 BUTYLEN GLYCOL **	V	0.208 +/− 0.097	0.143 +/− 0.055	0.174 +/− 0.070
GLYCEROL **	X	4.519 +/− 1.090	4.383 +/− 1.102	4.825 +/− 1.763
ETHYL ACETATE **	Y	0.069 +/− 0.025	0.066 +/− 0.018	0.072 +/− 0.034
ETHYL LACTATE **	Z	0.135 +/− 0.095	0.060 +/− 0.053	0.114 +/− 0.093
CAPROIC ACID ***	F6	0.201 +/− 0.077	0.184 +/− 0.072	0.168 +/− 0.090
CAPRYLIC ACID ***	F8	0.578 +/− 0.264	0.502 +/− 0.260	0.435 +/− 0.291
CAPRIC ACID ***	F10	0.173 +/− 0.097	0.113 +/− 0.095	0.111 +/− 0.113

1) EXPRESSED AS $\dfrac{\text{PEAK AREA COMPONENT}}{\text{PEAK AREA INTERNAL STANDARD}}$

*   METHYL BENZOATE
**   3-METHYL-2-BUTANOL
***   ENANTHIC ACID

rial data schows the existence of the correlation between objective (first) or indipendent variables and subjective (second) or dipendent variables.

This fact demonstrates that the analytical procedures described consenting to value qualitatively and quantitatively numerous costituents besides in analogue conditions to those perceived by the panelists and thus by the potential consumers they can also give a objective evaluation of wines quality.

V. REFERENCES

1. Schreier P., Drawert F. and Abraham K.O. - (1980) Lebensm. Wiss. und Technol. 13:318
2. Peynaud E. - "Le Goût du Vin", Dunod, Parigi, (1980)
3. Proc. of Symposium of Enology, S. Michele all'Adige (TN), (1980)
4. Proc. of International Symposium and poster presentation, University of Bristol, (1982)
5. Bertuccioli M., Daddi P. and Sensidoni A. - Proc. of International Symposium and poster presentation, University of Bristol, (1982)
6. Montedoro G. and Bertuccioli M. - Proc. of Symposium of Enology, S. Michele all'Adige (TN), (1980)
7. Williams A.A. - (1982) J. Inst. Brew, 43:88
8. Somers T.C. and Evans M.E. - (1974) I. Sci. Agric.,25:1369
9. Timberlake C.F. and Bridle P. - Proc. International Symposium and poster presentation, University of Bristol,(1982)
10. Montedoro G. and Miniati E. - (1976) S.& T.A.-NU, 6:177
11. Singleton U.L. - "Chemistry Winemaking", (A.D. Webb), Adv. Chem. Series n.137, New York, (1974)
12. Weurman C. and van Lunteren - (1967), Fortbildungskurs, 21
13. Badings H.T. - "Thesis", Wageningen, The netherlands,(1970)
14. Bertuccioli M. - (1982), Vini d'Italia, 137:62
15. Amerine M.A. and Ough C.S. - "Methods for analysis of must and wines", John Wiley and Sons Eds., New York, (1980)
16. Bertuccioli M. - (1982), Vini d'Italia, 138:149
17. Proc. of Symposium International "Vini spumanti", Salice Terme (PV), (1981)
18. Ayrapaa T. and Lindstrom I. - Proc. of "XVI European Brewery Convention", Elsevier Scientific Publ. Conf., (1977)
19. Nordström K. - (1964), J? Inst. Brew., 4:70

20. Forss A.D. - (1969), J. Agric. Food Chem., 17:681

21. Meilgard M.C. - (1975), M BAA Techn. Quarterly, 12:1510

22. Bertuccioli M. - in course of printing

23. Bertuccioli M. - Industrie delle Bevande, in course of printing

24. Montedoro G. - Proc. of "XXXVII Congresso Enotecnico Nazionale", Ostuni (BA), (1982)

CAPILLARY-CHROMATOGRAPHIC INVESTIGATIONS ON VARIOUS GRAPE VARIETIES

Adolf   Rapp

Werner   Knipser

Lorenz   Engel

Helene   Hastrich

Bundesforschungsanstalt für Rebenzüchtung Geilweilerhof

D - 6741 Siebeldingen

Grape breeding requires much time and labour and is therefore expensive. The development of a new variety requires a great number of examination procedures, spanning ca. 30 years, from cross breeding to registration. Therefore; efforts have been made to elaborate methods for early diagnoses which allow both evaluation and selection of a new variety already at the beginning of this process. Analytical, and thus objective, evaluation of the quality and the characterization of a variety depend largely on the individual constituent groups of grape berries or wines, e.g. sugar, acids, amino acids etc. In this connection, the aroma substances play a decisive role, since they affect one's sense organs intensively.

Instrumental Analysis of Foods
Volume 2

The first step in the analysis of aroma substances is
enrichment, that is the removal of water and ethyl
alcohol from the liquid sample. This is done to concentrate the aroma substances which are often present
in trace quantities. A proper enrichment must avoid
the development of artifacts (either by adding substances or from chemical changes of the substances
already present). Several such enrichment techniques
are described in the literature for wine or grape
juice which can be grouped either as gas (headspace)
extraction (1,2,3) or by liquid-liquid extraction
(1,3,4,5,6,7).

In gas extraction, the aroma substances are condensed
in cooling traps (1,2,3) or absorbed on solids such as
activated charcol (8,9,3), or polymers. The polymers
used are e.g. Tenax (10,11,12), Poropak Q (13,14) and
Chromosorb 105 (15,16,17) all of which have a minute
affinity for water and ethanol. However, it is possible
that besides water and ethanol, aroma-relevant polar
components can also be co-desorbed (17). Depending on
temperature and time of desorption, quantitative differences in the aroma components of the same initial
sample can arise (18). Moreover, in the case of chemical
desorption (15) and desorption using solvents, low molecular constituents of the polymers can render the
sample impure. Jennings (19) described the "Reflux-

trapping" method for the enrichment of the head-space components of ripening bananas. Because the extract contained water, the suitability of this method is rather limited with respect to its application to aqueous samples. By means of a new technique developed by the authors (20), careful and adequate concentration of trace components is achieved from the gas phase above aqueous and aqueous-alcoholic solutions. Using this method the aroma concentrates are obtained water-free by introducing the gas flow (N_2, CO_2) into Freon 11 (trichlorofluormethane). This enrichment technique has considerable advantages compared to the methods involving trapping with polymers because the extract can be stored indefinitely and portions can be analyzed at various times and in different ways (e.g. FID, GC-MS, sensorial evaluation). The volatile components introduced from the inert gas flow into the extraction apparatus are removed quantitatively (greater than 99%) (20). The extracts are water-free and do not contain any impurities. Moreover, artifact formation is practically impossible because of the mild conditions used (maximum temperature 23.8 $^{\circ}C$ = boiling point of Freon 11).

As a basis for quantitative investigations, liquid-liquid extraction is well suited for the reproducible enrichment of aroma substances. By selecting appro-

priate solvents or solvent mixtures (1,3,5,21,22,23),
a mild, complete, and artifact-free extraction of the
aroma substances can be achieved. A solvent largely
satisfying these criteria, which has been used with
advantage for the enrichment of aroma substances from
aqueous and aqueous-alcoholic solutions up to now, is
trichlorofluormethane (Freon 11) (1,3,6,24,7,25). Our
examinations on the extractability of several substances from homogenates of grape berries, grape juice, or
wine have shown that an extraction time of 20 hours
is sufficient (3,25).

The wine or grape concentrate, once obtained, can be
submitted to modern glass capillary chromatography for
separation into individual compounds. The results of
typical separations (Aromagram sections or "fingerprint" patterns - see Figure 1) show several hundred
individual compounds which are distributed among several classes of substituents and are largely responsible for the aroma of the wine or grape juice under
study. (In addition modern mass spectrometry can
identify most if not all of these compounds). The gas
chromatograph can quantitatively determine aroma-relevant compounds with in the ppb range. Although there
are only quantitative differences in the "fingerprint
patterns" of the varieties, these differences are
so distinct as to provide a basis for an analytical

characterization of a variety (3,25,26). Within the "fingerprint patterns" the amounts of certain com-

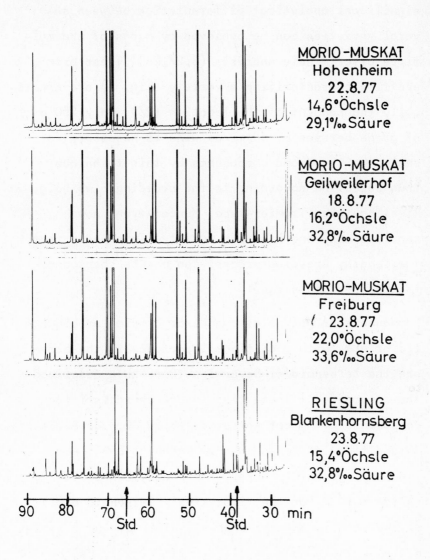

Fig. 1: Aromagram sections ("fingerprint patterns") of grape berries of the cvs. Morio-Muskat (different locations) and Riesling.

pounds or "key-substances" are characteristic of the
particular variety under study. As a result a highly
significant analytical differentation between several varieties can be achieved by means of the multiple discriminate analysis (3,27,28). These aroma
profiles characterizing a variety (Fig. 1) are practically independent of the location (27). Only 250 g
of grape berries or 50 ml of wine is necessary for
analyzing the aroma components by this technique;
thus selection according to the breeding aims (e.g.
Riesling-type, Tramier-type, Muscat-type) now
requires only one vine per variety. This means that
the lengthy breeding process can now be shortened
to the seedling stage.

Components or groups of compounds of similar biogene-
tic origin with a distinct importance to the aroma
are the terpenoides. Several authors have described
the occurence of these terpenes as characteristic
components of Muscat varieties (3,29,30,7,31,32,33,25).
In addition to the components known to date, e.g.
linalool, geraniol, nerol, hotrianol, linalooloxides,
we identified for the first time 2 terpenoid diols
(3,7-dimethyl-octa-1,5-dien-3,7-diol, I; and 3,7-di-
methyl-octa-1,7-dien-3,6-diol, II; Fig. 2) as constituents of the volatile components of grape berries
and wine (35,36). While compound I is the main com-

ponent of the volatile terpenoides in varieties with a fruity aroma e.g. "Scheurebe", compound II could only be identified in varieties having an intensive Muscat aroma.

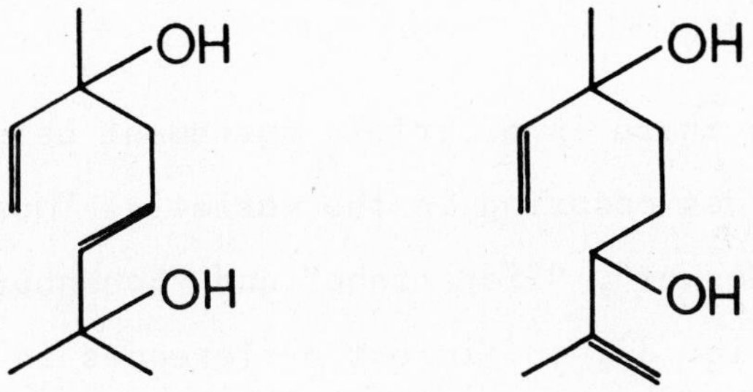

Fig. 2: Terpenoid diols identified in grape berries and wine (I: 3,7-dimethyl-octa-1,5-dien-3,7-diol; II: 3,7-dimethyl-octa-1,7-dien-3,6-diol).

There are clear differences in the "terpenoid profiles" between the varieties with Muscat-like aroma (Muscat-type"; the upper row in Fig. 3) and the varieties with a fruity and Riesling-like aroma ("Riesling-type"; lower row in Fig. 3) (37). The contents of the com-

pounds 11 and 12 (geraniol and trans-geranic acid) in
varieties of the "Muscat type" are significantly increased compared to varieties of the "Riesling-type".
The "terpenoid profile" of the "Riesling-type" is
mainly characterized by the compounds 3,5,7 (linalool,
trans-pyr.-linalooloxide, 3,7-dimethyl-octa-1,5-dien-
3,7-diol).

Although there is a certain agreement between the
terpenoides occuring in the varieties "Muskateller",
"Morio-Muskat", "Siegerrebe" and "Schönburger" ("Muskat-
type"; Fig. 3), sufficient differences in the amounts
of the 12 compounds within the terpenoid component
can be detected which permit an analytical differentiation of the varieties.

The aim to breed fungus-resistant grapevine varieties
by hybridizing American wild vines is of topical
interest. These varieties often have a distasteful
off-flavour (fox, grass, medicinal taste and other
flavours), unknown in European grapevine varieties
(V. vinifera). However, most of these off-flavours
have been eliminated in present-day hybrids (European x American cvs.) by an intensive selective breeding program. On the other hand, a marked strawberry-
like aroma note very often occurs which is not accepted
by the traditional wine connoisseur. Early and accurate
(analytical) identification of varieties with those

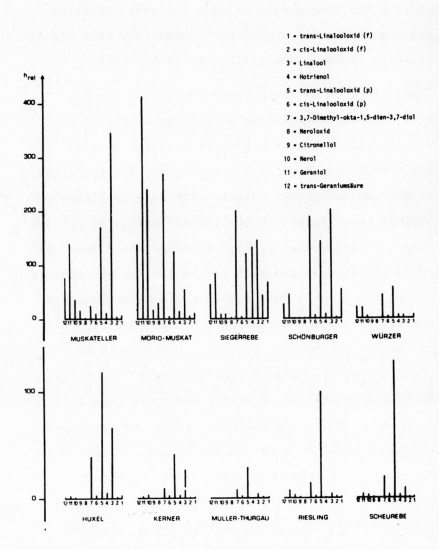

Fig. 3: Relative peak heights of terpenoid components ("terpenoid profiles") in grape
berries of European cultivars.

"negative" taste characteristics would considerably shorten the long period between the cross-breeding and the registration of a new variety and thus significantly increase the efficiency of breeding.

Identification of the compound or compounds which cause the strawberry-like aroma characteristic is very difficult because there are several hundred individual compounds in the aromagram. The technique of gas chromatography coupled with mass spectrometry (GC-MS) could result in the identification of all the compounds individually. Each compound could then be sensorially evaluated with regard to its importance to the off-flavor. This method would be tedious and unproductive for the most part because nearly all of the several hundred compounds which would be identified would have no contribution to the strawberry characteristic. A simple and rapid selection of those constituents which determine the aroma note can be effected by means of the "sniffing" technique. Applying this technique, those aroma components extracted from wine with a marked strawberry-like flavour were located in the "sniffing" chromatogram (Fig. 4; dotted field). By means of gas-chromatographic and mass-spectrometric investigations the peak causing the strawberry-like aroma note was identified as 2,5-dimethyl-4-hydroxy-2,3-dihydro-

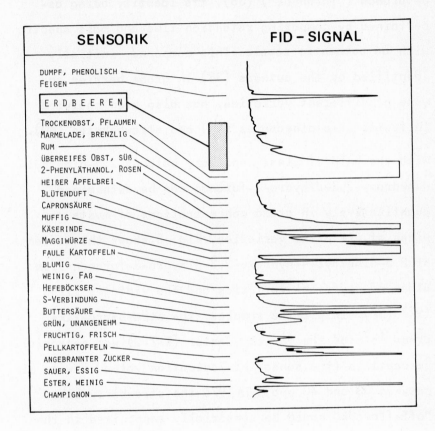

Fig. 4: Section of a "sniffing" chromatogram (25 th - 82 nd minute) of an aroma extract of the cv. Pollux (B-7-2). The strawberry-like aroma note typical of the variety is located in the dotted field.

3-furanon ("furaneol") (38), its identity being as-
certained by comparing retention time and mass spectra
with an authentic sample. This compound, initially
identified by the authors (38) in grape berries and
wine of different varieties, has also been analyzed
in fruits like pineapples (39) and strawberries (40).

With the help of glass capillary columns 2,5-dimethyl-
4-hydroxy-2,3-dihydro-3-furanon can be determined
quantitatively in aroma concentrations of musts and
wines of different varieties. Fig. 5 shows the contents
of 2,5-dimethyl-4-hydroxy-2,3-dihydro-3-furanon from
wines of different varieties and the vintages 1974-
1977 (37). In all the samples, the strawberry-like
aroma note of the variety Castor (B-7-2) was sensorially
perceptible (the sensorial indication value in wine is
between 30 and 40 ppb). In 1976 the strawberry-like
"off-flavour" could be sensorially recognized in the
cv. Pollux (B-16-8); in the other years, the content
of the furaneol was below the sensorial indication
value. Gas-chromatographic analysis however confirmed
that the furaneol occurred in the cv. Pollux in all
years (analytical indication value: 2ppb) (Fig. 5). As
a typical overmature component of the cv. Pollux, 2,5-
dimethyl-4-hydroxy-2,3-dihydro-3-furanon is not re-
cognizable by sensorial tests in bad vintage years;
however, it can be analyzed with the help of capillary

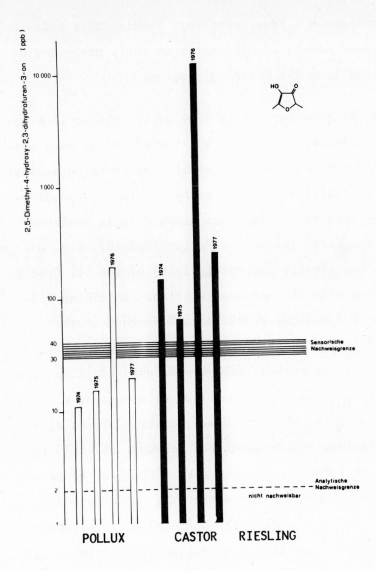

Fig. 5: Amounts of 2,5-dimethyl-4-hydroxy-2,3-dihydro-
3-furanon in different varieties and the vintages 1974-1977. Sensorial and analytical
threshold values of these are indicated.

chromatography. Thus, seedlings forming this undesirable component can be identified at an early stage and be excluded from the breeding program.

Since the strawberry-like flavour is unknown in European varieties (V. vinifera), investgations have been carried out in order to identify the crossing partner with the component (2,5-dimethyl-4-hydroxy-2,3-dihydro-3-furanon) causing this aroma note in resistant cvs. as Castor (B-7-2) or Pollux (B-6-18). Fig. 6 shows the genetic descent (origin) of the cv. Castor together with chromatogram sections containing 2,5-dimethyl-4-hydroxy-2,3-dihydro-3-furanon (marked peak). From figure 6 it can be seen that the furaneol (2,5-dimethyl-4-hydroxy-2,3-dihydro-3-furanon) is not present in berries of the American wild variety V. riparia or in V. vinifera Gamay, Oberlin 595 or V. vinifera Fosters' White Seedling. However, we could identify the furaneol in the cv. Vi 5861 (Fig. 6) as well as in the American wild variety V. labrusca and its descendents (e.g. Niagara, Buffalo) (Fig. 7). Based on these results it can be suggested that the origin of the cv. Vi 5861, called "Oberlin-595-Free-flowering" by the breeder, was due to crossfertilization with pollen from a variety containing V. labrusca genotypes. Further, as a result of this hybridization, progenies can occur which have both analytically measurable

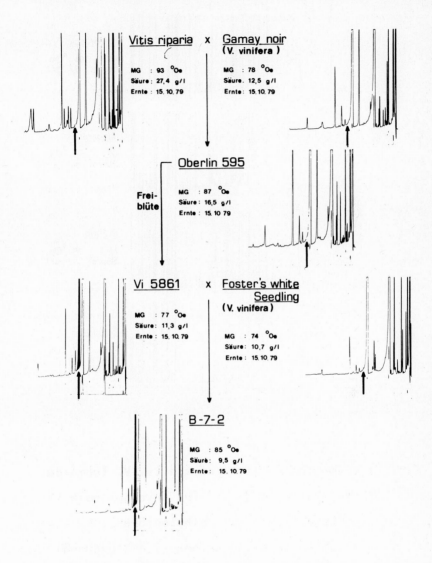

Fig. 6: Origin of the interspecific cv. B-7-2 (Castor).
Chromatogram sections of aroma extracts from
grape berries. Retention time for the furaneol
peak is marked with an arrow.

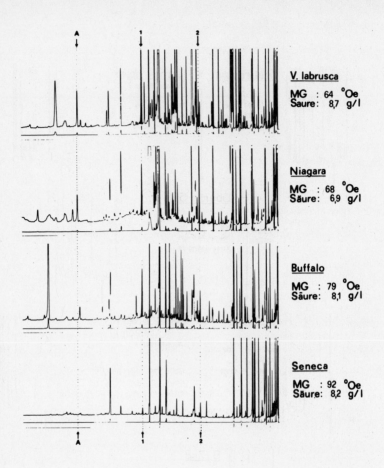

Fig. 7: Aromagram sections of the cv. V. labrusca
and its descendents. The furaneol peak is
marked. With 1. A: Methylanthranilat,
2: 2,5-Dimethyl-4-methoxy-2,3-dihydro-3-
furanon

contents of the furaneol and a sensorially perceptible
strawberry-like flavour.

The research on additional seedlings from Oberlin 595 Freiblüte showed that Furaneol in analytically traceable only in the seedling Vi 5861, while no furaneol was detected in the seedlings Vi 4704, Vi 4058, Vi 2932 and Vi 2703 which were also studied.

Because the gas chromatographic detection limit for Furaneol is very low, the analytical selection of now grape varieties in terms of Furaneol content is more accurate than a selection using sensory evaluation.

A 3-year study in the frome of the FDW (Forschungsring des Deutschen Weinbaues) yielded a parallel conclusion: Sensory evaluation of various other wines whose furaneol content was analytically measured clearly showed that a foreign flavor was clearly recognizbale at high concentrations, but that most of the tasters could not recognize the strawberry-like flavor at low concentrations (under 30 ppb).

The investigations show that it is not only possible to differentiate the individual varieties by analysis but also to correlate the analytical data with the sensorial evaluation. By these first and successful steps a basis has been provided for the application of the highly developed gas-chromatographic analytics for an early diagnosis in the field of grapevine breeding.

References

1. Rapp, A., Über Inhaltsstoffe von Traubenmosten
   und Weinen
   Diss. Univ. Mainz (1965)

2. Bertrand, A., Boidron, J.N. et Riberéau-Gayon, P.,
   Bull. Soc. Chim. France 5, 3149-3151 (1967)

3. Rapp, A., Hastrich, H., Engel, L. und Knipser, W.,
   Bull. OIV 53, 91-114 (1980)

4. Schreier, P., Drawert, F. und Junker, A.,
   Chem. Mikrobiol. Technol. Lebensm. 4, 154-157 (1976)

5. Rapp, A., Hövermann, W., Jecht, U., Franck, H. und
   Ullemeyer, H.,
   Chem. Ztg. 97, 29-36 (1973)

6. Rapp, A., Hastrich, H. und Engel, L.,
   Vitis 15, 29-36 (1976)

7. Hardy, P.J.,
   Phytochem. 9, 709-715 (1970)

8. Klimes, J. und Lamparsky, D.,
   In: G. Charalambous: Analysis of Foods and Bever-
   ages; Headspace-Techniques.
   Acad. Press, New York pp. 95-114 (1978)

9. Heinz, D.E., Sevenants, M.R. and Jennings, W.G.,
   J. Food Sci. 31, 63-68 (1966)

10. Jennings, W.G. and Filsoof, M.,
    J. Agric. Food Chem. 25, 440-445 (1977)

11. Noble, A.C.,
    In: G. Charalambous: Analysis of Foods and Bever-
    ages; Headspace-Techniques. Acad. Press New York
    pp. 203-228 (1978)

12. ter Heide, R. de Vries, P.J., Visser, J., Jaeger,
    R.P. and Timmer, R.,
    In: G. Charalambous: Analysis of Foods and Bever-
    ages; Headspace-Techniques.
    Acad. Press New York pp. 249-282 (1978)

13. Jennings, W.G., Wohleb, R. and Lewis, M.J.,
    J. Food Sci 37, 69-71 (1972)

14. Bertucioli, M. and Montedoro, G.,
    J. Sci. Food Agric. 25, 675-687 (1974)

15. Murray, K.E.
    J. Chromatogr. 135, 49-60 (1977)

16. Williams, P.J. and Strauss, C.R.,
    J. Inst. Brew. 83, 213-219 (1977)

17. Simpson, R.F.,
    Chromatographia 12, 733-736 (1979)

18. Wyllie, S.G., Alves, S., Filsoof, M. and Jennings,
    W.G.,
    In: G. Charalambous: Analysis of Foods and Bever-
    ages; Headspace-Techniques.
    Acad. Press New York pp. 1-16 (1978)

19. Jennings, W.G.,
    HRC & CC 2, 221-224 (1979)

20. Rapp, A. und Knipser, W.,
    Chromatographia 13, 698-702 (1980)

21. Stern, D.J., Lee, A., McFadden W.H. and Stevens,
    K.L.,
    J. Agric. Food Chem. 15, 1100-1103 (1967)

22. van Wyk, C.J., Webb, A.D. and Kepner, R.E.,
    J. Food Sci. 32, 660-664 (1967)

23. Prillinger, F. und Madner, A.,
    Mitt. Klosterneuburg 20, 202-205 (1970)

24. Stevens, K.L., Flath, R.A., Lee, A. and Stern, D.J.,
    J. Agric. Food Chem. 17, 1102-1106 (1969)

25. Rapp, A., Hastrich, H. und Engel, L.,
    Vitis 15, 183-192 (1976)

26. Rapp, A., Hastrich, H., Engel, L. und Knipser, W.,
    In: G. Charalambous: Flavor of Foods and Bever-
    ages. Acad. Press New York pp. 391-417 (1978)

27. Rapp, A. und Hastrich, H.,
    Vitis 17, 288-298 (1978)

28. Schreier, P., Drawert, F., Junker, A. und Rainer, L.,
    Mitt. Klosterneuburg 26, 225-234 (1976)

29. Terrier, A., Boidron, J.N. et Riberéau-Gayon, P.,
    C.R. Acad. Sc. (Paris) 275, 941-944 (1972)

30. Schreier, P., Drawert, F. und Junker, A.,
    Chem. Mikrobiol. Technol. Lebensm. 4, 154-157 (1976)

31. Wagner, R., Dirninger, N., Fuchs, V. et Bromer, A.,
    In: Internat. Symp. on the Quality of the vintage.
    Cape Town pp. 137-142 (1977)

32. Bayonove, C. et Cordonnier, R.,
    Ann. Techn. Agric. 19, 95-102 (1970)

33. Riberéau-Gayon, P., Boidron, J.N. et Terrier, A.,
    J. Agric. Food Chem. 23, 1042-1047 (1975)

34. Williams, P.J., Strauss, C.R. and Wilson, B.,
    Phytochem. 19, 1137-1139 (1980)

35. Rapp, A. und Knipser, W.,
    Vitis 18, 229-233 (1979)

36. Rapp, A., Knipser, W. und Engel, L.,
    Vitis 19, 226-229 (1980)

37. Rapp, A., Knipser, W., Engel, L. und Hastrich, H.,
    in Vorbereitung

38. Rapp, A., Knipser, W., Engel, L., Ullemeyer, H.
    und Heimann, W.,
    Vitis 19, 13-23 (1980)

39. Rodin, J.O., Himel, C.M., Silverstein, R.M. and
    Gortner, W.A.,
    J. Food Sci. 30, 280-285 (1965)

40. Buechi, G., Demole, E. and Thomas, A.F.,
    J. Org. Chem. 38, 123-125 (1973)

DIMETHYL DICARBONATE AS A BEVERAGE PRESERVATIVE

George Thoukis

Modesto, California

I. INTRODUCTION

Dimethyl dicarbonate (DMDC) is a homolog of diethyl dicarbonate (DEDC). The latter was successfully used in the United States as a yeast sterilant in various beverages, including wine, between 1963 until 1972 when the FDA cancelled the approval for use due to reasons which were largely political rather than scientific. Dimethyl dicarbonate has equal or better sterilant activity than diethyl dicarbonate against yeasts, some bacteria and molds on a weight for weight basis. It will inhibit yeast growth at levels ranging from 25 to 50 mg/L. It imparts no odor or taste to beverages when added at the 200-500 mg/L level. It hydrolyses readily in aqueous solutions to yield methanol and carbon dioxide as the primary products. Other products from side reactions are very minimal and they are of no consequence to human or animal health or safety. Ames tests and animal feeding studies conducted to date have proven negative with regards to teratogenesis or carcinogenesis. When added to wine, DMDC reacts with ethyl alcohol to form a small residue of ethyl methyl carbonate which can be very accurately measured by gas chromatography. Since the formation of ethyl methyl carbonate is linear with DMDC concentrations, its presence can be used to extrapolate the amounts of DMDC added to wines or other alcoholic beverages. DMDC is currently approved for use in several countries. An application for beverage use in the U.S. is now before the Food and Drug Administration.

II. GENERAL DISCUSSION

In 1961 Ough and Ingraham (1) proposed the use of diethyl dicarbonate (DEDC) as a wine sterilizing agent and in 1962 Thoukis and his co-workers (2) published experimental data on the reactivity of DEDC when added to wine and discussed the possible by-products which could be formed when certain substrates were present in wines. On the basis of the above-

mentioned publications and other data, the U.S. Food and Drug
Administration (FDA) in 1963 approved the use of DEDC as a
sterilant for wines. It was later approved for use in beer
and other beverages. DEDC functions as a "cold-pasteurizing"
agent whereby several enzyme systems are inactivated bringing
about yeast and bacterial death within minutes after addition
to beverages. Complete hydrolysis of DEDC to ethyl alcohol
and carbon dioxide takes place within a few hours following
its use and no detectable off flavors or aromas are imparted
to beverages to which it is added. DEDC reacts with compounds
containing an active hydrogen, such as amino acids or ammonia,
to form carboethoxy amino acids or ethyl carbamate respectively. Due to the very low concentration of free ammonia in
wines, the actual formation of ethyl carbamate in DEDC-
treated wines is negligible and it has been reported by Ough
(3) to be in the range of 0-10 parts per billion. Ethyl
carbamate is reported to be carcinogenic, and in view of the
Delaney clause which limits potential carcinogens in foods
sold in the United States to a level of zero tolerance the
use of DEDC in wines and other beverages was voluntarily
discontinued in 1972. In 1973 scientists at Logical International Corporation (4) proposed the use of dimethyl
dicarbonate (DMDC) which reportedly has equal or better anti-
yeast and antibacterial properties as a possible replacement
for DEDC. DMDC is a homolog of DEDC and it follows a similar
pattern of reactivity. In the presence of free ammonia DMDC
reacts to form methyl carbamate which is not a carcinogen.
It hydrolyses readily in aqueous solutions to yield methyl
alcohol and carbon dioxide; it imparts no off odor or off
taste to beverages to which it is added.

The structural formula of DMDC and some of the general
reactions it undergoes are shown in Figure 1. Keeping in
mind the general reactions listed in Figure 1, Ough (5) projected that if DMDC was added to wine at the level of 100
mg/L the following by-products would remain in wine:

methanol	48 mg/L
ethyl methyl carbonate	5 mg/L
methyl carbamate	0-0.007 mg/L
carbomethoxy derivatives	0-1 mg/L

The small concentrations of methanol which remain in wine
after DMDC hydrolysis is of no consequence from the organoleptic or health point of view since all wines contain small
amounts of methanol naturally. The presence of residual
ethyl methyl carbonate in DMDC-treated wines is of particular
importance. Stafford and Ough (6) showed that it is directly
proportional to the level of dimethyl dicarbonate added and

FIGURE 1

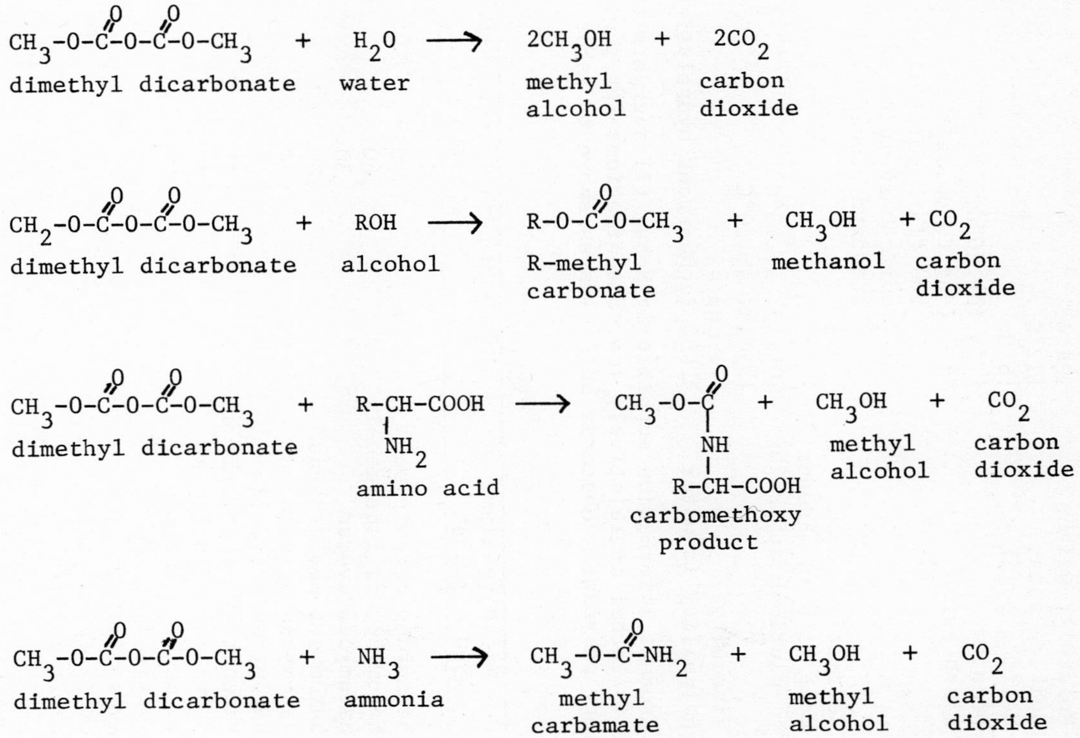

$$CH_3-O-\overset{O}{\underset{}{C}}-O-\overset{O}{\underset{}{C}}-O-CH_3 \; + \; H_2O \longrightarrow 2CH_3OH \; + \; 2CO_2$$

dimethyl dicarbonate   water   methyl alcohol   carbon dioxide

$$CH_2-O-\overset{O}{\underset{}{C}}-O-\overset{O}{\underset{}{C}}-O-CH_3 \; + \; ROH \longrightarrow R-O-\overset{O}{\underset{}{C}}-O-CH_3 \; + \; CH_3OH \; + \; CO_2$$

dimethyl dicarbonate   alcohol   R-methyl carbonate   methanol   carbon dioxide

$$CH_3-O-\overset{O}{\underset{}{C}}-O-\overset{O}{\underset{}{C}}-O-CH_3 \; + \; \underset{\underset{NH_2}{|}}{R-CH-COOH} \longrightarrow CH_3-O-\overset{O}{\underset{}{C}}-\underset{\underset{R-CH-COOH}{|}}{NH} \; + \; CH_3OH \; + \; CO_2$$

dimethyl dicarbonate   amino acid   carbomethoxy product   methyl alcohol   carbon dioxide

$$CH_3-O-\overset{O}{\underset{}{C}}-O-\overset{O}{\underset{}{C}}-O-CH_3 \; + \; NH_3 \longrightarrow CH_3-O-\overset{O}{\underset{}{C}}-NH_2 \; + \; CH_3OH \; + \; CO_2$$

dimethyl dicarbonate   ammonia   methyl carbamate   methyl alcohol   carbon dioxide

they developed a gas chromatographic method for the detection
and quantification of DMDC addition to alcoholic beverages
based on the amounts of ethyl methyl carbonate remaining.
   Mobay Chemical Corporation (7) reports the following properties for DMDC:

Molecular weight	134.45
Boiling point	172°C
Melting point	16°C
Density	1.25 (at 20°C)
Flash point	84°C (due to emission of CO_2, the flash point is diffi- cult to evaluate by standard procedures)
Refractive index	1.395
Half life in aqueous solutions at pH 2.8	40 min. at 10°C 15 min. at 20°C
Viscosity	2.1 cps
Solubility in water	3.65% (concurrent hydrolysis)

   Product information supplied by Mobay (7) suggests that
DMDC is lethal at relatively low concentrations for a wide
variety of yeasts, bacteria and molds as shown in Table 1:

TABLE 1

Microorganism 500 cells/ml	Minimum Lethal Concentration mg DMDC/L
Saccharomyces cerevisiae	40
Saccharomyces carlsbergensis	60
Saccharomyces uvarum	30
Saccharomyces pastorianus	100
Saccharomyces apiculatus	60
Saccharomyces oviformis	100
Zygosaccharomyces priorianus	75
Rhodotorula mucilaginosa	50
Pichia membranefaciens	40
Torulopsis stellata	65
Endomyces lactis	60
Kloeckera apiculata	40
Hansenula anomala	50
Acetobacter pastorianum	80
Lactobacterium buchneri	40
Penicillium glaucum	200
Byssochlamys fulva	100
Botrytis cinerea	100

   Medalen and Morenzoni (8) conducted experiments using
commercially available apple wine (ethanol 8.5% by volume;
total acidity 0.55%; pH 3.2; total SO_2 50 mg/L; reducing
sugar 7.0%) which they inoculated with 15, 6 and 3 yeast
cells per ml from an actively fermenting yeast starter of
Saccharomyces cerevisiae No. 449 and then treated with 0,
10, 20, 30, 40, 50, 60, 80, 100, 150 and 200 mg DMDC/L.
Their results which are presented in Table 2 indicated that
the lethal activity of DMDC is dependent on the number of
yeast cells/ml of wine and the concentration of DMDC added
to the contaminated wine. Even at 15 yeast cells/ml only 50
mg DMDC/L were needed to render the wine free of viable yeast
cells, as shown in Table 2.
   The hydrolytic decomposition rate of DMDC in simulated
wine model solution (10.0% ethanol by volume, buffered to
pH 3.1) has been studied by Christensen (9) using non-
dispersive infrared analysis. DMDC (50 µl) was dissolved in
5 ml of ethanol and immediately charged to a 500 ml gas
scrubbing bottle containing the simulated wine model solution. The solution was continually swept with nitrogen gas
at 500 ml/min into a Beckman Infrared Analyser to measure
the rate of CO_2 evolved. From this data which is shown in
Table 3, Christensen established a first order rate constant
of 0.033 min^{-1} for the hydrolytic decomposition of DMDC and
a half life of 21 min.

TABLE 2

| DMDC Added | Yeast Cells/ml | |
mg/L	Original Inoculation	After 8 weeks
0	6	TNTC ①
10	6	TNTC
20	6	TNTC
30	6	TNTC
40	6	0
50	16	0
60	3	0
80	3	0
100	16	0
150	16	0
200	16	0

① Too Numerous To Count (Spoilage)

TABLE 3

t(min)	$[CO_2]$	$\dfrac{a}{a-x_t}$	$\ln \dfrac{a}{a-x_t}$	$k_1 (min)^{-1}$
0	98.2	--		
10	79.1	1.24	0.216	0.0216
20	59.0	1.66	0.509	0.0255
30	42.0	2.34	0.849	0.0283
40	29.2	3.36	1.213	0.0303
50	19.7	4.98	1.610	0.0321
60	13.2	7.44	2.006	0.0334
70	9.8	10.02	2.305	0.0329
80	6.3	15.58	2.746	0.0343
90	4.5	21.82	3.083	0.0343
100	3.4	28.88	3.363	0.0336
110	2.8	35.07	3.557	0.0323
120	2.2	44.63	3.798	0.0317

$$k_1 = 0.033 \ min^{-1} \ (avg.) \qquad t\tfrac{1}{2} = \frac{\ln 2}{k_1} \qquad t\tfrac{1}{2} = 21 \ min$$

The rate of hydrolytic decomposition of DMDC in aqueous solutions was reported by Genth (10) to be greatly influenced by temperature at pH 3.2 as shown in Table 4.

This decomposition rate appears to be 3-4 times faster than that of DEDC, as reported by Thoukis et. al. (2). Consequently, in practical applications DMDC should be used at low enough temperature of substrate in order to provide greater contact time for effective antimicrobial action. Likewise, the degree of clarity and the number of offending microorganisms affect the antimicrobial activity of DMDC as shown in Table 5.

TABLE 4

Time (min)	% DMDC Remaining		
	10°C	20°C	30°C
15	78	50	30
30	60	25	7
45	46	11	2
60	36	5	0
120	13	0	0
180	5	0	0
240	2	0	0
300	0	0	0

TABLE 5

Fermentative Yeast (cells/ml)	Lethal DMDC Concentration (mg/L)
8	30
70	60
680	100
6,500	200

III. CONCLUSION

Dimethyl dicarbonate is a strong antimicrobial compound which can be used to protect alcoholic and non-alcoholic beverages from yeast, bacterial or mold contamination when used at levels below 100 mg/L just prior to packaging.

It hydrolyses quite readily to carbon dioxide and methyl alcohol to leave the treated beverage free of DMDC a few hours after addition.

In alcoholic beverages DMDC reacts with ethyl alcohol to yield a small amount of ethyl methyl carbonate which can be measured readily by gas chromatographic techniques to provide quality control and legal safeguards.

When ammonia is present in DMDC-treated beverages less than 10 ppb of methyl carbamate is formed. Methyl carbamate is not carcinogenic and it poses no known health hazards to humans or animals.

Regulatory approval of DMDC for use in alcoholic and other beverages is strongly recommended on the basis of its safety, efficiency of antimicrobial activity and absence of residue hours after it is used.

REFERENCES

1. Ough, C.S. and Ingraham, J.L. (1961). Am. J. Enol. Vitic. 12, 149.
2. Thoukis, G., Bouthilet, R.J., Ueda, M., and Caputi, A., Jr. (1962). Am. J. Enol. Vitic. 13, 105.
3. Ough, C.S. (1976). J. Agric. Food Chem. 27, 328.
4. Anon. (1973). Introducing DMPC t.m. A New Cold Sterilant. Logica International Corp., Milwaukee, Wisconsin.
5. Ough, C.S. (1975). Am. J. Enol. Vitic. 26, 130.
6. Stafford, P.A. and Ough, C.S. (1976). Am. J. Enol. Vitic. 27, 7.

7. Anon. (1982). Product Information "Velcorin." Mobay
    Chemical Corp., Pittsburg, Pennsylvania.
8. Medalen, A. and Morenzoni, R. (1976). Unpublished data
    and personal communication.
9. Christensen, E. (1976). Unpublished data.
10. Genth, H. (1978). Report to Wine Institute Technical
    Committee, San Francisco, California.

CONTAMINATION ASSESSMENT OF COFFEE SEEDS AFTER ACCIDENTAL SPILLAGE OF LEAD-ACID BATTERY CONTENTS - A CASE STUDY

M.K.C. SRIDHAR

G. KASI VISWANATH[1]

S.C. PILLAI[2]

Department of Preventive and Social Medicine, University of Ibadan, Ibadan, Nigeria.

I. INTRODUCTION

It is not uncommon that at times, foodstuffs get damaged or seem to have been damaged during transport from place to place. These apparent damages may lead to wastage and also to heavy economic losses to the transport or to insurance companies. These extra costs may be saved by the application of scientific methods. This paper reports an incident when a truck carrying 120 bags (weighing 50kg each) of good quality coffee seeds met with an accident where some sulphuric acid from car batteries spilled over them resulting in discoloration of the seeds. The customer rejected the whole batch and demanded compensation from the transporting company. The transporting company approached us for a scientific investigation to find out whether the coffee seeds are fit for human consumption or not. An account of this investigation is presented here.

1. NCITR/CNTE, School of Engineering and Applied Science, University of California, Los Angeles, U.S.A.

2. CIERS Research and Consultancy Private Limited, 340 Sampige Road, Malleswaram, Bangalore, 560003, India.

Instrumental Analysis of Foods
Volume 2

II.   MATERIAL AND METHODS

A. Materials

From each of the 120 bags, 250 g of the samples were taken representing all parts of the bag. In addition, one sample of the same variety of the coffee seeds were collected from the coffee Board (customer) and six samples were purchased from the local market which act as uncontaminated control samples.

B   Methods

Samples of the coffee were washed in distilled water and the washings were examined for pH value, acidity, alkalinity, sulphate, chloride, hardness and lead. In the preparation of aqueous washings of the coffee seeds, different quantities of water were used for washing a given weight of control sample of coffee seeds to ascertain the maximum quantity of water required for bringing the different constituents into aqueous solution. After ascertaining the maximum quantity of water required for washing, the maximum period of washing (by shaking) of the seeds with water was worked out to obtain the different constituents in aqueous solution. It was found that 20g of seeds in 200 ml distilled water shaken for 30 minutes on a rotary shaker at 200 r.p.m. gave maximum leaching of the constituents. Thus, the samples were subsequently treated optimally and were analysed.

The pH was determined by using the Elico pH meter with glass electrode. The acidity, alkalinity, sulphate, chloride, hardness and lead determined according to the standard methods (1,2,3). The results of analysis of the aqueous washings are expressed as mg of the constituent per 100 g seeds.

III.   RESULTS AND DISCUSSION

A. Appearance of the aqueous washings

As the coffee seeds were shaken in distilled

water, the seeds imparted a slight pale brownish or yellowish tinge. When the shaking of the seeds in water was prolonged for various periodson mechanical rotary shaker for 5,10,15,20,25 and 30 minutes, the color of the aqueous washings became increasingly intense and developed into a brown color. All the 120 test samples and the control samples (uncontaminated) imparted the same color to the water.

B. Appearance of the aqueous washings of the control coffee seeds pretreated with various concentratious of sulphuric acid and dried

Ten grams samples of control coffee seeds were immersed for 5 minutes in 10 ml dilute sulphuric acid of following concentrations: 0.1, 0.2, 0.3, 0.4, 0.5, 0.6, 0.7, 0.8, 0.9, 1, 2, 3, 4, 5, 10, 15, 20, 25, 30, 35, 40, and 50 per cent. After 5 minutes the seeds were dried on a filter paper fold, then each lot was placed in a conical flask and was shaken with 100 ml distilled water on the shaker for 30 minutes. The supernant liquids in each of the lots were practically colourless.

If, for example, the acid from fully charged batteries (about 38 per cent; 1.280 sp. gravity) from a truck fell over the coffee seeds, the aqueous washings of the matter on the surfaces of the seeds would be colourless. The results show that the quantity of acid (dilute sulphuric acid from a battery) that fell over the coffee seeds under examination was inconsiderable or negligible. It is an interesting point that treatment of control samples of coffee seeds even with 0.1 per cent dilute sulphuric acid eliminated the colour of the aqueous washings of the seeds.

C. Chemical analysis of aqueous washings of the coffee seeds

1. Test samples. The washings of the 120 samples of coffee seeds received were examined for pH value, acidity, alkalinity, sulphate, chloride, hardness and lead.

The minimum, maximum, and the mean values are given
in Table 1. In addition, the total number of the
test samples that gave values below the mean value,
and those that gave above the mean value are given
in Table II. The results indicate that the
aqueous washings of the different samples of seeds
are more or less similar in their composition and
that the variations are not significant.

  2. <u>Control samples</u>. The control samples
obtained from the coffee Board and the local market
were treated in the same way as the test samples
and were analysed. The results (Table III) indicate
that there is considerable variation.

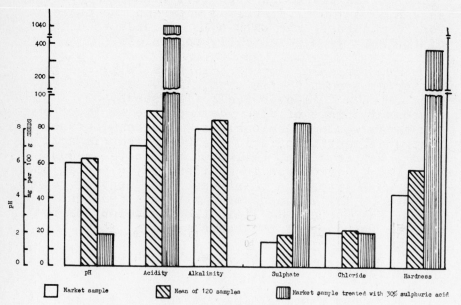

Figure 1. RESULTS OF ANALYSIS OF AQUEOUS WASHINGS OF A CONTROL OR MARKET SAMPLE OF COFFEE SEEDS, THE TEST SAMPLES OF
COFFEE SEEDS AND THE CONTROL SAMPLE OF COFFEE SEEDS TREATED WITH 30 PER CENT SULPHURIC ACID

TABLE I

The minimum, maximum and mean values for each item of analysis

Item of analysis	Minimum Value	Maximum Value	Mean Value
p^H	6.10	6.50	6.27
Acidity, mg/100 gm	60	130	90.59
Alkalinity, mg/100 gm	50	120	84.90
Sulphate, mg/100 gm	6.4	32.60	18.46
Chloride, mg/100 gm	10	30	20.58
Hardness, mg/100 gm	39	78	56.66
Lead, mg/100g	Nil	Nil	Nil

TABLE II: The numbers of the test samples of coffee seeds having the values for the different items of analysis below and above the mean value

	Mean Value	Numbers having the values below the MEAN VALUE						Numbers having the values above the MEAN VALUE				
pH	6.27	1 to 31	36	42	44	45		32 to 35	37 to 41	43		
		46 49	51 52	54	56			47 48	50 53	55 58		
		57 59	60 91	to 96	98			61 62	60 97 (90/50)	100		
		99 102	to 108 110 (70)	to 120				101 109				
Acidity	90.6	1 to 15	21 31	33	35			16 to 20	22 to 30	32		
		36 38	39 41	42	43			34 37	40 44	47 to 50		
		45 46	51 52	53	57			54 55	56 58	59 60		
		61 62	64 66	67	69			65 68	71 77	97 100		
		70 72	to 76 78	to 96				101 103	104 105	107 to 111		
		98 99	102 106 112 (69)	113 116				114 115 117 (51)	to 120			

468

TABLE II CONTD.

Mean Value		Numbers having the values below the Mean Value	Numbers having the values above the Mean Value
Alkalinity	84.9	2 to 12 14 15 32 34	1 13 16 to 33 36
		35 38 41 45 54 56	37 39 40 42 44
		57 58 75 90 to 109	55 59 to 74 76 to 89
		111 to 119 (61)	110 120
			(59)
Sulphate	18.5	5 7 9 14 17 19 20	1 to 6 8 10 to 13
		21 22 24 to 31 33 34	15 16 18 23 32 35
		38 to 51 53 55 57 58	36 37 52 54 56 60
		59 61 to 75 80 88 89	76 to 79 81 to 87 91
		90 106 107 109 111 112	108 110 113 119 120
		114 to 118 (67)	(53)

TABLE II CONTD.

Mean Value	Numbers having the values below the Mean Value	Numbers having the values above the Mean Value
Chloride 20.6	1 to 4, 6 to 7, 11, 16 to 22, 24 to 29, 31 to 42, 44 to 57, 59, 62, 65, 69, 71 to 74, 77, 82, 84 to 85, 89, 92 to 93, 96 to 97, 99 to 113, 115 to 120 (84)	5, 8 to 10, 12 to 15, 23, 30, 43, 58, 60 to 61, 63 to 64, 66 to 68, 70, 75 to 76, 78 to 81, 83, 86 to 88, 90 to 91, 94 to 95, 98, 114 (36)
Hardness 56.7	2 to 3, 5, 8 to 9, 16, 18 to 19, 21 to 23, 31 to 38, 40 to 44, 46 to 49, 51 to 53, 56 to 58, 62 to 64, 66 to 74, 78, 80, 82 to 84, 86 to 88, 96 to 98, 100 to 101, 103 to 104, 106 to 107, 112, 116 to 118 (67)	1, 4, 6 to 7, 10 to 15, 17, 20, 24 to 30, 39, 45, 50, 54 to 55, 59 to 61, 65, 75 to 77, 79, 81, 85, 89 to 95, 99, 102, 105, 108 to 111, 113 to 115, 119 to 120 (53)

TABLE III: Results of analysis of the aqueous washings of control "uncontaminated" samples of coffee seeds

Coffee seeds (control samples) obtained from:	pH	Acidity	Alkalinity	Sulphate	Chloride	Hardness
		mg per 100 gram seeds				
The Coffee Board	6.05	70	80	14.4	20	42
Local market (mean values for 6 samples)	6.40	100	110	45.5	15	133

D. Examination of the aqueous washings of
coffee seeds after immersing them in dilute
sulphuric acid.

Twenty gram samples of the control coffee
seeds from the local market were immersed in 25ml
of 30 per cent sulphuric acid for 5 minutes and
then removed and dried over a filter paper on
the laboratory bench. These seeds were washed
with distilled water, as indicated, and the
washings analysed. The comparison of the results
of control and the test samples (Figure 1) revealed
that those when immersed in acid showed (a) very
low pH (1.85); (b) very high acidity (1040mg per
100g); (c) no alkalinity; and (d) very high
hardness due to hydrogen ions. None of the test
samples showed up any of these features.

E. The extent of contamination of the aqueous
    washings of the test samples.

The results of analysis of the aqueous
washings of the test samples of coffee seeds (mean
of 120 samples), the control samples obtained from
the Coffee Board and the local market, and the
sample of coffee seeds experimentally "contaminated"
with 30 per cent sulphuric acid are compared in
Table IV which also includes the permissible values
for drinking water according to different standards
(4,5,6,7). The results confirm that the level of
contamination in the test samples is within the
permissible limits set by the different Organisations.
The washings obtained from acid treated samples do
not pass the standards.

TABLE IV: Values for the aqueous washings of the test samples of coffee seeds, control samples of coffee seeds and of the control sample No.2 treated with 30% sulphuric acid before washing as compared with the values for water for drinking as recommended by different authori-ties

	pH	Acidity	Alkalinity	Sulphate	Chloride	Hardness	Lead
			mg per 100 gram seeds				
Mean value for the 120 samples	6.27	90.59	84.90	18.46	20.58	56.66	Nil
Control samples No.1 (from the Coffee Board)	6.05	70.00	80.00	14.40	20.00	42.00	Nil
Control sample No.2 (from the local market)	6.40	100.00	110.00	45.5	15.00	133.00	Nil
The control sample No.1 treated with 30% H_2SO_4	1.85	1040.00	Nil	83.5	20.00	362.00	Nil
ICMR Standards Highest desirable	7 to 8.5	*	*	200.00	250.00	300.00	-
Permissible	6.5 to 9.2	*	*	400.00	1000.00	600.00	0.1

TABLE IV CONTD.

	PH	Acidity	Alkalinity	Sulphate mg per 100 gram	Chloride Seeds	Hardness	Lead
European Standards							
Highest desirable	-	*	*	250.00	200.00	100 to 500	-
Permissible	-	*	*	-	-	-	0.1
International Standards							
Highest desirable	7 to 8.5	*	*	200.00	200.00	100.00	-
Permissible	6.5 to 9.2	*	*	400.00	600.00	500.00	0.1
WHO Standards							
Highest desirable	7 to 8.5	*	*	200.00	200.00	500.00	-
Permissible	6.5 to 9.2	*	*	400.00	600.00	1000.00+	0.1

* Standards have not been fixed.

+ Magnesium sulphate + Sodium sulphate.

It may also be indicated in this connection that
for the water used for food canning and freezing
the following values have been recommended (7)
pH, = 7.5; alkalinity, 30 to 250 mg/l; hardness,
25 to 75 mg/l (for legumes); 100 to 200 mg/l
(for fruits and vegetables); and 200 to 400 mg/l
(for peas).

F. Experiments on germination

   1. Effect of pretreatment with sulphuric acid.
From each of the 120 test samples, the controls
treaed for 5 minutes with varying per cent of
sulphuric acid (1,5,10,15,20,25,30,40 and 50 per
cent) and dried, 10 seeds were taken into test
tubes, added 10 ml of distilled water and allowed
to stay for 48 hours during which period the seeds
started germinating. All the seeds germinated
indicating that contact with sulphuric acid up to
50 per cent did not have any deleterious effect
on germination. In one experiment the pre-treatment
of the coffee seeds was prolonged to 2 hours and
even then the germination occured.

   2. Effect fof temperature on the germination.
One set each of acid treated coffee seeds from
seeds from control sample, market samples and the
test samples was heated to boiling in distilled
water (which took about 5 minutes) and left for
germination. It was most interesting that all the
seeds germinated. Even 50 per cent acid treated
samples boiled for 10 minutes germinated showing
that the coffee seeds have a hard and highly
resistant seed coat.

G. Comparative taste of coffee prepared from the
    test samples and control sample.

   It was considered necessary to brew coffee
from the test samples and the control samples after
roasting and powdering. The coffee prepared from
test samples showed a pH value of 5.90 and the
coffee samples from controls showed a pN value of
5.75.

Other items could not be performed as the brewed
coffee had colour which was not removable by the
conventional methods.

The comparative "cup taste" of the coffee
prepared from those samples and the controls showed
similarity as revealed by human volunteers.
Observation over a period of 5 weeks also showed
that there was no ill effect as a result of
consumption of the coffee prepared from the
representative test samples.

These and other circumstantial evidence
revealed that the coffee seeds were not significantly affected as a result of acid spillage and
the coffee was fit for human consumption.

IV.   SUMMARY

A truck load of coffee seeds were rejected
unfit for consumption as a result of accidental
spillage of acid from batteries which were being
carried along. Chemical analysis of the aqueous
washings of the coffee seeds, experiments
conducted by deliberately spilling the seeds with
sulphuric acid, germination pattern was insignificant and the coffee seeds were fit for consumption.

V.   REFERENCES

1.   Indian Council of Medical Reseach (1963),
     Manual of Methods for the Examination of
     water, sewage and Industrial Wastes, special
     Report series No.47, New Delhi.

2.   American Public Health Association (1976),
     Standard Methods for the Examination of Water
     and Wastewater, Fourteen, Edition.

3.   Association of Official Analytical Chemiss,
     (1975), Official Methods of Analysis of the
     Association of Official Analytical Chemists,
     Twelfth Edition.

4.   Indian Council of Medical Research (1962)
     Manual of Standards of quality for Drinking
     Water Supplies, Special Report series No.44,
     page 10.

5. World Health Organisation (1970), European
    standards for drinking water, 2nd Edition,
    Geneva.

6. World Health Organisation (1971), International
    standards for drinking water,third edition,
    Geneva.

7. American Water Works Association, (1971), Water
    Quality and Treatment - A handbook of Public water
    supplies Mc Graw - Hill Book company, New York,
    Third Edition, pp.27-31.

ANALYTICAL INSTRUMENTATION IN MEASUREMENT
AND ASSESSMENT OF QUALITY PARAMETERS
IN CEREALS AND THEIR PRODUCTS

Vassiliki Pattakou

Chemistry and Technology Laboratory
Cereals Institute
Thessaloniki, Greece

E. Voudouris

University of Ioannina
Greece

I. INTRODUCTION

It is known for long that the quality of cereals and cereal products is designated and assessed by a number of features in the raw materials processes and eventually the final products.

The initial chemical composition, some measurable characteristics and the added ingredients have proved to be significant for the quality of the end-products.

Starting from the evaluation and selection of the grains during marketing, up to the preparation of the baked goods and in front of the great differentiation in quality requirements, a remarkable number of tests and analytical procedures are applied, aiming at the assessment of genetic factors, milling behavior and suitability and baking potential.

Cereal scientists today have been endeavoring in revealing the real importance of quality factors, by means of analytical procedures and instrumentation designed to meet the requirements of modern analysis and especially the specificity, accuracy and sensitivity (1-6).

In this paper some instrumental assessments of basic quality factors are presented and also physicochemical changes of wheat flour properties, occuring by the use of food additives, to correct defects and insufficiencies for a better end-product.

479

II. WHEAT QUALITY FACTORS

Two basic groups of factors are considered today of great importance in testing the quality of wheat and wheat flour.

The first group comprises the factors connected with the gluten proteins and their functional properties in the dough system and are measured by means of the rheological tests. The term "strength" is used to characterize the physical properties and many workers have shown that the strength of flour and dough is the result of the proteins structural formation (7,8). The protein structure of course has genetic sources, but further undergoes changes as a result of diversified environmental conditions and cultivation procedures (soil, climate, fertilizing, rainfalls, drought, etc.)

The protein quality and quantity is of basic interest to the cereal technologist in the assessment of the overall quality and the most important correlation to the final products.

The second group of quality factors deals with the starch and its behavior and properties. The enzymatic activities and particularly the diastatic power and gasing power responsible for the ability of CO_2 production during the baking process are mentioned.

The assessment of these two basic groups of quality factors in the raw material, as well as the improvement conferred by the use of additives in the flours, e.g. dry gluten, oxidizing agents, malt flour, enzymes, organic acids, emulsifiers, is achieved today with the aid of especially devised instruments, which beside other advantages are equipped with recording units to graphically visualize all the changes taking place during the processes of making the final products.

Today, therefore, the cereal scientist and technologist does not have to resort to complicated chemical procedures for basic quality control and further assessment of the results for actions to be taken, but just to observe and explain physical changes in terms of components and ingredients, their properties and interactions.

III. ASSESSMENT OF GLUTEN STRENGTH

A. *Farinograph*

The figure and shpae of farinographic curves produce a reliable indication of wheat flour strength (Fig. 1). A weak flour gives curve with poor elasticity, short stability time, quick break down. In contrast, a strong flour is accompanied with long development and stability time and delayed breakdown.

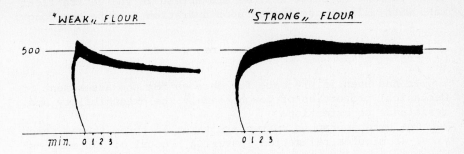

Fig. 1. Examples of normal farinograms of "weak" and "strong" flour.

Water absorption capacity is also a useful piece of information derived by the faringraph. This property is closely related to the quality of the gluten forming proteins of wheat in their capacity to form cohesive bonds in the dough system.

The Interruption farniogram with rest period of one hour depicts fairly well the weakening and breakdown of gluten with time (Fig. 2). This weakening, when kept in low leverl, is considered normal, attributed to the mechanical breakdown of the dough (cases a-c). When appearing pronounced, it is the result of degradation effect of proteolytic enzymes on gluten (cases d-f).

The use of interruption farnigram is widely used in Greece as a valuable tool to intimate and segregate heavily infested wheat lots, by a species of insects called *Pentatomidae*. These bugs bite the wheat kernels on the ear just before maturation, in order to be fed, thus injecting proteolytic enzymes via

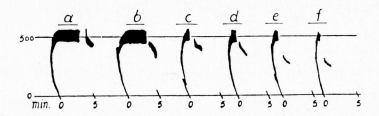

Fig. 2. Interruption faringrams after one hour rest. Graphs a, b, c, normal drop off. Graphs d, e, f, high proteolytic activity.

their saliva and these enzymes are spread in the entire flour
mass with milling, activated upon hydration to make the dough.

B. Extensograph

It has been in practice for decades for the assessment of
mechanical properties of dough, namely the extensibility and
resistance to extention.

The extensographic curves are taken in three stages, each
after 45 minutes relaxation, covering a total of 135 minutes
which is normal baking process time. They give a measure for
the suitability of wheats to derive flour for various applications.

The extensographic curves of Fig. 3 make clear deferences
in resistance to extension by the height of the curves and the
area under them. Case *a* shows a flour of medium strength, suitable for ordinary bread making; Case *b* with high extensibility,
suitable for production of puff pastry goods (Filo Kroustas is
a typical Greek speciality) and Case *c* with high resistance to
extension, desirable for luxury breads and confectionery goods
involving a number of ingredients in rather high percentages,
as sugar, shortening, eggs, flavors, etc., that require strong
and tough gluten.

The Extensograph is the outstanding instrument in the testing and assessment of the action of oxidative agents and ingredients, aimed to improve and rectify rheological properties

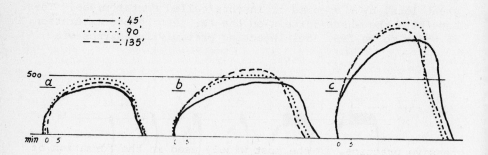

*Fig. 3. Extensograms of flours with different strength
and their suitability for basic applications. Case b - bread;
case b - Filo (thin dough sheets); case c - toast.*

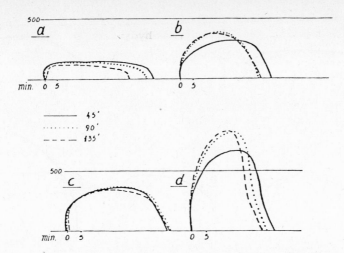

*Fig. 4. Changes of dough rheological properties with 0.1%
addition of citric acid in the flour of two wheat varieties.
1) Variety Vergina, a = untreated, b = improved. 2) Variety
Siete Cerros, c = untreated, d = improved.*

of the dough, originating from varietal deficiencies or cultivation abnormalities and damages.

Apart from the widely known improving ability of L-ascorbic
acid, extensively used due to low effective dosages, we know
that considerable rheological improvement is imparted to the
dough by organic acids such as citric, tartaric and lactic in
the presence of NaCl (9-11).

The improving effect of 0.1% addition of citric acid in the
flour of two quality different wheat varieties is depicted in
Fig. 4. a-Vergina is a low quality Greek variety and c - S.
Gerros a good quality Mexican variety. Both wheat samples were
milled in Brabender senior experimental mill, so that to ascertain that the only variable is the quality of the treated substrate.

The addition of vital dry gluten in the flour, although expensive practice, is the most widely used in the formation of
desired mechanical properties of the dough. The achievement of
tailor made characteristics desirable for the great variety of
baked goods is affected by the carefully determined percentage
of added gluten in combination with its rheological behavior.
The latter is a function of a number of features widely varying
caused of raw material origin, processing technology and typical
specifications. All these parameters and characteristics can
be successfully assed by means of the extensograph (12,13).

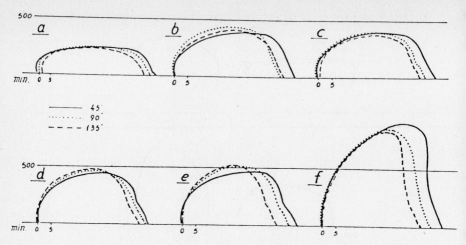

Fig. 5.   Changes in the extensographic curves with addition
of dry gluten.   a, untreated flour; b-e, 97% flour + 3% dry
gluten of different origins; f, 90% flour + 10% dry gluten
(case b).

Differences in the quality of dry gluten and the improving
effect on book baking flour are shown by the curves of Fig. 5.

The production of bread and baked goods from composite flours
of high protein level, popular in the last years, has been
achieved with incorporation in the wheat flour of flours and
products from other sources, such as soy flour. The need of
improved compatibility of these components towards workable systems and eventually good customer acceptance of the final products, dictated the use of minor ingredients such as lipids and
emulsifiers. The action of these emulsifiers lies in the formation of complex structures between starch and proteins resulting in better bread making properties, expressed with increased
mixing time, better dry dough surface and higher bread volume
(14-18).

The curves of Fig. 6 show the improvement in extensographic
energy and peak height of two surfactant preparations in medium
strength flour.

IV.   ASSESSMENT OF STARCH BEHAVIOR - ENZYMATIC ACTIVITY

The quality control of wheat flour is integrated only after
the determination of the diastatic power. The most used analytical procedures for this purpose are the amylograph test, a-
and b-amylase activity, gassing power and Hagberg test (falling number) (19-21).

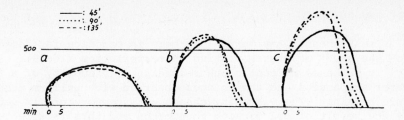

Fig. 6. Addition of two different composite bread improvers b,c.

A. *Falling Number Test*

The Hagberg test has internationally been adopted as the official standard method of ICC (International Cereal Chemistry), of AACC (American Association of Cereal Chemists) and of ISO (International Standardization Organization)(22-24).

The method is based on the ability of a-amylase to liquefy the starch gel made up during heating of a water flour suspension under standard conditions. The enzyme strength being a combination of concentration and activity is measured in terms of time in seconds required by the special stirrer to drop a determined distance in the flour gel.

The falling number test is the simplest, fastest, and reasonably sensitive one in determining the activity of commercial preparations added to the flours to optimize the diastatic power. The same test is rendered valuable in indicating the initiation and stage of wheat sprouting, which when exceeds tolerated limits results to non-machinable doughs from soft wheats and low quality pasta products from durum (25).

Table I shows the F.N. changes of flour possessing medium diastatic ability, after addition of a-amylase commercial preparations in varying levels.

TABLE I. The F.N. Values (Sec.) After the Addition of Commercial Preparations

Untreated flour		403
Bacterial amylase I	0.02%	243
Bacterial amylase I	0.05%	177
Bacterial amylase II	0.05%	156
Malt flour	0.2%	305

B. Amylograph

The Brabender Amylograph has been in practice for decades
for the study of starch pasting characteristics (26-29).

It is based on continuous recording of viscosity of a flour-
water suspension and the changes observed with uniform temperature raising. The viscosity change with temperature is naturally the result of gel forming phenomenon and this is affected by
the action of starch degrading enzymes a- and b-amylases. The
amylogram curves reveal this action by measuring of time needed for the gel formation as well as the peak value of the viscosity, after which liquefaction starts.

Figure 7 shows the graphs of three experimentally milled
flours, a, b and c, with corresponding low, medium and high
diastatic activity, the last due to high proportion of sprouted
kernels.

The amylograph is considered as the most reliable instrument for the adjustment of additions of starch liquefying enzyme
preparations.

Figure 8 shows the graphs of three cases of commercial enzyme additions, of bacterial and fungal source and malt flour
each in two different levels in flour (A) of very poor diastatic
power.

C. Fermentograph

In the manufacture of yeast leavened products the production
of CO_2 is the agent governing the aeration and volume of the
baked goods, both in the fermentation and the oven stages. It
is essential that its production rate is adjusted and kept in
agreement with the gas retention ability of the dough and gluten
matrix.

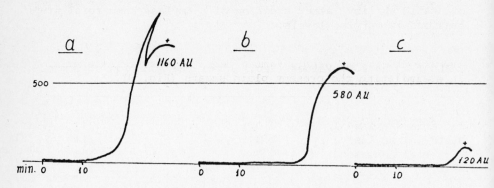

*Fig. 7. Amylograms of three wheats with widely varying
diastatic power. a, low; b, medium; c, high (sprouted wheat).*

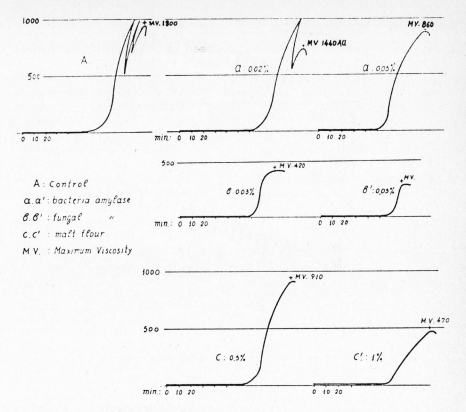

Fig. 8. Changes in amylographic characteristics of a low diastatic flour (A) with varying levels of added amylolytic preparations.

The recorded CO_2 production, with the aid of Brabender Fermentograph, every hour and in total of four, gives valuable information on the raw material potential and the changes that can be effected with the addition of agents acting either on the starch, or the yeast stimulating the yeast-cell function in assimilating the present plain sugars (Fig. 9).

V.   INSTRUMENTAL METHODS FOR PREDICTION OF BREAD QUALITY

The baking test has remained one of the most objective procedures in controlling the raw material, as well as the adjustments of flour blends for production of various products. In large industrial bakeries employing continuous processes, the stable flour baking properties are of great importance.

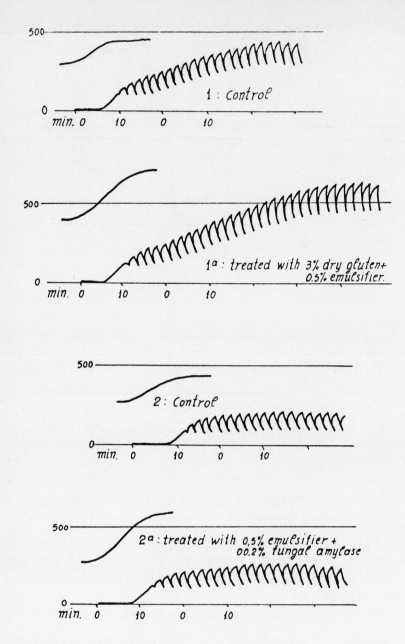

Fig. 10. Effect of various additives on flour baking properties with medium (1) or low (2) strength. Improvement of dough elasticity, stability and volume (1a). Volume improvement (2a).

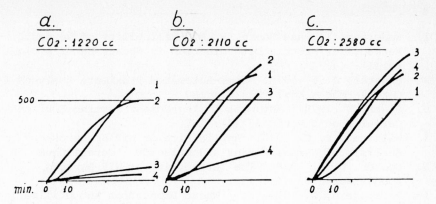

Fig. 9. Fermentographic CO_2 production in four hours. a, untreated flour; b, with addition of commercial a-amylase preparation I; c, with addition of commercial a-amylase preparation II.

The baking test though presents basic difficulties in practice, chiefly the long time needed for its performance and inefficiency in reliable adjustment in a great number of products.

The most important information to be taken are the dough behavior in the fermentation and oven stages and the characteristics of the finished products, the volume prevailing.

The new instruments developed by Brabender, the Maturograph and Ovenriserecorder are able to register all the phases of a baking test without having to resort to its performance.

The curbs of Fig. 10 taken on the Maturograph and Ovenriserecorder with two flours before and after treatment with various ingredients, make it possible to predetermine the proper baking conditions and optimal final characteristics.

VI.   CONCLUSIONS

The brief presentation of the basic and most applied tests and instrumental methods in assessing the quality parameters of cereals and their products and the changes observed and measured after technological treatments, reveals their importance. Cereal chemists and technologists in all parts of the world use successfully these procedures in quality screening of breeding programmes to perform the thousands of tests required for that, as well as in milling and baking operations for the best possible final products.

REFERENCES

1. Pomeranz, Y., and Moore, R. B. Reliability of several
    Methods for Protein Determination in Wheat, Baker's Dig.
    49 (1): 44.
2. Shuster, J. C. 1978. Semi-Automated Proximate Analyses.
    Cereal Food's World 23 (4): 180.
3. Sietz, W. 1978. Advances in Ingredient Testing Equipment.
    Cereal Food's World 23 (4): 187.
4. Williams, P. C. 1975. Application of NIR Spectroscopy to
    Analysis of Cereal Grains and Oilseeds. Cereal Chem. 52:561.
5. Hooton, D. E. 1978. The Versatility of NIR Devices. Cereal
    Food's World 23 (4): 176.
6. Bodenstein, B. 1974. In "Wheat Production and Utilization"
    AVI Publishing CO., Westport, Connecticut. p. 366-383.
7. Tipples, K. H., Preston, K. R., and Kilborn, R. H. 1982.
    "Strength" as related to Wheat and Flour Quality. Baker's
    Dig. 56 (6): 16.
8. Bloksma, A. H. Basic Considerations of Dough Properties.
    In "Wheat Chemistry and Technology" AACC St. Paul. Minnesota,
    1964.
9. Maher Galai, A., Variano-Marston, E., and Johnson, J. A.
    1978. Rheological Dough Properties as affected by Organic
    Acids and Salt. Cereal Chem. 55 (5): 683.
10. Valtadoros, A. 1964. Neuere Untersuchungen über Wanzen-
    weizen. Getreide u.Mehl 3: 25.
11. Stephan, H. 1965. Möglichkeiten zur Verbesserung der Back-
    fähigkeit von Mehl aus Wanzenstichweizen. Brot u. Gebäck
    9: 172.
12. Sarkki, M. E. 1981. Quality of Wheat Gluten from different
    Wheat Varieties and Manufacturing Methods. In Charalambous/
    Inglett: The Quality of Foods and Beverages, Academic Press.
    Vol. 1: 289.
13. Pattakou, V. 1982. Possibility of vital dry Gluten (and
    starch) Production from Greek Wheat (in Greek) Agricultural
    Research 6: 129.
14. Finney, K. F., and Shorgen, M. D. 1971. Surfactants supple-
    ment each other make foreing proteins compatible in bread-
    making. Baker's Dig. 45 (1): 40.
15. Chung, O. K. and Tsen, C. C. 1975. Functional Properties
    of Surfactants in Breadmaking. I. Role of surfactants in
    relation to flour constituents in a dough system. Cereal
    Chem. 52: 832.
16. Tsen, C. C., Banck, L. J., and Hoover, W. J. 1975. Using
    surfactants to improve the quality of cookies made from
    hard wheat flours. Cereal Chem. 52: 629.
17. Landfried, B. W. 1977. Surfactants used by bread makers.
    Cereal Food's World 22 (8): 338.

18. Rusch, D. 1981. Emulsifiers: Uses in Cereal and Bakery
     Foods. Cereal Food's World 26 (3): 111.
19. Hagberg, S. A. 1960. A rapid method for determining alpha-
     Amylase activity. Cereal Chem. 37: 218.
20. Perten, H. 1964. Application of the Falling Number Method
     for Evaluation a-Amylase Activity. Cereal Chem. 41: 127.
21. Malhot, W. C. 1980. The falling Number Method. Cereal
     Food's World 26 (10): 584.
22. ICC-Standard No. 107. "Determination of the Falling Num-
     ber" (according Hagberg-Perten) as a measure of the degree
     of a-amylase activity in grain and flour.
23. AACC Improved Methods (Method 56-81A). The Association.
     St. Paul, Minnesota. 1962.
24. ISO 3093-1982. TC34/Sc 4 Cereals-Determination of Falling
     Number.
25. Matsuo, R. R., Dexter, J. E., and MacGregor, A. W. 1982.
     Effect of Spouted Damage on Durum Wheat and Spaghetti Qual-
     ity. Cereal Chem. 59 (6): 468.
26. Anker, C. A., and Geddes, W. F. 1944. Gelatinization Stud-
     ies upon Wheat and Other Starches with the Amylograph.
     Cereal Chem. 21: 360.
27. Loska, D. 1956. The Heating Characteristics of an Amylo-
     graph. Cereal Chem. 33: 266.
28. Pomeranz, Y., and Shellenberger, J. A. 1962. Starch-lique-
     fying activity of alpha-amylase. Cereal Chem. 39: 327.
29. Yasunaga, T., Bushuk, W., and Irvine, G. N. 1968. Gelaniti-
     zation of starch during bread-baking. Cereal Chem. 3: 269.

SIMULTANEOUS DETERMINATION OF LEAD AND CADMIUM IN SODIUM CHLORIDE BY SOME TECHNIQUES OF ANODIC STRIPPING VOLTAMMETRY

E. Casassas, M. Esteban and C. Ariño

Department of Analytical Chemistry
Facultat de Quimica
Universitat de Barcelona
Barcelona, Spain

SUMMARY

An electroanalytical method for the fast simultaneous determination of Pb(II) and Cd(II) in common salt is described. Techniques used are mainly Direct Current Anodic Stripping Voltammetry (ASV), Differential Pulse ASV and Fundamental Altern Current ASV. The techniques and their results are compared. The determination can be made by calibration line method or by addition-standard method. The latter being the better one. The different instrumental and experimental parameters are discussed. DPASV and AC_1ASV are the better working techniques because of their higher sensibility and their better resolution when the concentration ratio (Pb:Cd) is high. The determination limits of Pb(II) and Cd(II) are given. Copper does not interfere. The application of Normal Pulse ASV and Second Harmonic Altern Current ASV is also discussed.

Instrumental Analysis of Foods
Volume 2

INTRODUCTION

The generalized interest observed in the past years in relation to the determination of traces and ultra-traces of metal-
lic elements in foods is also shown in connection with common
salt. As a result of this situation a large number of research
papers appeared and several programs for development of standard
reference methods have been established. Among them a collaborative study of the European Committee for the Study of Salt (E.C.
S.S.)-Joint F.A.O.-W.H.O. Food Standards Programme. Codex Ali-
mentarius Commission. Codex Committee on Food Additives (C.C.F.A.
for the validation of a standard reference spectrophotometric
method for As and AAS methods for Hg, Pb and Cd in common salt
(1,2,3,4) is to be mentioned. The literature shows that most of
the studies about heavy metal traces determination in sodium
chloride (for food seasoning or for industrial use) propose AAS
methods. The effectivity of the modern electroanalytical techniques has not been fully developed yet, and it is in this sense
that this paper is addressed.

DC polarography was first used in the analysis of common salt
for the determination of its contents in NaCl (5), Pb (6), V (7),
Fe (8), Cu and Zn (9), and Pb, Cu and Zn (10). Some micromethods
were developed for Pb and Cu in salt (11), for Cu(I) and Cu(II)
in ionic cristalls as NaCl (12), for Pb, Cd and Ag in eutectic
mixtures of NaCl and KCl (13), for Cu, Pb, Zn and Cd in salt (14)
(including a comparison of the polarographic results for Pb with
the spectrophotometric results), and for NO_2^- and NO_3^- in water,
fertilizers and sodium chloride (15).

Differential pulse polarography (DPP) has been used by De
Galan et al. (16) for the determination of Cu, Pb, Tl, Cd, Zn,
Ni, Co, Fe, Mn, Sb, Bi, Sn, Mo and As in high purity NaCl. The
detection limits in each case are given in this work.

More recently, the most used electroanalytical techniques
has been anodic stripping voltammetry (ASV) applied at its beginning by Vinogradova to the determination of Cd in support
electrolytes of NaCl between others (17), or of Cu and Pb in salt
(18), and by Kemula to the determination of Cu, Pb and Zn in high
purity salts (19). Lately, it was applied to the determination
of Cu in NaCl after its extraction as a 8-hydroxyquinolinato complex (20), or as a thiocyanato complex (21); and also to the determination of Sb in common salt (22).

Several modern techniques of potential sweep for ASV were
applied by Atley and Florence (23) in a study (similar in some
aspects to the present one) about Cd(II) and Pb(II) determination in 0.1 M HCl and in several environmental samples (not in
sodium chloride). The most complete studies published on Pb(II)
and Cd(II) determination by ASV in foods (but not in salt) (24,
25) and in raw agricultural crops (26), are respectively from

Gajan and Satzger, both from Food and Drug Administration, which
follow in general terms former work by Holak (27) and by the same
authors (45) and include a collaborative inter-laboratory study.

The recommended method is differential pulse ASV or DC sweep
ASV. The background electrolyte is a pH 4.3 acetic acid-acetate
buffer prepared from the nitric acid solution of the ashed samples.
Thus, the methods cannot be directly applied to common salt.

More recently, Adelaju et al. (28) propose an ASV method for
Se, Cu, Pb and Cd determination in biological materials. Brihaye
et al. (29) perform a Pb, Cd and Cu determination in pure water
and in "pure" sodium chloride using a rotating ring-disc glassy
carbon electrode, but the results obtained for Pb and Cd in the
"pure" sample of sodium chloride are much lower than those obtained for a Merck Suprapur sample in the E.C.S.S.-C.C.F.A (3,4)
and E.C.S.S. (30) inter-laboratory studies referred before.

Among the difficulties encountered in trace and ultra-trace
analysis the most important are perhaps, apart from those derived
from the availability of a reliable reagent blank, those derived
from the need of avoiding the contamination of samples, reagents
and solutions handled and standard reference materials, even
though working in "clean" desks or "clean" laboratories (31).
All this enhances the interest in the development of a determination for Pb(II) and Cd(II) in salt by a single stage method. This
method will be possible, in principle, taking advantage of the
selectivity inherent to ASV technique, especially through the use
of the several stripping modes available in modern laboratories,
even in those not excessively sophisticated.

As we have seen, several voltammetric methods for Pb, or Cd,
or both, determination have been described for natural or industrial samples, using several different background electrolytes,
including in some cases complexing reagents but no information
is found about the analysis of traces of these elements in sodium chloride, in which the nature of the background electrolyte
is imposed by the sample's own nature. Chloride interference in
ASV, reported by several authors (32, 33), giving rise to a background noise enhancement, is unlikely to be encountered if too
positive electrode potentials are avoided, i.e., if anodic polar-
ization of the electrode, which leads to calomel formation, is
avoided.

EXPERIMENTAL

Apparatus

METROHM Polarecord E 506 Interfaced to a Stand 505, using a
three-electrode system with a H.M.D.E. (Metrohm EA 290) as the
working electrode. Reference electrode was a Ag/AgCl electrode

(Metrohm EA 441/5), and the auxiliary electrode was a platinum electrode (Metrohm 285).

The cell was located in a reproducible position on a constant-speed magnetic stirrer (Metrohm E 504) used to rotate a 2 cm Teflon-coated stirrer bar. All measurements were made at $25°C \pm 0.5°C$.

The electrodeposition time and the rest period were measured with an Omega precision chronometer.

All glassware was washed for 24 hours in nitric acid-sulfuric acid (1:1), rinsed with deionized water, washed three times with doubly-distilled water and finally steam-washed at least one with doubly-distilled deionized water just before use.

Reagents

All solutions were prepared in water that was first demineralized then distilled twice over permanganate.

Hydrochloric acid, nitric acid: Suparapur Merck. All other chemicals: Merck reagent grade.

In order to remove oxygen from the cell solution high purity nitrogen was used further purified through vanadous perchlorate solution and equilibrated with the background electrolyte solution. The minimum deareation time used was twenty minutes.

Preparation of Samples

Sample solution ($NaCl$ 1 mol L^{-1}). Weigh, to nearest 0.01 g, 11.7 ± 0.05 g of the salt sample. Dissolve it in water and add 0.5 ml of Suprapur hydrochloric acid. Heat and keep boiling moderately during some minutes. Allow to cool and transfer quantitatively into a 200 ml volumetric flask. Dilute and mix.

Reference Solution

Prepared from vacuum salt (which contained 0.001 mg Cd Kg^{-1} and 0.025 mg Pb Kg^{-1} as determined by AAS). This one was the best quality NaCl found in regard to Cd(II) and Pb(II) contents (3,4).

Lead, Cadmium and Copper Standard Solutions

A stock solution of lead (100 mg L^{-1}) was prepared by dissolving 160 mg of $Pb(NO_3)_2$ in 2 ml of nitric acid (1:1) and diluting to 1000 ml with water.

A stock solution of cadmium (1000 mg L^{-1}) was prepared by dissolving 1000 mg cadmium metal powder in 10 ml nitric acid (3:1) and diluting to 1000 ml with water.

Standard lead solutions, standard cadmium solutions and standard copper solutions (all of them at the 1 ppm and 0.1 ppm level) were prepared daily by diluting aliquots of the respective stock solution.

Analytical Procedure and Experimental Technique

A 20 ml aliquot of the sample solution is added to the polarographic cell and deoxygenated with nitrogen stream for at least 20 min. During the preelectrolysis and the potential scan, nitrogen is passed over the solution. The working electrode area is adjusted to 2.20 ± 0.05 mm^2 (corresponding to 4-division turn of the micrometer).

The pre-electrolysis is performed for 180 s at a applied plating potential of -1.0 v (vs Ag/AgCl), with stirring of the solution. After a rest period of 60 s, without stirring and with the solution held at the deposition potential, the function selector of the instrument is switched to the stripping mode, and the stripping curve is recorded at 10 mV s^{-1}.

The several stripping modes used in the experiments, with their main settings, were the following:

```
-- Direct current mode (DC):        ------------
-- Differential pulse mode (DP):    Pulse amplitude = 50 mV
-- Fundamental a.c mode (AC₁):      Modulation amplitude =
                                      10 mV
                                    Frequency = 75 Hz
                                    Phase angle = 0°

-- Normal pulse mode (NP):          Pulse base voltage =
                                      60 mV
                                    Pulse base voltage =
                                      300 mV

-- Second harmonic a.c mode
     (AC₂):                         Modulation amplitude =
                                      10 mV
                                    Frequency = 37.5 Hz
                                    Phase angle = 0°
```

For best resolution when the amount of the firstly stripped element is much greater than that of the element stripped in the second place, it is recommended (only in DC mode) to stop the potential sweep at an intermediate value until the current drops to a stationary minimum value. Usually a 30 s stop-time suffices. During this time the sensitivity switch can be adjusted to the best setting for the peak measurement of the second element.

Quantitation of the results can be obtained from calibration curves prepared as described later from the reference solution fortified with known amounts of lead ion and/or cadmium ion. However, because of the proved presence of these metal ions in the sodium chloride taken as reference material, it is recommended, for very small amounts of them in the sample, to perform quantitation by the standard-addition technique, as follows: after the stripping potential sweep, a 0.10-0.20 ml volume of the appropriate standard lead solution or standard cadmium solution (or both) is added into the polaro graphic cell, and the full analytical procedure is performed again. Three different additions to original level are performed. The amount of lead and cadmium in the cell is calculated from the linear regression coefficients (24, 25) of the standard addition data.

RESULTS AND DISCUSSION

Discussion of the Instrumental Parameters

For voltametric studies of trace levels of metals ions, the use of a superimposed differential pulse modulation or an a.c. modulation with phase selective detection increase the sensitivity over conventional DC methods (34, 35). In differential pulse voltammetry, the peak current is proportional to modulation (pulse) amplitude (ΔE). For this work an intermediate value of ΔE (50 mV) was chosen because at high ΔE values peak widths are increased and poorer resolution results. The use of a superimposed differential pulse waveform at the H.M.D.E. requires a slow scan rate. Batley and Florence (23) used 2 mV s^{-1}. For this work, 10 mV s^{-1} was chosen as a compromise between a reasonably fast analysis and accuracy. Scan rates faster than 10 mV s^{-1} lead to smaller and broader peaks.

The height of the stripping peak using a.c. modulation is a function of both the modulation amplitude and frequency (35, 36). In Polarecord E 506 the frequency is 75 Hz for fundamental a.c. mode and 37.5 Hz for second harmonic a.c. mode. These low frequencies suffice to render available a large range of instrumental sensitivity. To improve resolution it is desirable to use the lowest available modulation amplitude. In this work peak resolution was good using 10 mV as a.c. modulation amplitude, and a scan rate of 10 mV s^{-1}.

Experiments with DC, DP and AC_1 Stripping Modes

The calibration curves have been obtained from vacuum salt reference solutions fortified with Cd(II) alone in the concen-

tration range from 0.5 µg L^{-1} to 1 mg L^{-1}, or with Pb(II) alone
in the range from 5 µg L^{-1} to 1 mg L^{-1}. Table 1 and/or Fig. 1
and Fig. 2 show the results obtained, taken in different ranges
of concentrations. In both cases, the behavior is lineal in all
the ranges studied. In Table 1 is given too the sensitivity,
slope of the calibration curve $I_p = f(C)$, (37), for each one of
the three stripping modes. The use of the DP or AC_1 modes provides a 8-10 times greater sensitivity than that from the DC
mode.

A similar study has been done from the reference solution
fortified with Cd(II) and Pb(II) together, in order to establish the intermetallic effects and the resolution of each stripping mode. We studied solutions with concentration ratios Pb:Cd
from 2000/1 down to 1/200, and the results for each ion were good
in the full range studied (Table 2). However, in the presence
of large Cd(II) excess, in order to obtain good results with the
DC mode, it is essential to stop the potential sweep at an intermediate value between $E_{p\ Cd}$ and $E_{p\ Pb}$ (Fig. 3 and 4). If the
Cd(II) excess is not so large, stopping the potential sweep can
be avoided. Anyhow, the DP and AC_1 modes give better resolution.
This one depends on the width at the half-height ($W_{1/2}$) of the
dissolution peaks. As Table 2 shows, DP and AC_1 stripping modes
give much sharper peaks than DC mode. Calibration curves obtained with both Pb(II) and Cd(II) present (included too in Fig. 1
and Fig. 2), show that the results obtained for one of them are
not influenced by the presence of the other one, and that the
sensitivities of the procedures are not modified (Table 1).

Experimentally it has been shown that equally good results
are obtained whether the standard-addition method is separately
applied to Pb(II) and to Cd(II) in two independent aliquots, or
it is used for their joint determination in a single aliquot
(Table 3). The joint-determination method has been adopted because of it is faster, more convenient, and more free from contamination dangers.

It has been shown that standard-addition method yields more
exact and more precise results than the calibration-curves method.

Quantitation Limits

When the calibration curves are used, the quantitation limit
is obtained when the sample signal equals the reference balnk
signal plus 3 times the standard deviation of this last signal.
As shown in Table 4, from the statistical study of the reference
blank, the following quantitation limits are obtained:

-- For cadmium: By DC mode 0.25 µg L^{-1} (in the cell solution)
                 or 4.25 µg Kg^{-1} (in the salt sample); by DP
                 mode 0.15 µg L^{-1} or 2.6 µg Kg^{-1}; by AC_1 mode
                 0.11 µg L^{-1} or 1.9 µg Kg^{-1}.

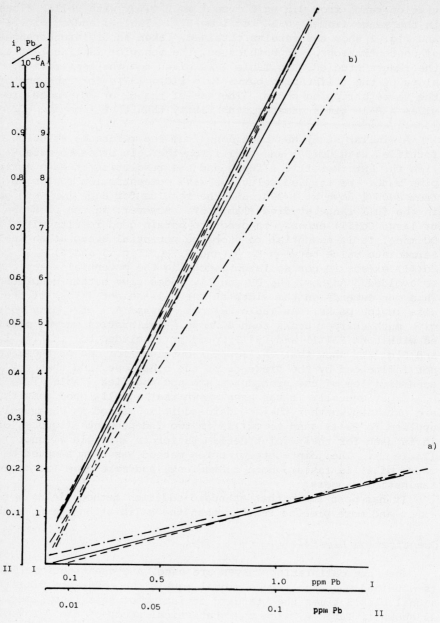

Fig. 1. Calibration curves for Pb(II)(alone or in mixture with Cd(II) for the three stripping modes in all concentration ranges studied. a) DC mode; b) DP and AC$_1$ modes. (——[Pb] >0.1 ppm, Fortified with Cd; --- [Pb]<0.1 ppm, Fortified with Cd; —·—·— Pb alone).

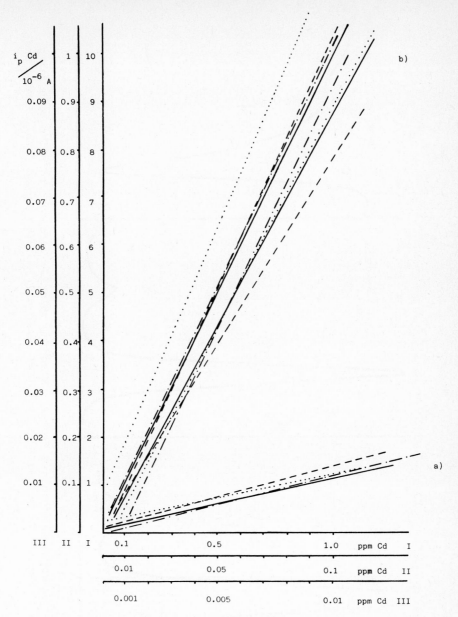

Fig. 2. Calibration curves for Cd(II)(alone or in mixture with Pb(II) for the three stripping modes in all concentration ranges studied. a) DC mode; b) DP and AC_1 modes. (———[Cd]>0.1 ppm, Fortified with Pb; —·—·—[Cd]<0.1 ppm, fortified with Pb; ---[Cd]<0.01 ppm, Fortified with Pb; . . . . . Cd alone).

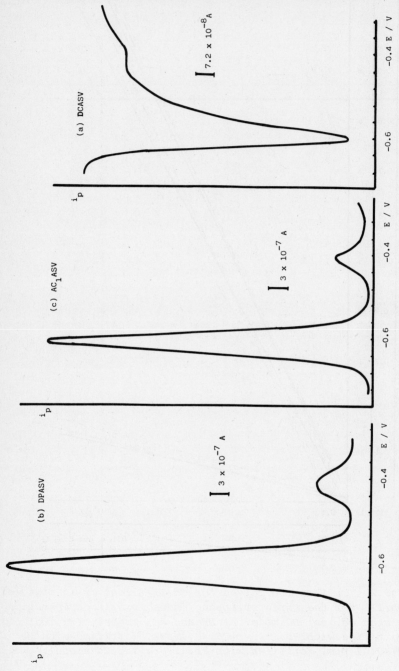

Fig. 3. Voltammograms obtained from a 1 M sodium chloride reference solution containing [Pb]= 0.05 mg L^{-1} and [Cd]= 0.5 mg L^{-1}; [Pb]/[Cd] ratio = 1/10. (a) DC mode; (b) DP mode and (c) AC_1 mode.

502

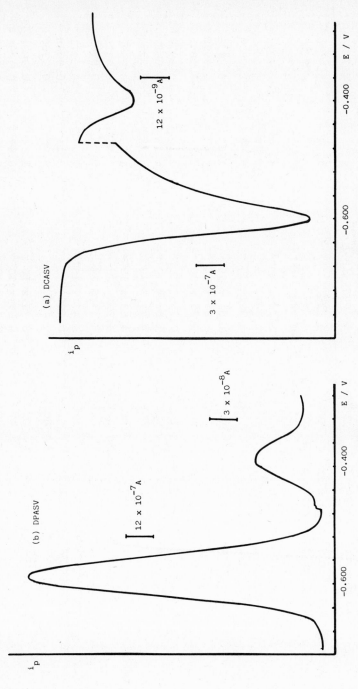

Fig. 4. Voltammograms obtained from a 1 M sodium chloride reference solution containing [Pb]= 0.005 mg L⁻¹ and [Cd]= 1 mg L⁻¹; [Pb]/[Cd] ratio = 1/200: (a) DC mode (at −0.476 V potential stop and change in the sensitivity knob are made); (b) DP mode (at −0.470 V a change in the sensitivity knob is made).

TABLE 1. Calibration Curves for Pb(II), Cd(II) and Mixtures of Both, in Different Concentration Ranges. Sensitivity, Σ, (μA $\mu g^{-1} cm^{-3}$) and Intercept in I_p Axis for the Three Stripping Modes Under Comparison

Metal	Concentration range	Stripping mode	Intercept (μA)	Slope (Σ)	Regression
CADMIUM	0.1 mg L^{-1} -1 mg L^{-1} (No fortified with PB(II))	DC	0.235	1.02	0.9890
		DP	0.797	11.79	0.9989
		AC_1	-0.240	9.32	0.9903
	0.1 mg L^{-1} -1 mg L^{-1} (Fortified with Pb(II))	DC	1.77×10^{-2}	1.14	0.9993
		DP	0.147	10.05	0.9962
		AC_1	-1.59×10^{-2}	8.94	0.9994
	5 μg L^{-1} -0.1 mg L^{-1} (Fortified with Pb(II))	DC	1.31×10^{-3}	1.17	0.9956
		DP	6.65×10^{-4}	10.47	0.9998
		AC_1	-6.75×10^{-2}	10.17	0.9845
	0.5 μg L^{-1} -0.01 mg L^{-1} (With 1 mg L^{-1} Pb(II))	DC	7.25×10^{-4}	1.35	0.9977
		DP	-1.66×10^{-4}	10.33	0.9965
		AC_1	2.55×10^{-3}	7.72	0.9959
LEAD	0.1 mg L^{-1} -1 mg L^{-1} (No fortified with Cd(II))	DC	4.63×10^{-2}	1.28	0.9998
		DP	0.320	7.80	0.9990
		AC_1	9.32×10^{-2}	9.94	0.9981
	0.1 mg L^{-1} -1 mg L^{-1} (Fortified with Cd(JJ))	DC	8.47×10^{-3}	1.25	0.9971
		DP	0.480	9.11	0.9988
		AC_1	0.434	9.70	0.9960
	5 μg L^{-1} -0.1 mg L^{-1}	DC	-8.70×10^{-3}	1.34	0.9638
		DP	6.65×10^{-4}	10.47	0.9998
		AC_1	-1.25×10^{-2}	10.25	0.9956

TABLE 2. Resolution Obtained with the Different Stripping Modes[a]

Metal		D. C. mode	D. P. mode	A.C$_l$. mode
Pb	E_p	-0.403 v	-0.390 v	-0.414 v
	$W_{1/2}$	76 mV	72 mV	50 mV
	I_p	1.75 µA	9.48 µA	12.81 µA
Cd	E_p	-0.608 V	-0.600 V	-0.624 V
	$W_{1/2}$	56 mV	56 mV	50 mV
	I_p	9.6×10^{-4} µA	5×10^{-3} µA	7.5×10^{-3} µA
$\Delta E_p = E_{p\ Pb} - E_{p\ Cd}$		205 mV	210 mV	210 mV
$I_{p\ Pb}/I_{p\ Cd}$		1882	1896	1708

[a] *1 M NaCl reference solution containing 1 mg Pb L^{-1} and 0.55 µg Cd L^{-1}. Pb:Cd ratio = 1818.*

TABLE 3. Results Obtained by the Standard Addition Method in the Determination of Pb(II) and Cd(II) in Two Common Salt Samples[a]

Sample	Determination	D. C. mode Pb	Cd	D. P. mode Pb	Cd	A. C_1 mode Pb	Cd
A	Simultaneous	0.414	0.010	0.394	0.010	0.409	0.011
	Independent	0.412	0.010	0.414	0.010	0.410	0.010
B	Simultaneous	0.328	0.025	0.364	0.026	0.371	0.026
	Independent	0.329	0.023	0.347	0.026	0.390	0.026

[a] Concentrations in mg Metal $\times$ Kg^{-1} salt. Samples A and B are respectively the samples nr 11 and nr 12 from Table 8.

TABLE 4. Statistical Study of the "Reagent Blank" Signal[a]

	DC mode	DP mode	AC_1 mode
Pb(II)	$\overline{X} = 0.0334$	$\overline{X} = 0.0303$	$\overline{X} = 0.0261$
	$SD = 0.0121$	$SD = 0.0057$	$SD = 0.0105$
	$\overline{X} + 3SD = 0.0699$	$\overline{X} + 3SD = 0.0475$	$\overline{X} + 3SD = 0.0576$
Cd(II)	$\overline{X} = 0.0010$	$\overline{X} = 0.0008$	$\overline{X} = 0.0006$
	$SD = 0.0010$	$SD = 0.0005$	$SD = 0.0004$
	$\overline{X} + 3SD = 0.0042$	$\overline{X} + 3SD = 0.0025$	$\overline{X} + 3SD = 0.0018$

[a] $\overline{X}$ = Mean from 20 independent replicates in mg Kg; SD = Standard deviation; $\overline{X} + 3SD$ = Quantitation limits take for the calibration curve finish).

-- For lead: By DC mode 4.1 µg L^{-1} (in the cell solu-
                      tion) or 70 µg Kg^{-1} (in the salt sample);
                      by DP mode 2.8 µg L^{-1} or 48 µg Kg^{-1}; by
                      AC_1 mode 3.4 µg L^{-1} or 58 µg Kg^{-1}.

As shown before, the vacuum salt used as reference material in the preparation of the blanks contains some lead and cadmium. Use of the standard-addition method of measurement avoids using these blanks and allows to reach very much lower quantitation limits. In the standard-addition method as used here no reference material is used other than the pure metal salts used as addition, and the limit only depends:

a) on the ability to distinguish the sample signal from the base line of the voltammetric curve, and this in turn depends on the standard deviation due exclusively to instrumental noise.

b) on the ability to distinguish the standard-added sample signal from the original sample signal, which, if the addition is properly chosen, depends too on the instrumental noise.

However, because no sodium chloride sample has been available to us with a Cd(II) or Pb(II) content lower than that of the vacuum salt described before, the criterion a) could not be fulfilled and a quantitation limit based only in criterion b) was established.

From sets of 10 experiments performed at Cd(II) and Pb(II) levels much lower than those given above, it has been shown than by the standard addition method the quantitation limits are now:

-- For cadmium: By DC, DP and AC_1 modes lower than 0.06 µg
                      L^{-1} (in the cell solution) or 1.00 µg Kg^{-1}
                      (in the salt sample)

-- For lead: By DC mode lower than 0.95 µg L^{-1} (in the
                      cell solution) or 16 µg Kg^{-1} (in the salt
                      sample); by DP and AC_1 modes lower than
                      0.85 µg L^{-1} or 14 µg Kg^{-1}.

Interferences

Since copper is the only element which led to accidental contamination problems in some uses of salt (40), its potential interference is the only one studied here. It has been known for at least twenty years that Cu(II) content in sodium chloride trade samples is lower than 0.1 mg Kg^{-1} (41), and usually is about 0.01 mg Kg^{-1}. At this last concentration level, Cu(II) does not interfere in Pb(II) or Cd(II) determination by ASV using anyone of the stripping modes. However, at somewhat higher levels, for instance 0.022 mg Cu(II) Kg^{-1} salt, does not interfere if the stripping mode used is DC- or DP-mode. Using DP stripping mode it is even possible to perform the simultaneous determination of all three elements, Pb, Cd and Cu (Fig. 5). With all stripping modes, however, unusual width of the Cu-peak are obtained (Table 5 and Fig. 5). At these concentration levels, using AC_1 mode, Cu(II) ion does interfere in Pb(II) and Cd(II) determination. No further study has been made about the reason of this interference. It is probably due to the formation by Cu(II) of chlorocomplexes such as $CuCl_2^-$, which, according to Siebert and Hume (42), can be adsorbed over the electrode surface.

Precision and Accuracy

The precision was evaluated by calculating the relative standard deviations of the peak heights for Cd(II) and Pb(II) from ten successive voltammograms in a sea salt sample. The relative standard deviations of a single determination are shown in Table 6.

The accuracy of the method was assessed by determining PB(II) and Cd(II) in salt samples prepared by the E.C.S.S.-C.C.F.A., for an inter-laboratory study on AAS and spectrophotometric reference methods for Hg, Pb, Cd and As (2,3,4,1) and a linear sweep ASV reference method for Pb determinations (30).

Study by NP Mode and AC_2 Mode

Although these stripping modes have not been so fully worked as the former ones, the results obtained (Fig. 6) from several samples studied by AC_2 mode (43) are equivalent in sensitivity and resolution to those obtained with the other stripping modes.

On the contrary, NP mode shows strong limitations, at least for the samples studied.

Applications

Lead and cadmium contents in a number of sodium chloride samples from different origin, quality and use have been determined by the ASV methods here proposed (Table 8). Among these, there are three samples that were used as standard samples in the E.C.S.S.-C.C.F.A. collaborative study mentioned earlier, three samples of chemical reagent for laboratory use of different brands and qualities, two samples of table salt (sea salt origin) from the supermarket, one industrial salt from a salted fish factory and, finally, seven samples of Mediterranean and Atlantic sea salt from different Spanish saltworks, whose charac-

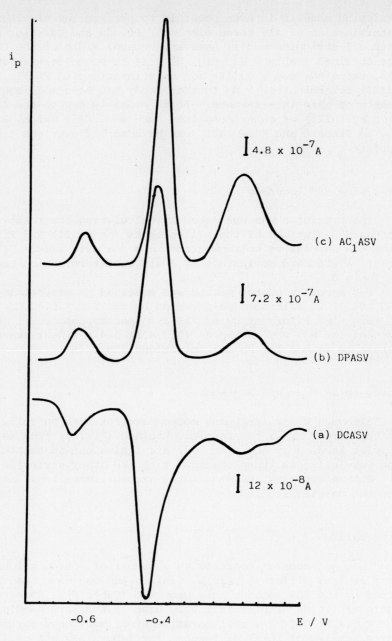

Fig. 5. Voltammograms obtained from a 1M sodium chloride reference solution containing Cu(II)(0.16 mg L⁻¹), Pb(II)(1 mg L⁻¹) and Cd(II)(0.2 mg L⁻¹). (a) DC mode; (b) DP mode and (c) AC₁ mode.

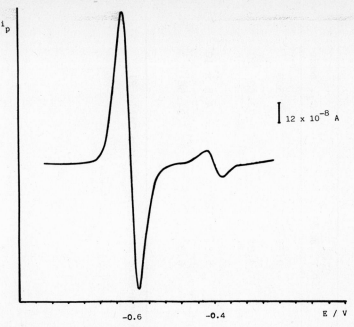

Fig. 6. AC_2 Voltammogram from 1 M sodium chloride reference solution containing [Pb] = 0.05 mg L^{-1} and [Cd] = 0.5 mg L^{-1}; [Pb]/[Cd] ratio = 1/10.

TABLE 5. Half-Height Width of Cd(II), Pb(II) and Cu(II) Dissolution Peaks Obtained from a 1 M NaCl Solution Containing the Three Metal Ions[a]

Stripping mode	Cd	Pb	Cu
D.C.	68 mV	65 mV	--
D.P.	64 mV	64 mV	100 mV
A.C$_1$.	54 mV	54 mV	109 mV

[a]0.011 mg Cd, 0.057 mg Pb and 0.021 mg Cu by Kg salt.

TABLE 6. Relative Standard Deviations (R.S.D.) of the Mean Peak Heights for Cd(II) and Pb(II)[a]

	D.C. mode	D.P. mode	A.C_1. mode
Concentration of Pb (mg x L^{-1})	0.516	0.523	0.509
R.S.D.	4.5	4.85	3.63
Concentration of Cd (mg x L^{-1})	5.65 x 10^{-3}	5.55 x 10^{-3}	4.5 x 10^{-3}
R.S.D.	38	6.3	34.56

[a]$R.S.D. = SD \times 100 / \bar{x}.$ (10 replicates of the sea-salt sample nr 1 from Table 8)

512

TABLE 7. Accuracy of Pb(II) and Cd(II) Determination[a]

Sample	Metal	DC mode	DP mode	AC_1 mode	Inter-laboratory study (E.C.S.S.-C.C.F.A.) AAS	(E.C.S.S.) DCASV
SEA SALT (sample nr 9)[b]	Pb	0.804	0.806	0.810	0.808	0.804
	Cd	0.010	0.009	0.011	0.011	---
ROCK SALT (sample nr 10)[b]	Pb	0.057	0.058	0.055	0.045	0.058
	Cd	0.002	0.002	0.002	0.002	---
VACUUM SALT (sample nr 8)[b]	Pb	0.033	0.030	0.026	0.025	0.032
	Cd	0.001	0.001	0.002	0.001	---

[a] Results in mg kg^{-1} salt.

[b] From Table 6.

TABLE 8. Cd(II) and Pb(II) Contents in Several Sodium Chloride Samples[a]

Sample	Cd(II) DC mode	DP mode	AC$_1$ mode	Pb(II) DC mode	DP mode	AC$_1$ mode
1 Sea Salt nr 1	0.0057	0.0055	0.0056	0.499	0.505	0.496
2 Sea Salt nr 2	0.0027	0.0026	0.0019	0.362	0.376	0.369
3 Sea Salt nr 3	0.0045	0.0052	0.0050	0.419	0.426	0.390
4 Sea Salt nr 4	0.0035	0.0045	0.0047	0.481	0.441	0.422
5 Sea Salt nr 5	0.0075	0.0064	0.0071	0.290	0.317	0.318
6 Sea Salt nr 6	0.0046	0.0043	0.0044	0.221	0.221	0.250
7 Sea Salt nr 7	0.0104	0.0092	0.0092	0.490	0.440	0.459
8 Vacuum Salt (E.C.S.S.)	0.0017	0.0010	0.0025	0.033	0.030	0.026
9 Sea Salt (E.C.S.S.)	0.0100	0.0090	0.0110	0.804	0.806	0.810
10 Rock Salt (E.C.S.S.)	0.0025	0.0023	0.0020	0.057	0.058	0.055
11 Table Salt (paper bag)	0.0100	0.0090	0.0110	0.486	0.448	0.496
12 Table Salt (plastic bag)	0.0250	0.0260	0.0260	0.328	0.364	0.371
13 Industrial Salt	0.007	0.0083	0.0082	0.845	0.886	0.873
14 Sodium Chloride, Suprapur Merck, Lot: 2353414	0.0025	0.0023	0.0020	0.037	0.033	0.034
15 Sodium Chloride, Pro-anal. Merck, Lot: 0119905	0.0063	0.0058	0.0057	0.123	0.127	0.126
16 Sodium Chloride, Reagent Probus, Lot: 50569	0.0105	0.0116	0.0095	0.353	0.385	0.331

[a]Results in mg Kg^{-1}.

TABLE 9. Chemical Analysis of Six of Sea-Salt Samples Analysed for Cd(II) and Pb(II)[a]

	Sample nr. 1	Sample nr. 2	Sample nr. 3	Sample nr. 4	Sample nr. 5	Sample nr. 6	Ref.[b]
Calcium	0.143%	0.215%	0.143%	0.175%	0.163%	0.100%	1
Magnesium	0.114%	0.118%	0.174%	0.181%	0.017%	0.053%	1
Potassium	0.055%	0.053%	0.066%	0.088%	0.036%	0.044%	2
Sodium (calculated)	37.352%	37.538%	36.366%	36.853%	37.405%	36.968%	-
Sulfate	0.495%	0.681%	0.603%	0.710%	0.435%	0.294%	3
Bromide	0.033%	0.034%	0.035%	0.043%	0.032%	0.033%	4
Chloride	57.856%	58.139%	56.436%	57.201%	57.715%	57.146%	5
Matters insoluble in water	0.051%	0.058%	0.014%	0.016%	0.026%	0.013%	6
Loss of mass at 110°C	3.731%	2.971%	5.893%	4.790%	4.066%	5.212%	7
Sodium chloride content (dry basis)	98.780%	98.514%	98.487%	98.303%	99.199%	99.260%	8
Iron (mg Kg^{-1})	7.5	6.0	5.0	7.5	4.0	5.0	9
Copper (mg Kg^{-1})	0.010	0.017	0.021	0.010	0.007	0.010	10
Nitrogen (mg Kg^{-1})	2	1	2	1	1	1	11

teristics and whole chemical analysis, for six of them, are
given in Table 9 (41).

All results for lead are within the range 0.221-0.886 mg Pb
Kg^{-1} salt except for the sample of rock salt (0.057 mg Kg^{-1}),
the one of vacuum salt (0.030 mg Kg^{-1}) and two of the sodium
chloride reagent studied (pro analysi: 0.123 mg Kg^{-1} and suprapur: 0.035 mg Kg^{-1}).

Results for cadmium are in the range 0.0025-0.0105 mg Cd Kg^{-1}
salt, except for the sample of vacuum salt (0.0020 mg Kg^{-1}) and
rock salt (0.0022 mg Kg^{-1}).

It is noteworthy to comment about the results obtained for
the two common table salt samples, traded by the same firm, one
of them packed in paper, the other in plastic bag. The sample
packed in plastic has a normal Pb(II) content, but its Cd(II)
content is higher than twice the Cd(II) content of the salt
packed in paper. It must be noted that on the outside of the
plastic bag there is the brand name printed in big letters of
a red color, and since sometimes the red printing-inks contain
cadmium pigments, the high Cd(II) content perhaps can be assigned
to contamination from this origin, as indicated by Preda et al
(44) in similar cases.

Footnotes to Table 9 (p. 515).

[a]*Results for Cd(II) and Pb(II) in Table 8.*

[b]*(1): UNE Standard 34-204-81, equiv. to ISO Standard 2482-73;
(2): UNE Standard 34-208-81, equiv. to ECSS/SC Standard 183-1979;
(3): ISO Standard 2480-1972; (4): ECSS/SC Standard 182-1979;
(5): UNE Standard 34-205-81, equiv. to ISO Standard 2481-1973;
(6): UNE Standard 34-202-81, equiv. to ISO Standard 2479-1972;
(7): UNE Standard 34-203-81, equiv. to ISO Standard 2483-1973;
(8): R.D. 1424-1983, 27th April; B.O.E. nr 130, 1983.06.01, pp
15261-15264, Art. 22.12, equiv. to Report of 16th Session of the
C.C.F.A., The Hague 22nd-28th March 1983 ALINORM 83-12A, Appen-
dix XI, Section 8.2; (9): UNE Standard 34-209-81, equiv. to
ECSS/SC Standard 118-1976; (10): ECSS/SC Standard 144-1977;
(11): Merck Standards method, modified by E. Casassas, J. Bar-
bosa and L. Garcia (in press).*

ACKNOWLEDGMENTS

The authors thank Dr. J. M. Ráfols, from Unión Salinera de España S.A., for his cooperation and comments.

REFERENCES

(1): European Committee for the Study of Salt, Doc. ECSS/SC nr 311 (1982).
(2): Ibid Doc ECSS/SC nr 312 (1982).
(3): Ibid Doc ECSS/SC nr 313 (1982).
(4): Ibid Doc ECSS/SC nr 314 (1982)
(5): H. Y. Li and C. S. Shen, *Chemistry 1*, 3-12 (1956).
(6): K. Sugihara and T. Saito, *Japan Analyst 4*, 27-30 (1955).
(7): E. Muto and H. Oka, *Study of Polarography 2*, 195-196 (1954), C. A. *50*, 7006 (1955).
(8): K. Sugihara and T. Saito, *Japan Analyst 7*, 1-4 (1958).
(9): K. Sugihara and T. Saito, *Japan Analyst 7*, 139-142 (1958).
(10): K. Sugihara and T. Saito, *Japan Analyst 9*, 202-205 (1958).
(11): K. G. Berezina, L. V. Volkova and E. I. Gun'ko, *Zavodsk. Lab. 31*, 656-657 (1965).
(12): J. Trnka and K. Funk, *Talanta 16*, 1587-1590 (1969).
(13): H. Goto and F. Maeda, *J. Chem. Soc. Japan 90*, 787-790 (1969).
(14): V. Fano, *Microchem. J. 14*, 511-518 (1969).
(15): M. I. Montenegro, M. J. Cruz and M. L. Matias, *Rev. port. Quim. 13*, 217-221 (1972).
(16): L. De Galan, C. Erkelens, C. Jongerius, W. Maertens and C. I. Mooring, *Z. Anal. Chem. 264*, 173-176 (1973).
(17): E. N. Vinogradova, A. I. Kamenev and N. V. Lisenkova, *Zavodsk. Lab. 31*, 1180-1182 (1965).
(18): A. I. Kamenev and E. N. Vinogradova. *Zh. Analit. Khim. 20*, 1064-1068 (1965).
(19): W. Kemula and S. Sacha, *Microchem. J. 11*, 62-72 (1966).
(20): L. A. Chernova, T. V. Martynenko, E. A. Zakharova and Y. A. Karbainov. *Zavodsk. Lab. 31*, 1180-1182 (1965).
(21): E. A. Zakharova and L. A. Chernova. *Zh. Analit. Khim. 29*, 881-885 (1974).
(22): E. N. Karbainov and S. N. Karbainova. *Uzv. Tomsk. Politekh. Inst. 254*, 110-113 (1975). Referat. Zh. Khim. 19GD, 1976(12) nr 12G 20.
(23): G. E. Batley and T. M. Florence. *J. Electroanal. Chem. 55*, 23-43 (1974).
(24): R. J. Gajan, S. G. Capar, C. A. Subjoc and M. Sanders. *J. Assoc. Off. Anal. Chem. 65*, 970-977 (1982).

(25): S. G. Capar, R. J. Gajam, E. Madzsar, R. H. Albert, M. Sanders and J. Zyren. *J. Assoc. Off. Anal. Chem. 65*, 978-986 (1982).

(26): R. D. Satzger, C. S. Clow, E. Bonnin and F. L. Fricke. *J. Assoc. Off. Anal. Chem. 65*, 987-991 (1982).

(27): W. Holak. *J. Assoc. Off. Anal. Chem. 65*, 777-780 (1975).

(28): S. B. Adeloju, A. M. Bond and H. C. Hughes. *Anal. Chim. Acta 148*, 59-69 (1983).

(29): G. Brihaye, G. Gillain and G. Duyckaerts. *Anal. Chim. Acta 148*, 51-57 (1983).

(30): European Committee for the Study of Salt. Doc ECSS/SC nr 315 (1982).

(31): E. Wänninen. "Trace analysis and the contamination problem" in "Euroanalysis IV, Reviews on Analytical Chemistry" L. Niinistö (Ed) pag. 157. Akadémiai Kiadó, Budapest

(32): (1982).

(32): T. M. Florence. *Anal. Chim. Acta 119*, 217 (1980).

(33): J. Wang. *Environ. Sci. Technol. 16*, 104-109 (A)(1982).

(34): T. R. Copeland, J. H. Christie, R. A. Osteryoung and R. K. Skogerboe. *Anal. Chem. 45*, 2171 (1973).

(35): E. D. Moorhead and P. H. Davis. *Anal. Chem. 45*, 2178 (1973)

(36): A. M. Bond. *Anal. Chem. 44*, 315 (1972).

(37): H. Kaiser. *Pure Appl. Chem. 34*, 35 (1973).

(38): G. Brihaye and G. Duyckaerts. *Anal. Chim. Acta 143*, 111-120 (1982).

(39): G. Brihaye and G. Duyckaerts. *Anal. Chim. Acta 146*, 37-43 (1983).

(40): J. Obiols and J. M. Rafols. *Afinidad 25*, 23-27 (1968).

(41): Dr. J. M. Rafols. Private communication.

(42): R. J. Sieberg and D. N. Hume. *Anal. Chim. Acta 123*, 335 (1981).

(43): M. Stulikova and F. Vydra. *J. Electroanal. Chem. 42*, 127 (1973).

(44): N. Preda, L. Popa, M. Aricsan and E. Bordas. *Igiena 27*, 151-156 (1978).

(45): J. W. Jones, R. J. Gajan, K. W. Boyer and J. A. Fiorino. *J. Assoc. Off. Anal. Chem. 60*, 826-832 (1977).

EFFECTS OF STORAGE DURATION OF FRESH ASPARAGUS ON THE QUALITY OF THE RESULTING CANNED PRODUCT

G. D. Karaoulanis

Food Technology Institute
Athens, Greece

E. D. Paneras

Food Science and Technology Dept.
University of Thessaloniki
Greece

SUMMARY

The raw material consisted exclusively of the white parts (under ground stalks) of Asparagus (Asparagus officinalis L.) Darbonne N°4 variety. The Asparagus stalks were stored at 0°C (±0.5) and RH 95% for a period of 30 days.

Two samples were taken every 5 days which were evaluated for their qualitative characteristics (color, fibrousness, rot, diameter, pH, acid, ascorbic acid, organoleptic test, etc.) and then were peeled and canned.

The qualitative evaluation of the Asparagus performed after different periods of cold storage and after their canning showed that the samples which were stored for a long period were slightly inferior compared to those stored for a short time.

Therefore, it is deduced that Asparagus of the above variety stored under the above conditions for 30 days can be canned without serious effect on their quality and constitute a suitable raw material for canning.

Instrumental Analysis of Foods
Volume 2

519

INTRODUCTION

Asparagus is marketed as fresh or processed, canned or frozen.

Important quality characteristics of asparagus for fresh or processed use are the color, flavor, tenderness and fiber content (1, 12).

Various investigators as Kramer et al., (10, 11, 12, 13), Wiley et al., (27) and others developed objective methods for evaluating the quality of fresh and processed asparagus.

Other investigators including Scott and Kramer (22) Kramer (15), Dame et al., (4,5,6) studied the physiological changes which occurred in asparagus after harvest and the effects of the post-harvest storage and other variables on the processed stalks of asparagus.

Several investigators as Lipton (18) Lougheed et al., (19, 20) and others have examined the effects of storage of fresh asparagus in low temperatures and under modified atmospheres.

However the effects of extended low temperature storage of asparagus to be subsequently canned with the intent of extending the processing season of the factories has not been investigated until now.

The present study is of special interest for Greece where harvesting of asparagus occurs in a short period which results in difficulties in absorption of the produced quantity of asparagus in the limited number of processing plants.

The purpose of this study was to investigate the effects of extended cold storage on the quality characteristics of fresh asparagus in relation to the quality of the resulting canned product. Any possibility of increasing the storage time of raw asparagus without harming the quality of the processed product will lengthen the period of utilization of the plants and will offer to the growers better prices for their crop.

MATERIALS AND METHODS

For the present study only white asparagus, Darbonne N°4 variety, grown in the fields of the Thessaloniki area was used.

The asparagus stalks when harvested had white color with a very slight violet color at their tips.

The stalks after harvesting were thoroughly washed, trim= med to a length of approximately 17 to 22 cm placed carefully in plastic bags and shipped air mail inside heavy cartons to the Laboratory in Athens where they were received the day of harvesting.

Upon arrival the asparagus was separated into shallow perforated plastic boxes containing 2.5 to 3 Kg stalks each. The boxes were placed in a cold room, jacket type, where the temperature was maintained at $0^{\circ}C \pm 0.5^{\circ}C$ and the relative humidity at 95%.

Before placing the asparagus in the cold room (0 days) and every 5 days thereafter up to 30 days total storage time samples were taken out of the cold room for evaluating the quality characteristics of the raw asparagus which was then canned and subsequently studied for quality.

The measurements performed in the raw asparagus included diameter, fibrousness, soluble solids (Brix), pH, weight loss, ascorbic acid, color and appearance of rot. Upon removal of the stalks from the cold room they were separated into three sizes on the basis of their diameter: Large (L) with diameter 17 to 21 mm, Medium (M) 13 to 17 mm and Small (S) 9 to 13 mm. The measurements of diameter were done on the separated unpeeled as well as on the stalks after hand peeling. Fibrousness was also measured in both unpeeled and peeled stalks. The measurements of soluble solids, pH, weight loss, ascorbic acid and color were performed on the unpeeled stalks after the appropriate preparation for each test.

Upon completion of the above tests the peeled stalks of each size were cut to a length of 15 cm rinsed and placed tips up in tall 1 Kg cans used for asparagus canning. Then the spears were blanched inside the cans with steam, drained and immediately hot 2% brine (95° to $98^{\circ}C$) with 0.3% citric acid was added to the cans.

After closing, the cans were sterilized at $115^{\circ}C$ for 23 minutes, cooled and stored at room temperature for three months. Then the following measurements were made on the canned product: vacuum, color, soluble solids, pH of liquor, pH of spears, titratable acidity and organoleptic evaluation.

Fibrousness which is the most important factor for determining the edible quality of fresh or processed asparagus was measured with an apple pressure tester modified specifically for asparagus by placing at the end of the plunger a special stainless steel blade of 0.017 in thickness. The measurements were made on the asparagus stalks 15 cm from the tip. Ascorbic acid content was determined by the National Canners Association photometric method (21) using metaphosphoric acid to acidify the samples. Results were computed to mgms per 100 gr of weight.

Color of the external surface of the asparagus, which directly influences consumer preference, was determined with a Hunter colorimeter. The stalks were cut in 2 cm pieces and examined.

The instrument color values were: 1) L (darknes-lightness), (2) a (+ a redness, - a greenness) and 3) b (+ b yellowness, - b blueness). Before each measurement the instrument was standardized with a prototype yellow plaque having values L + 83.2, a = -4.7, and b = + 26.4.

Soluble solids were determined in degrees Brix with a refractometer after homogenization of the samples. Titratable acidity on homogenized samples is expressed as % citric acid.

Organopleptic examination of the canned spears included visual evaluation of color, fibrousness and flavor. The organoleptic evaluation was performed by a six member panel of our staff. The panel was asked to evaluate the samples with the following five point scales.

For color a score of 5 indicated white, 4 yellowish, 3 light yellow, 2 yellow-light yellow and 1 yellow.

For fibrousness a score of 5 indicated free from fiber, 4 traces of fiber, 3 limited fibrous, 2 borderline edible and 1 woody.

For flavor a score of 5 indicated excellent, 4 good, 3 fair, 2 poor and 1 off-flavor.

The experiment was repeated three times in three consecutive growing seasons (3 years). The cold storage duration for the 1st and 3rd year was 30 days while for the 2nd year was 25 days. Every measurement was made on two samples and the average is presented in the Tables.

RESULTS AND DISCUSSION

The results of the investigation are presented in Tables 1 through 6. Tables 1, 2 and 3 show the results of the measurements of values of the raw products for the 1st, 2nd, and 3rd year of the study. Tables 4, 5 and 6 present the results of the measurements of values of the canned products for the 1st, 2nd and 3rd year.

I. BEFORE CANNING

a. *Loss of weight*

As shown in the Tables 1, 2 and 3 the loss of weight during storage was very small in all three years (repetitions) the exception being the asparagus stored for 30 days where the weight loss amounted from 2.20 to 3.26% which is considered within acceptable limits. No significant differences were observed among the large, medium and small sizes of asparagus for the same storage period.

b. *Pressure*

The pressure which represents fibrousness as it was expected was greater in the non-peeled than in the peeled stalks. During storage some small fluctuations were observed in the pressure with a small tendency for increase as time of storage increased. The same phenomenon was observed with both peeled and unpeeled stalks.

c. Soluble solids (Brix)

A constant decrease in soluble solids was observed during storage.

d. Ascorbic acid

The content of ascorbic acid (vitamin C) decreased continuously during storage and at the end of storage time (30 days) the ascorbic acid had dropped to approximately 1/3 of the initial quantity. Because of the insignificant variations of ascorbic acid values in the three diameters of asparagus stalks the results in Tables 1, 2 and 3 are averages of the values obtained.

e. Color

There was considerable change in the color of the stalks during storage. The values of L showed a slight decrease while the a value which initially was negative changed to positive numbers after 20 days of storage. The b value showed a small decrease during storage. Above measurements indicated that the color of the asparagus stalks darkens continuously as storage time progresses.

f. pH

There were slight changes in the pH of the asparagus stalks during storage. There were no significant differences among the stalks of various diameters stored for the same period of time.

g. Appearance of rot

Sporadic molding in a small number of stalks was observed at the tips after 20 days of storage. At the end of the storage period (30 days) a slight molding was present on approximately 15% of the stalks.

TABLE 1. Effect of Storage on Raw Asparagus (1st Year)

Days of storage		Diameter (mm)		Pressure (lbs) (Fibrousness)		Brix	pH	Loss of weight (%)	Ascorbic acid (mg/100 gm)	Color values			
		Unpeeled	Peeled	Unpeeled	Peeled					L	a	b	a/b
0	L	20.10	19.40	8.00	6.35	5.10	5.50	0	41.9	64.1	-3.1	18.2	-0.170
	M	14.05	13.60	15.90	4.65	5.00	5.25	0					
	S	10.20	9.80	4.50	4.10	5.00	5.53	0					
5	L	20.80	20.20	8.00	6.40	5.00	5.65	0.45	31.0	63.3	-2.9	18.0	-0.161
	M	14.30	14.00	5.10	5.00	5.00	5.62	0.40					
	S	11.00	8.80	4.70	4.05	4.70	5.63	0.45					
10	L	19.80	19.60	8.20	7.28	4.70	5.68	0.86	27.0	62.9	-2.3	17.5	-0.131
	M	14.80	14.10	6.10	5.80	4.50	5.65	0.80					
	S	10.50	9.50	5.60	5.30	4.20	5.60	0.81					
15	L	17.90	17.70	8.35	7.30	4.20	5.65	0.83	19.5	62.0	-1.4	17.0	-0.082
	M	13.30	13.10	6.25	5.70	4.20	5.65	0.73					
	S	8.00	7.70	5.40	4.80	4.00	5.60	0.85					
20	L	16.90	16.50	8.20	8.10	4.00	5.82	1.15	17.2	61.1	-1.2	17.1	-0.071
	M	13.60	13.40	6.90	5.80	4.00	5.80	1.13					
	S	10.10	9.70	6.00	5.90	4.10	5.70	1.10					
25	L	20.00	19.30	8.50	7.14	4.00	4.00	1.15	15.8	61.3	-1.1	17.2	-0.069
	M	14.00	12.40	6.90	5.85	4.05	4.00	1.25					
	S	9.09	8.60	6.37	4.90	3.90	4.00	1.10					
30	L	18.60	18.00	8.80	7.37	3.90	3.90	2.50	13.9	62.1	-1.2	16.9	-0.071
	M	14.40	13.00	6.95	5.80	3.85	3.90	2.30					
	S	9.75	9.30	6.50	5.10	3.85	3.90	2.20					

TABLE 2. Effect of Storage on Raw Asparagus (2nd Year)

Days of storage		Diameter (mm) Unpeeled	Diameter (mm) Peeled	Pressure (lbs) (Fibrousness) Unpeeled	Pressure (lbs) (Fibrousness) Peeled	Brix	pH	Loss of weight (%)	Ascorbic Acid (mg/100 gm)	Color values L	Color values a	Color values b	Color values a/b
0	L	18.90	17.00	9.50	7.05	3.65	5.66	0	42.4	63.4	-2.4	17.4	-0.161
	M	14.00	13.74	8.04	6.20	3.60	5.55	0					
	S	9.90	9.70	8.05	6.20	3.60	5.56	0					
5	L	21.00	18.25	9.22	6.09	3.55	5.58	0.30	33.0	62.7	-2.6	17.4	-0.149
	M	13.42	12.75	7.97	5.88	3.55	5.58	0.35					
	S	10.25	8.90	8.00	5.69	3.50	5.58	0.40					
10	L	18.55	15.35	9.10	6.07	3.50	5.50	0.69	21.0	62.2	-2.2	17.0	-0.129
	M	13.50	12.67	8.38	6.84	3.55	5.30	0.59					
	S	10.57	8.67	8.15	6.59	3.50	5.40	0.70					
15	L	18.60	15.12	8.79	5.78	3.40	5.42	1.20	18.0	61.2	-0.9	22.4	-0.040
	M	12.84	11.30	8.70	5.98	3.30	5.40	1.25					
	S	10.45	8.25	7.94	6.22	3.40	5.45	1.35					
20	L	17.70	17.50	9.40	7.25	3.40	5.52	2.42	16.2	63.1	+2.2	17.8	+0.0123
	M	12.88	12.44	7.97	6.79	3.45	5.52	2.32					
	S	9.19	9.00	6.22	6.10	3.45	5.52	2.40					
25	L	17.60	16.34	9.49	6.71	3.00	5.31	3.26	14.7	62.3	+1.4	15.7	+0.089
	M	13.57	11.42	8.67	5.34	3.05	5.29	3.06					
	S	8.30	7.96	7.45	5.23	3.05	5.35	3.16					
30	L	--	--	--	--	--	--	--	--	--	--	--	--
	M	--	--	--	--	--	--	--					
	S	--	--	--	--	--	--	--					

TABLE 3. Effect of Storage on Raw Asparagus (3rd Year)

Days of storage		Diameter (mm) Unpeeled	Peeled	Pressure (lbs) (Fibrousness) Unpeeled	Peeled	Brix	pH	Loss of weight (%)	Ascorbic acid (mg/100 gm)	Color values L	a	b	a/b
0	L	20.1	19.5	10.09	6.68	4.15	5.20	0					
	M	15.30	11.70	7.16	5.04	4.17	5.20	0	42.3	62.0	-2.6	17.1	-0.152
	S	10.20	7.32	6.64	3.82	4.13	5.15	0					
5	L	18.85	17.65	9.18	6.75	4.40	4.90	0.51					
	M	13.80	12.55	7.22	5.69	4.40	4.85	0.50	32.1	62.2	-2.4	17.0	-0.141
	S	9.30	7.40	6.74	4.72	4.41	4.90	0.01					
10	L	17.45	15.80	9.38	6.20	4.00	4.82	1.15					
	M	14.04	11.55	7.18	4.52	4.20	4.80	1.10	26.4	62.2	-0.5	16.1	-0.031
	S	9.6	8.50	6.89	4.81	4.15	4.79	1.10					
15	L	20.2	19.20	9.05	6.02	3.90	4.82	1.82					
	M	13.90	11.30	7.94	4.45	3.95	4.75	1.85	20.2	61.0	-0.9	15.0	-0.036
	S	9.70	6.80	6.59	4.70	3.90	4.50	1.88					
20	L	18.40	16.40	9.80	6.36	3.60	4.80	1.95					
	M	13.60	11.40	7.40	4.88	3.65	4.82	1.96	17.9	64.5	+2.7	17.0	-0.154
	S	9.10	7.35	6.29	4.09	3.60	4.60	1.95					
25	L	19.50	17.35	9.15	6.22	3.60	5.90	1.98					
	M	13.50	11.80	7.95	3.95	3.60	5.90	1.88	16.0	63.8	+1.3	24.3	+0.053
	S	10.10	9.30	6.24	4.52	3.60	5.90	1.98					
30	L	21.45	18.80	9.83	6.48	3.60	5.81	3.08					
	M	14.30	10.80	7.75	4.54	3.60	5.80	2.98	14.2	61.4	+0.1	15.3	+0.006
	S	10.30	7.30	6.90	4.75	3.53	5.80	2.88					

II.   AFTER CANNING

a.   Vacuum of cans

As shown in the Tables 4, 5, and 6 the vacuum in the cans
was satisfactory which indicated that the experimental canning
of the asparagus in the laboratory was done in a proper manner.

b.   Soluble solids (Brix)

A decrease in soluble solids of the spears was observed
after canning. The soluble solids of the spears canned after
various times of cold storage fluctuated with a trend to decrease as storage time prior to canning increased. Similar
values of soluble solids with those of the spears were found
in the liquor of the cans.

c.   pH and titratable acidity

A small decrease in the pH of the liquor as well as in the
pH of the canned stalks is observed as storage time of the raw
asparagus prior to canning increases. A small increase in the
titratable acidity of the canned stalks is observed correspondingly.

d.   Ascorbic acid

A considerable decrease of the ascorbic acid content of
the canned stalks is observed as the storage time prior to canning increases. However due to insignificant variations the
ascorbic acid content of the canned stalks of various sizes for
the same period of storage the average values are presented in
Tables 4, 5 and 6.

e.   Color

There were no considerable change in the color of the
peeled canned stalks which showed that the effects of storage
time on the color of the stalks is manifested on the skin of
the asparagus and not internally.

TABLE 4. Effect of Storage on the Resulting Canned Asparagus (1st Year)

Storage in days prior to canning		Vacuum in Hg	Brix		pH		Tritatable acidity %	Ascorbic acid mg 100 gm	Color values				Organoleptic evaluation (1)		
			Liquor	Spears	Liquor	Spears			L	a	b	a/b	Color	Fibrous-ness	Flavor
0	L	17.00	3.90	3.90	5.15	5.10	0.55								
	M	16.80	3.80	3.80	5.10	5.08	0.55	35.10	72.9	-2.9	23.2	-0.125	4.30[a]	4.50[a]	3.90[a]
	S	17.50	3.75	3.80	4.85	4.65	0.50								
5	L	14.00	3.50	3.57	5.14	5.40	0.50								
	M	19.00	3.30	3.60	4.78	5.00	0.40	22.30	72.6	-2.7	22.6	-0.119	4.20[a]	4.40[a]	3.90[a]
	S	21.00	3.20	3.50	5.07	5.05	0.50								
10	L	14.20	3.60	3.60	4.85	5.15	0.55								
	M	15.80	3.60	3.80	5.30	5.10	0.55	16.00	72.5	-2.6	22.4	-0.115	4.20[a]	4.40[a]	3.80[a]
	S	16.20	3.60	3.80	5.15	5.10	0.65								
15	L	16.40	3.95	4.05	4.88	5.18	0.60								
	M	17.05	3.75	3.80	4.83	5.02	0.55	13.10	72.5	-2.3	24.5	-0.093	4.10[a]	4.40[a]	3.80[a]
	S	20.20	3.20	3.40	4.55	4.92	0.50								
20	L	11.30	3.47	3.60	5.00	4.95	0.57								
	M	18.20	3.13	3.30	4.97	4.93	0.55	11.80	73.0	-2.1	23.1	-0.090	3.50[b]	4.00[ab]	3.50[ab]
	S	15.30	3.20	3.30	4.80	4.78	0.54								
25	L	16.20	3.10	3.10	4.99	5.03	0.58								
	M	21.80	3.00	3.10	4.99	5.01	0.50	10.30	72.2	-1.6	22.9	-0.009	3.40[b]	3.60[b]	3.30[b]
	S	19.80	2.95	3.00	5.00	5.07	0.50								
30	L	15.30	3.70	3.90	4.97	4.88	0.60								
	M	17.90	3.60	3.80	4.98	5.05	0.55	9.90	72.0	-1.6	23.0	-0.069	3.30[b]	3.50[b]	3.20[b]
	S	17.50	3.80	3.40	4.61	4.70	0.50								

(1) The scores for color, fibrousness and flavor with the same letter exponent indicate no significant difference (Duncan's Multiple Range Test).

TABLE 5. Effect of Storage on the Resulting Canned Asparagus (2nd Year)

Storage in days prior to canning	Vacuum in Hg	Brix		pH		Tritatable acidity %	Ascorbic acid mg 100 gm	Color values				Organoleptic evaluation (1)		
		Liquor	Spears	Liquor	Spears			L	a	b	a/b	Color	Fibrousness	Flavor
0 L	17.50	3.25	3.35	5.10	4.98	0.47	35.90	73.7	-3.0	23.3	-0.128	4.60[a]	4.20[a]	4.00[a]
M	18.15	3.20	3.45	5.00	4.97	0.45								
S	17.00	3.25	3.30	5.00	4.95	0.50								
5 L	17.70	3.50	3.50	4.90	4.88	0.50	23.70	72.8	-2.6	22.7	-0.114	4.15[a]	4.15[a]	4.00[a]
M	14.50	3.45	3.70	4.94	4.82	0.42								
S	16.00	3.48	3.50	4.90	4.87	0.45								
10 L	16.10	3.42	3.45	5.00	5.05	0.50	16.90	73.1	-2.4	22.4	-0.107	4.35[a]	4.15[a]	3.90[a]
M	16.50	3.40	3.65	5.02	4.88	0.50								
S	15.50	3.39	3.40	5.10	5.00	0.50								
15 L	12.60	3.20	3.20	4.24	4.24	0.40	13.80	72.1	-2.3	22.1	-0.103	4.30[a]	3.70[b]	3.90[a]
M	14.75	3.10	3.20	4.08	3.85	0.48								
S	15.75	3.15	3.20	4.50	4.20	0.45								
20 L	16.05	3.40	3.55	3.85	3.90	0.62	10.90	71.8	-2.4	21.7	-0.110	3.90[ab]	3.60[b]	3.30[b]
M	19.40	3.10	3.25	3.85	3.82	0.62								
S	17.10	3.45	3.50	3.90	3.85	0.60								
25 L	16.95	3.10	3.89	3.89	3.85	0.65	10.10	72.1	-1.9	22.1	-0.085	3.40[b]	3.50[b]	3.20[b]
M	19.40	3.40	3.85	3.85	3.78	0.65								
S	16.80	3.30	3.92	3.92	3.90	0.67								

(1) The scores for color, fibrousness and flavor with the same letter exponent indicate no significant difference (Duncan's Multiple Range Test).

TABLE 6. Effect of Storage on the Resulting Canned Asparagus (3rd Year)

Storage in days prior to canning		Vacuum in Hg	Brix		pH		Tritatable acidity %	Ascorbic acid mg 100 gm	Color values				Organoleptic evaluation (1)		
			Liquor	Spears	Liquor	Spears			L	a	b	a/b	Color	Fibrousness	Flavor
0	L	18.0	3.69	3.80	4.92	4.96	0.75								
	M	16.5	3.95	3.65	4.85	4.87	0.72	36.10	73.8	-2.8	23.1	-0.121	4.50[a]	4.10[a]	3.80[a]
	S	15.0	3.31	3.45	4.77	4.77	0.73								
5	L	15.75	3.60	3.95	4.61	4.62	0.88								
	M	14.50	3.32	3.65	4.73	4.74	0.75	24.30	72.90	-2.4	22.7	-0.105	4.20[a]	4.10[a]	3.70[a]
	S	11.40	2.95	3.20	4.51	4.45	0.60								
10	L	16.0	3.41	3.58	4.85	4.82	0.70								
	M	17.80	3.45	3.75	4.82	4.80	0.68	17.20	72.60	-2.60	22.4	-0.115	4.10[a]	4.00[a]	3.70[a]
	S	16.00	3.35	3.52	4.62	4.60	0.68								
15	L	16.00	3.40	3.60	4.68	4.68	0.78								
	M	18.25	3.10	3.50	4.55	4.50	0.72	14.0	73.60	-2.8	20.4	-0.137	4.10[a]	4.00[a]	3.60[a]
	S	16.50	3.00	3.30	4.28	4.25	0.68								
20	L	16.50	3.35	3.75	4.66	4.68	0.68								
	M	17.00	3.30	3.45	4.74	4.75	0.75	11.50	73.70	-2.2	22.4	-0.098	4.00[a]	3.50[b]	3.60[a]
	S	15.00	2.88	3.05	4.38	4.26	0.60								
25	L	14.00	3.32	3.32	4.84	4.84	0.72								
	M	16.75	3.18	3.25	4.76	4.74	0.78	10.10	70.6	-2.2	21.0	-0.105	3.40[b]	3.30[b]	3.10[b]
	S	17.00	2.90	3.05	4.88	4.86	0.52								
30	L	18.00	3.40	3.55	4.95	4.96	0.55								
	M	15.50	3.00	3.15	4.95	4.46	0.55	10.10	72.1	1.5	22.6	-0.066	3.30[b]	3.30[b]	3.00[b]
	S	15.00	3.10	3.10	4.90	4.50	0.58								

(1) The scores for color, fibrousness and flavor with the same letter exponent indicate no significant difference (Duncan's Multiple Range Test).

f. Organoleptic evaluation

There were insignificant changes in the asparagus after storage for 5, 10 and 15 days for the 1st and 2nd year and after storage of 20 days for the 3rd year of the study. After storage of 25 and 30 days a decrease is observed in the organoleptic characteristics of the canned stalks however the product was still suitable for commercial utilization.

CONCLUSIONS

From the present investigation which was repeated for three years we can draw the following conclusions:

1. The stalks of white asparagus, Darbonne N^O4 variety, can be stored at 0^OC and RH 95% for at least 30 days.

2. In this period there is a decline of the quality mainly in the color of the skin, tenderness and ascorbic acid content.

3. During cold storage and after subsequent canning there is a decline of the quality of the canned product especially in the ascorbic acid content.

4. The data obtained showed that white asparagus Darbonne N^O4 variety can be stored for 20 days at $0^OC \pm 0.5^OC$ and RH 95% and then canned giving a good quality canned asparagus.

The storage of 20 days can be extended for 10 more days with some lowering of the quality of the resulting canned product.

REFERENCES

1. Backinger, G. D., Kramer, A., Decker, W. R., and Sidwell, A. P. Application of work measurement to the determination of fibrousness in asparagus. *Food Technology 11*, 583-585 (1957).

2. Berger, W. R., Harvey, J. M., Stewart, J. K. and Ceponis, M. J. California asparagus. U. S. Dept. of Agr. Marketing Res. Rept. N^O 428.

3. Dame, C. Jr., Chichester, C. O. and Marsh, G. L. Studies of processed all-green asparagus. I. Quantitative analysis of soluble compounds with respect to strain and harvest variables and their distribution within the asparagus spear. *Food Research 22*, 658 (1957).

4. Dame, C. Jr., Chichester, C. D. and Marsh, G. L. Studies of processed all-green asparagus. II. The effect of post-harvest storage and blanching variables upon the chemical

composition of processed asparagus. *Food Research 22*, 673 (1957).

5. Dame, C. Jr., Chichester, C. D. and Marsh, G. L. Studies of processed all-green asparagus III. Qualitative and quantitative studies of non volatile organic acids by chromatographic techniques. *Food Research 24*, 20-27 (1959).

6. Dame, C. Jr., Chichester, C. D. and Marsh, G. L. Studies of processed all-green asparagus. IV. Studies on the influence of tin on the solubility of rutin and on the concentration of rutin present in the brines of asparagus processed in glass and in containers. *Food Research 24*, 28-36 (1959).

7. Davis, R. B., Guyer, R. B., Daly, J. J. and Johnson, H. T. Control of rutin discoloration in canned asparagus. *Food Technology 14*, 212-217 (1961).

8. Decker, R. W., Yeatman, J. N., Kramer, A. and Sidwell, A. P. Modification of the shear-press for electrical indicating and recording. *Food Technology 11*, 343-347 (1957).

9. Ellison, H. J. Commercial growing of asparagus. U.S.D.A. Farmers Bull. N° 2232 (1975).

10. Kramer, A., Haut, C. I., Scott, L. E. and Eide, L. Objective methods for measuring quality factors of raw canned and frozen asparagus. *Proc. Am. Soc. Hort. Sci. 50*, 411-425 (1949).

11. Kramer, A., Haut, I. C. and Scott, L. E. Quick recorders of fibrousness in asparagus. *Food Industries 21*, 1075-77 (1949).

12. Kramer, A., Haut, I. C. and Scott, L. E. Objective methods for measuring quality factors of raw canned and frozen asparagus. *Proc. Am. Soc. Hort. Sci. 53*, 411-425 (1949).

13. Kramer, A., Burkhardt, G. T. and Rodgers, H. P. The shear-press, a device for measuring food quality. *Canner 112*, 34-36 (1951).

14. Kramer, A., Aamlid, K., Guyer, R. B. and Rodgers, H. P. New shear-press predicts quality of canned lima beans. *Food Eng. 23*, 112-113 (1951).

15. Kramer, A. Mandatory standards program of quality factors for frozen asparagus and peas. An industry approach. III. Evaluation of results. *Food Technology 8*, 468-470 (1954).

16. Kramer, A., Kornetsky, A., Elehway, N., Steinmetz, G. and Morin, E. L. A procedure for sampling and grading raw green asparagus. *Food Technology 10*, 212-214 (1956).

17. Kramer, A., Wiley, R. C. and Twingg, R. A. The measurement of fibrousness of asparagus. *Proc. Amer. Soc. Hort. Sci. 76*, 382-388 (1960).

18. Lipton, W. J. Post-harvest of asparagus spears to high carbon dioxide and low oxygen atmospheres. *Proc. Am. Soc. Hort. Sci. 86*, 347-356 (1965).

19. Lougheed, E. C. and Dewey, D. H. Factors affecting the tenderizing effect of modified atmospheres on asparagus spears during storage. *Proc. Am. Soc. Hort. Sci. 89,* 336-345 (1966).

20. Lougheed, E. C. and Franklin, E. W. Influence of carbon dioxide on the uptake of water by asparagus. Nature N° 5003, 1313 (1965).

21. National Canners Assoc. A Laboratory Manual for Food Canners and Processors. The Avi Publishing Company, Inc. Vol. 2 (1968).

22. Scott, L. and Kramer, A. Physiological changes in asparagus after harvest. *Proc. Am. Soc. Hort. Sci. 54,* 357-366 (1949).

23. Smith, H. R. and Kramer, A. The fiber content of canned green asparagus. *Canner 104,* 14-16 (1947).

24. Schmith, H. P. Mandatory standards program on quality factors for frozen asparagus and peas. An industry approach. I. Purpose and organization of program. *Food Technology 8,* 463-465.

25. U. S. Grades for canned asparagus. Mimeo May 1, 1945. U.S.D.A.

26. U. S. Standards for grades of fresh asparagus. April 1, 1966. U.S.D.A.

27. Wiley, R. C., Elehway, N. and Kramer, A. The shear-press. An instrument for measuring the quality of foods. IV. Application to asparagus. *Food Technology 10,* 439-443 (1956).

28. Wilder, J. C. and Samuels, C. E. Mandatory standards program on quality factors for frozen asparagus and peas. An Industry approach. II. Selection of official analytical procedures. *Food Technology 8,* 465-467 (1954).

INDEX